D1496275

EMI Troubleshooting Techniques

Other Reference Books of Interest by McGraw-Hill

Handbooks

BENSON ▪ *Television Engineering Handbook*
CHEN ▪ *Fuzzy Logic and Neural Network Handbook*
CHRISTIANSEN ▪ *Electronics Engineers' Handbook,* 4/e
COOMBS ▪ *Printed Circuits Handbook,* 4/e
DI GIACOMO ▪ *Digital Bus Handbook*
DI GIACOMO ▪ *VLSI Handbook*
HARPER ▪ *Electronic Packaging and Interconnection Handbook,* 2/e
JURAN AND GRYNA ▪ *Juran's Quality Control Handbook*
JURGEN ▪ *Digital Consumer Electronics Handbook*
OSA ▪ *Handbook of Optics,* 2/e
RORABAUGH ▪ *Digital Filter Designers' Handbook,* 2/e
SERGENT AND HARPER ▪ *Hybrid Microelectronic Handbook*
WAYNANT ▪ *Electro-Optics Handbook*
WILLIAMS AND TAYLOR ▪ *Electronic Filter Design Handbook*
ZOMAYA ▪ *Parallel and Distributed Computing Handbook*

Other

ANTOGNETTI AND MASSOBRIO ▪ *Semiconductor Device Modeling with SPICE*
BEST ▪ *Phase-Lock Loops,* 3/e
DEVADES ▪ *Logic Synthesis*
GRAEME ▪ *Photodiode Amplifiers Op Amp Solutions*
GRAEME ▪ *Optimizing Op Amp Performance*
HECHT ▪ *The Laser Guidebook*
JOHNSON ▪ *LabVIEW Graphical Programming,* 2/e
KIELKOWSKI ▪ *Inside SPICE*
PILLAGE ▪ *Electronic Circuit and System Simulation Methods*
RHEM ▪ *Oscillator Design Computer Simulation,* 2/e
SANDLER ▪ *SMPS Simulation with SPICE*
SCLATOR ▪ *McGraw-Hill Electronics Dictionary,* 6/e
SEALS ▪ *Programmable Logic*
SMITH ▪ *Thin-Film Deposition*
STEARNO ▪ *Flexible Printed Circuits*
SZE ▪ *VLSI Technology*
TSUI ▪ *LSI/VLSI Testability Design*
VAN ZANT ▪ *Microchip Fabrication,* 3/e
WOBSCHALL ▪ *Circuit Design for Electronic Instrumentation*
WYATT ▪ *Electro-Optical System Design*

EMI Troubleshooting Techniques

Michel Mardiguian

Boston, Massachusetts Burr Ridge, Illinois
Dubuque, Iowa Madison, Wisconsin New York, New York
San Francisco, California St. Louis, Missouri

Library of Congress Cataloging-in-Publication Data

Mardiguian, Michel.
EMI troubleshooting techniques / Michel Mardiguian.
p. cm.
ISBN 0-07-134418-7
1. Electromagnetic interference. 2. Electromagnetic compatibility. I. Title.
TK7867.2.M27 1999
621.382'24—dc21 99-33555
CIP

McGraw-Hill

A Division of The McGraw-Hill Companies

2 3 4 5 6 7 8 9 BKM BKM 9 0 9 8 7 6 5 4 3 2 1 0

ISBN 0-07-134418-7

The sponsoring editor for this book was Scott Grillo and the production supervisor was Pamela Pelton. It was set in New Century Schoolbook by J. K. Eckert & Company, Inc.

Contents

Preface

As long as the scientific community and the engineering world have existed, they generally have worked in cooperation to achieve what is their *raison d' être* and their endeavor: to search, discover, and realize things that perform better, faster, or more economically. In the world of electronics, this translates into designing systems that, from the ground up, are supposed to chirp like birds in paradise. But often, despite our plans, all hell breaks loose, and an EMC engineer is called upon to provide a remedy. This too-typical situation may be one of the most exciting and challenging aspects of EMC: instead of spending most of their time on the design of things that will work, EMC engineers and technicians punch in days and weeks of troubleshooting on things that do not work.

This, however, is not to say that they are perverse people who like nothing more than the dark side of a project. We have developed an impressive amount of literature, software tools, and math models to help each other in building the dream: a system in which EMC has been ingrained, as have other elements of performance. We love to predict field coupling, shielding performance, crosstalk, EMI scenarios, EMC budgets and margins, and finally validating the whole architecture by mock-up tests that prove that we were right—with enviable accuracy. But the hard fact does remain: we spend an enormous amount of time shoveling coal in the boiler rooms. This author being no exception, I realized that, while dozens of excellent books are now available on EMI control, analysis, coupling, measurement, etc., in at least ten languages, it seems that none to date has been fully devoted to the orchestration of one mischievous art: troubleshooting EMI problems. This book is designed to fill that gap.

Michel Mardiguian

Preface

Introduction

Electromagnetic interference (EMI), radio frequency interference (RFI), and electrostatic discharge (ESD) are complex problems, involving many interactive mechanisms. No human brain can envision at a glance all the possibilities and limitations of the possible solutions.

Having spent 20 years studying, measuring, and solving EMI, RFI, and ESD problems, the author has drawn from experience a set of troubleshooting routines and fixes that avoid reinventing the wheel at each occurrence of a failure, whether it occurs during product development or in the field. The result is a simple guide, laid out in the general structure of a checklist, that provides substantial assistance.

This book on EMI troubleshooting allows the reader (who may be a development engineer, field technician, test engineer, or hobbyist) to select and apply efficiently the best component(s) for solving an electromagnetic interference problem. It will equip the EMI/RFI firefighter with a quick and comprehensive set of techniques and parts with traceable characteristics for the diagnosis and correction of most electrical noise and interference problems. For convenience, a list of recommended tools, accessories, and components is provided so the reader can easily assemble and maintain an inventory of the most commonly used EMI-fix items.

The hardware solutions described in the book are organized in a manner similar to a medical prescription, containing

1. The name and a description of the fix (the "medication")
2. The indications associated with specific, recognizable symptoms (the "disease")
3. The prescription and installation hints (how many pills, before or after meals)

4. The limitations of the fix (medication not efficient if you have diabetes, should not be taken during pregnancy)

The problems and solutions, as described, can be easily handled by an EMC engineer or any electronic engineer or technician with a decent RF background. They are applicable in the field or in a test lab where the equipment under test (EUT) is failing the EMC specification, or in the engineering lab for straightening up a prototype.

With reference to standard lab equipment such as an oscilloscope and a spectrum analyzer, the manual provides the EMI firefighter with needed tools to detect, diagnose, and correct most interference problems.

A final word of caution: This is not a treatise on electromagnetism. Many of the suggested shortcuts are crude and might raise the eyebrows of a rigorous theoretician. However, the methods described never violate the laws of electricity; they simply have been devised for optimum practicality and efficiency.

The manufacturers listed at the end of appropriate chapters are simply mentioned as examples—as being those with whom the author has had the most experience and contacts. This does not imply, in any way, that other vendors and products are of inferior quality.

About the Author

Michel Mardiguian, born in Paris in 1941, is a graduate electrical engineer with BSEE and MSEE degrees. He served with the French Air Force and subsequently worked as an engineer with Marcel Dassault Aviation (1965 to 1968). In 1968, he moved to the IBM Research and Development Laboratory near Nice, France, and worked in the packaging of modems and digital PBXs.

Mr. Mardiguian started his career in EMC in 1974, becoming the local IBM EMC specialist, with frequent contacts with his U.S. counterparts at IBM/Kingston, U.S.A., acquiring the ABCs of EMC analysis and testing. From 1976 to 1980, he was also the mandated French delegate to the CISPR Working Group on computer RFI, participating actively in the development of what was to become CISPR Publ. 22, the root document for FCC 15-J and European EN 55 022.

In 1980, he joined Don White Consultants (later renamed Interference Control Technologies, Inc.) in Gainesville, Virginia, becoming Director of Training, then Vice President/Engineering Director, a position he kept until 1990. As such, he helped to increase the market for EMC seminars, himself teaching more than 160 classes in the United States and worldwide.

Since 1990, Michel Mardiguian has established himself as a private consultant in France, teaching EMI/RFI/ESD classes and working constantly on consulting tasks ranging from EMC design to "firefighting." One of his major involvements has been EMC in the TransManche Link, with his colleagues from Interference Technology International.

He has written seven other widely sold handbooks and co-authored two others with Don White. He has published 21 papers at IEEE and Zurich EMC symposia and various conferences, in addition to 7 articles in engineering magazines.

Chapter

1

Introduction: Brush-Up on Some EMI/EMC Basics

This chapter recapitulates some basic topics that are probably more thoroughly covered in one or several books in your library. But these books will not be with you when you are troubleshooting EMI problems, whereas this manual could be a regular companion in your fix-up work.

1.1 Terminology

Generally, *electromagnetic interference* (EMI) occurs when an electrical disturbance from either a natural phenomenon (e.g., *electrostatic discharge* [ESD], lightning, and so on) or an electrical or electronic equipment causes an undesired response in another equipment. *Electromagnetic compatibility* (EMC) is just the opposite of EMI; that is, EMC is said to exist when no equipment or system causes EMI to other equipments or systems.

EMI comes in all degrees, from barely perceptible to overwhelming. It could be graded into three levels: nuisance, intermediate, and catastrophic. Nuisance EMI is illustrated when an electric shaver or food mixer causes disturbance to radio or TV services. The nuisance usually lasts for only a few minutes and causes no damage.

Catastrophic EMI exists, for example, when radiation from a radar triggers ordnance on an aircraft carrier or when two airliners collide during approach due to garbled radio messages from the tower. Loss of life or extensive damage is usually involved.

Most EMI situations fall in between; i.e., they are intermediate EMI. Examples include radiation from portable transmitters causing interference to computers, ESD disturbing cash registers or banking terminals, or lightning upsetting automatic test equipment (ATE) and industrial process controls

1.2 EMI/EMC Glossary

The following terms are common in this book and in the general EMC community. It would be good for the reader to become familiar with them.

absorption loss That part of shielding effectiveness (SE) dealing with energy absorption through a metal barrier.

antenna factor A calibration term equal to the unknown electric field strength divided by the measured voltage during an EMI measurement.

aperture leakage the compromise in shielding effectiveness (SE) resulting from holes, slits, slots, and the like used for windows, cooling openings, and joints of metal boxes, such that EMI can get in or out.

bandwidth The frequency interval between the upper and lower 3 dB down response of a receiver.

bond A temporary or permanent mating of two metallic parts through a low-impedance interface (the bond).

broadband (BB) EMI Electrical disturbances whose frequency spectrum covers several octaves or decades in the frequency spectrum, or exceeds the receiver bandwidth.

common mode (CM) As applied to two or more wires, all currents flowing therein with the same polarity.

common-mode rejection ratio (CMRR) A measure of the immunity of an op amp to CM voltages.

corner frequency That frequency for which the slope on a Fourier transform or Bode plot changes, usually by 20 dB/decade.

coupling path The conducted or radiated path by which interfering energy gets from a source to a victim.

cross modulation Energy from one transmitter that causes the modulation to change on a received signal from another transmitter.

crosstalk The ratio, expressed in decibels, of the coupled voltage onto a victim cable to the voltage on the nearby culprit cable.

current probe An EMI sensor that clamps onto a current-carrying wire, cable, or strap to measure intentional or interference current.

differential mode (DM) Voltages or currents on a wire pair that are of opposite polarity.

DIL Dual-in-line style of circuit package.

dropout A short power interrupt that may last from a fraction of a cycle up to several cycles.

E^3, or EEE Electromagnetic environmental effects, an umbrella term used to cover EMC, EMI, RFI, EMP, ESD, RADHAZ, lightning, and the like.

electric field A radiated wave's potential gradient in volts per meter (V/m).

electrical gasket A compressible bond used between two mating metal members to secure a low-impedance path between them.

EMC Electromagnetic compatibility, the conditions under which an ensemble of several equipments do not interfere with each other, nor with their environment.

EMI Electromagnetic interference (exactly the opposite of EMC).

EMP Electromagnetic pulse.

ESD Electrostatic discharge.

EUT Equipment under test.

far fields Radiated fields where the source distance to the point measurement is greater than about 1/6 wavelength. Also called *plane waves.*

ferrite Powdered magnetic material in form of beads, rods, and rings used to absorb EMI on wires and cables.

field strength The radiated voltage or current per meter corresponding to electric or magnetic fields.

filter A device to block the flow of EMI while passing the desired 50/60/400 Hz or signal frequencies.

Fourier envelope The frequency-domain envelope of the Fourier transform or Fourier series of a time-domain function.

ground loop A potential EMI condition formed when two or more equipments are interconnected and tied to a common ground (signal, chassis, structure, earth) for safety, power return, or other purposes.

impulse bandwidth For an instrument, the difference between two frequencies corresponding to the ratio of the maximum amplitude divided by the area under the frequency response curve; approximately the 6 dB bandwidth.

impulse noise A transient disturbance, usually repetitive, that produces in-phase components up to first corner frequency of the Fourier spectrum.

intermodulation (IM) and IM interference EMI produced by integer multiples of sums and differences of two or more signals mixing in a nonlinear junction.

isolation transformer A transformer with a high primary-to-secondary isolation device used in power mains or signal links to break ground loops.

LISN Line impedance stabilization network. A device inserted between the power mains and a test item to permit repeatable conducted EMI measurements.

magnetic field A radiated wave's current gradient, expressed in amperes per meter (A/m).

narrowband (NB) EMI Interference whose emission bandwidth is less than the bandwidth of the EMI measuring receiver or spectrum analyzer.

near field In radiated fields, an EMI point source distance of less than about 1/6 wavelength.

NEMP Nuclear electromagnetic pulse.

noise immunity level The voltage threshold in digital logic families above which a logic zero may be sensed as a one and vice versa—sometimes termed *noise margin,* although the two are not exactly equivalent.

optical isolator An electro-optical device offering a high galvanic isolation, inserted in signal lines to block common-mode current (generally used for digital or on/off-type signals).

plane waves See *far field.*

power conditioning Reduction of EMI pollution on power mains by inserting filters, isolators, regulators, or an uninterruptible power supply (UPS).

power density Radiated power flux flow divided by the observing area, in watts per square meter (W/m^2).

RADHAZ Military code name for radiation hazards to humans, animals, ordnance, or fuels.

reflection loss The part of shielding effectiveness that is due to energy reflection from impedance mismatch between incident field and metal barrier.

RFI Radio frequency interference.

sag A sudden voltage drop on the power mains.

sensitivity For bandwidth-limited white noise in analog devices, the point at which S = N. Below this value, no or negligible response is exhibited by the device.

shielding effectiveness (SE) The ratio of field strengths before and after installing a shield. This consists of absorption and reflection losses.

skin depth The calculated metal layer thickness through which 63 percent of the surface current flows.

SMT Surface mount technology. Leadless style of component package.

spectrum analyzer A special receiver designed to scan, intercept, and display the amplitude of a signal versus frequency.

spurious response Undesired responses of superheterodyne receivers resulting from mixing of harmonics of local oscillator with a signal.

surge A sudden voltage increase on the power mains.

TEMPEST An unclassified code word for potentially compromising emanations from electrical and electronic equipments that process classified data.

transfer impedance (Z_t) A figure of merit of the quality of cable shield performance—the ratio of the coupled voltage to the surface current, in ohms per meter (Ω/m).

UPS Uninterruptible power supply. A power source whose output power continues even when the mains supply disappears.

1.3 Units of Measurement: the Bel

Many things in physics, as well as in real life, are relative (i.e., a ratio of something to some reference). Thus, such things can be represented on an exponential or logarithmic scale.

The Bel was originally defined as the logarithm (log to the base of 10) of the ratio of two power levels:

$$\text{Bel} = \text{B} = \log\left(\frac{P_2}{P_1}\right) \tag{1.1}$$

where P_1 = original or reference power in watts
P_2 = measured power in watts after a change

The Bel did not provide a fine enough gradation measure for many types of measurements, so the decibel was soon adopted:

$$\text{decibel} = \text{dB} = 10\log\left(\frac{P_2}{P_1}\right) \tag{1.2}$$

Quite often, data are measured in units of voltage, current, field strength, or similar units. Substituting $P = V_2/Z$ into Eq. (1.2) yields

$$\text{dB} = 20\log\left(\frac{V_2}{V_1}\right) + 10\log\left(\frac{Z_1}{Z_2}\right) \tag{1.3}$$

If the impedances Z_1 and Z_2 are equal, Eq. (1.3) becomes

$$\text{dB} = 20\log\left(\frac{V_2}{V_1}\right) = 20\log\left(\frac{I_2}{I_1}\right) \tag{1.4}$$

Equations (1.2) and (1.4) are tabulated in Tables 1.1a and 1.1b. Corresponding negative dB values are obtained by reciprocating any of the ratios.

Table 1.1a Decibels vs. Amplitude and Power Ratio

Logarithms		
log[ab] = log[a] + log[b]		
log[a/b] = log[a] – log[b]		
log[1/a] = –log[a]		
$\log[a^n] = n \log[a]$		
dB Quick Math		
Linear:	dB:	
multiply	add	
divide	subtract	
10 × 10 = 100	10 + 10 = 20	
exponent	multiply	
Amplitude (V, I, E,...) factor	**Power factor**	**Decibels**
× 1.12	× 1.25	+1
× 1.25	× 1.6	+3
× 1.4	× 2.0	+3
× 1.58	× 2.5	+4
× 2.0	× 4.0	+6
× 2.5	× 6.25	+8
× 3.16	× 10	+10
× 4.0	× 16	+12
× 5.0	× 25	+14
× 6.3	× 40	+16
× 10	× 100	+20
× 100	× 10,000	+40
× 1000	× 1,000,000	+60

The decibel values of voltage or current may be obtained by substituting 1 V or 1 A into Eq. (1.4):

$$\text{dBV} = 20\log(\text{V}) \text{ or dBI} = 20\log(\text{I}) \tag{1.5}$$

Voltages, current, or field strength may be obtained by taking the antilog ($\log^{-1}$) of any of the corresponding values in Eq. (1.5):

$$\text{V} = \log^{-1}\left(\frac{\text{dBV}}{20}\right) \text{ or I} = \log^{-1}\left(\frac{\text{dBI}}{20}\right) \tag{1.6}$$

Table 1.1b Practical Conversion Examples, EMI Units

Decibels	
Power	Voltage
dB = 10 log (P_{1meas}/P_{2ref})	dB = 20 log (V_{1meas}/V_{2ref})
Units	Units
dBW → P_2 = 1 watt	dBV → V_2 = 1 volt
dBm → P_2 = 1 milliwatt	dBμV → V_2 = 1 μV
dBm/MHz	dBμV/MHz

Examples:

(a) Convert 50 W to dBW
(10 × 10)/2 W = 10 log(10) + 10 log(10) – 10 log(2) = 10 + 10 – 3 = 17 dBW

(b) Change reference to milliwatts (dBm)
1 W = 10^3 mW, so
0 dBW = 30 dBm, hence
17 dBW ≡ 30 dBM + 17 = 47 dBm

(c) Convert 5 mV to dBμV
5 mV = 5.10^3 μV = (14 dB + 60 dB) above 1 μV = 74 dBμV

(d) Convert dBm to dBμV into 50 Ω
VdBμV = PdBm + 107
e.g., –64 dBM into 50Ω = –64 + 107 = 43 dBμV

Table 1.1a may also be used for antilogs by looking up the dB values and reading out the corresponding voltage or current.

For narrowband EMI, where only a single spectral line (i.e., a sine wave signal) is present in the receiver's (or victim's) bandwidth, the magnitude of EMI signals can be expressed in the following forms:

Voltages:	V or dB above 1 V (dBV) or dB above 1 μV (dBμV)
Currents:	A, μA, or dB above 1 μA (dBμA)
Power:	W, mW, or dB above 1 mW (dBm)
E-fields:	V/m, μV/m, or dBμV/m
H-fields:	T, A/m, μA/m, or dBμA/m (Strictly, the tesla is the unit of magnetic induction.)
Radiated power density:	W/m², mW/cm², or dBm/cm²

For broadband EMI, where many spectral lines (i.e., harmonics) are adding up in the receiver (or victim) bandwidth, the magnitude of the received interference is normalized to a unity bandwidth:

Voltages:	dBμV/kHz or dBμV/MHz
Currents:	dBμA/kHz or dBμA/MHz
E-fields:	dBμV/m/kHz or dBμV/m/MHz
H-fields:	dBμA/m/kHz or dBμA/m/MHz

There are many ways to distinguish between NB and BB EMI conditions. A simple (and fail-proof) one is as follows. Given the passband (Bp) or the 3 dB bandwidth of the receiver (or victim's input stage), and F_0, the EMI source's fundamental frequency, the interference will be

$$\text{BB if } Bp > F_0$$
$$\text{NB if } Bp < F_0$$

1.4 Time-to-Frequency Conversions

Quite often, the repetitive signals causing EMI are known by their time domain waveform. Furthermore, single-pulse threats such as lightning, ESD, power line surges, and so on are always characterized by a waveform. On the other hand, EMC specifications are defined in the frequency domain, as are the performances of filters, shielding materials, and many EMC components. So, one way or another, time waveforms need to be translated into frequency, or vice versa.

Using Fourier theory, any periodic signal can be expressed in terms of a series of sine and cosine signals, at frequencies that are multiple integers of the basic repetition frequency. However, since the EMI domain covers from hertz to gigahertz, it would take a significant time to perform a rigorous Fourier analysis of the amplitudes of every harmonic from the fundamental to the thousandth or ten thousandth term.

Fortunately, simplifying assumptions can be made that allow for quick calculations with acceptable results, using the maximum envelope representation. The routine has two steps:

1. Draw the waveform and period of the time-varying signal.
2. Perform the time-to-frequency conversion, which yields the broad domain of the occupied spectrum.

This can be precious, even when based on quick, offhand calculations, in evaluating which types of fixes will work. Applying some simple coupling coefficients (see Appendix B), this will also give an idea of the required dB level of reduction.

Step 1 is relatively easy using an oscilloscope. Make sure that its bandwidth is large enough to display correctly the rise time (t_r) without appreciable distortion. Use BW ≥ $0.35/t_r$ as an absolute minimum, corresponding to the 3 dB BW, i.e., causing a small smoothing of the rise front.*

Then, outline approximately the waveform by one of the following shapes:

A. Trapezoidal Waveform

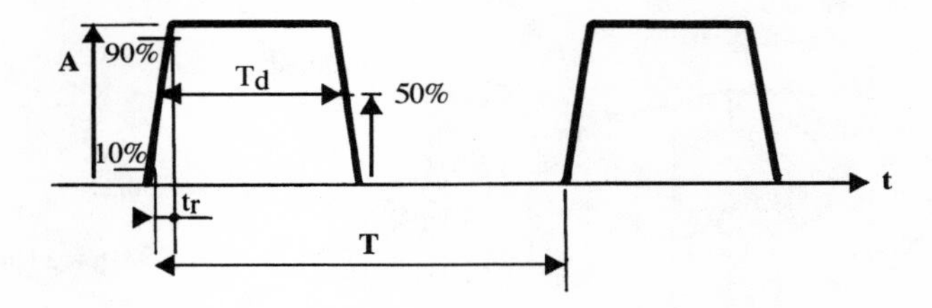

B. Triangular Waveform

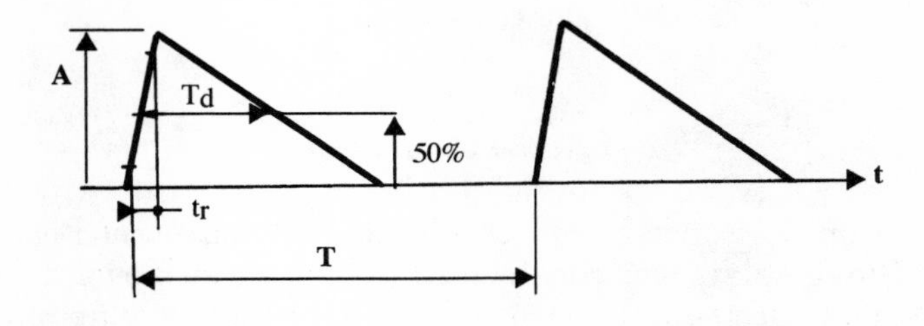

---

* *Note:* This is a kind of iterative process because, in order to determine t_r, you need to measure it, but you don't know if the instrument is adequate for the task. The way out of this vicious circle is the following:

1. Always try to use an oscilloscope with at least 100 MHz bandwidth. This will allow you to measure, with minimum distortion, any rise time t_r that is ≥ $0.35/100.10^6$, or 3.5 ns.
2. Then, when measuring an unknown signal,
   - If you read $t_r > 3.5$ ns, you can trust this value. The approximation error decreases as rise times get longer. In fact, for an actual 3.5 ns rise time, this scope will show

     $$t_r\,(\text{displayed}) = [(t_{r\,\text{scope}})^2 + (t_{r\,\text{true}})^2]^{1/2} = \sqrt{3.5^2 + 3.5^2} \approx 5 \text{ ns}$$

   - If you read 3.5 ns (the instrument cannot show less than this), you should change to a scope with a higher bandwidth (200 MHz instruments are quite common as of this writing). Do not try using the ×10 magnifier; it is a viewing magnifier, not a bandwidth expander.

Step 2 consists of drawing the envelope of the maximum harmonic amplitudes by computing:

- The first corner frequency, $1/\pi T_d$.
- The second corner frequency, $1/\pi t_r$.

Note that if $t_r \neq t_f$, only the smallest of the two is considered. For a periodic waveform, the Fourier spectrum is composed of a series of discrete frequency components, comprising the fundamental ($F_0 = 1/T$) and all integer multiples of F_0. The units of $2A \times T_d/T$, the envelope amplitude at $F = 0$, are volts, microvolts, amps, and so on.

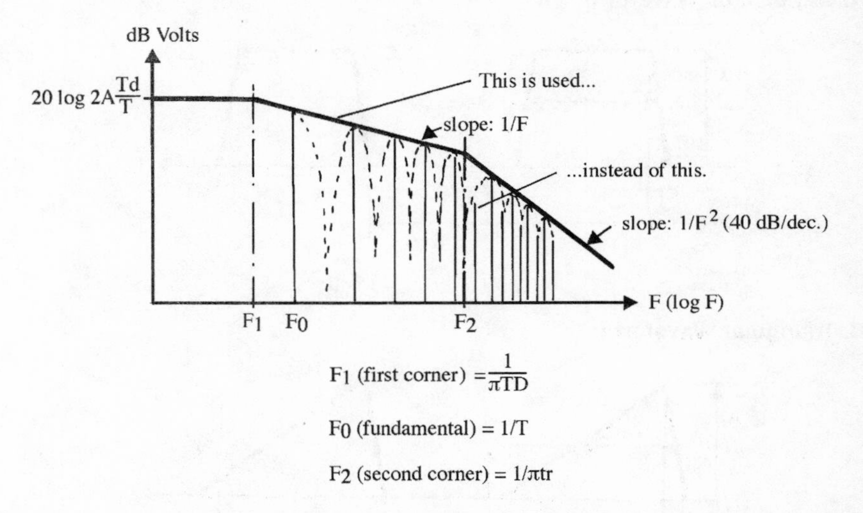

To speed up the construction profile, the nomogram of Fig. 1.1 can be used. The template was constructed to help a quick overview of the total occupied spectrum, as well as the amplitude of maximum harmonics, at any frequency.

Example of Using the Nomogram

Consider the clock signal with the characteristics of Fig. 1.2. The first thing is to identify the key frequencies for the Fourier envelope:

- Fundamental, $1/T = 20$ MHz.
- First corner frequency, $F_1 = 1/(\pi \times 25 \times 10^{-9}) = 12$ MHz. (Note that this is a construction point for the envelope, not a harmonic.)
- Second corner frequency, $1/\pi t_r = 64$ MHz. (This frequency is critical, showing that the smaller the rise time, the higher the spectrum occupancy.)

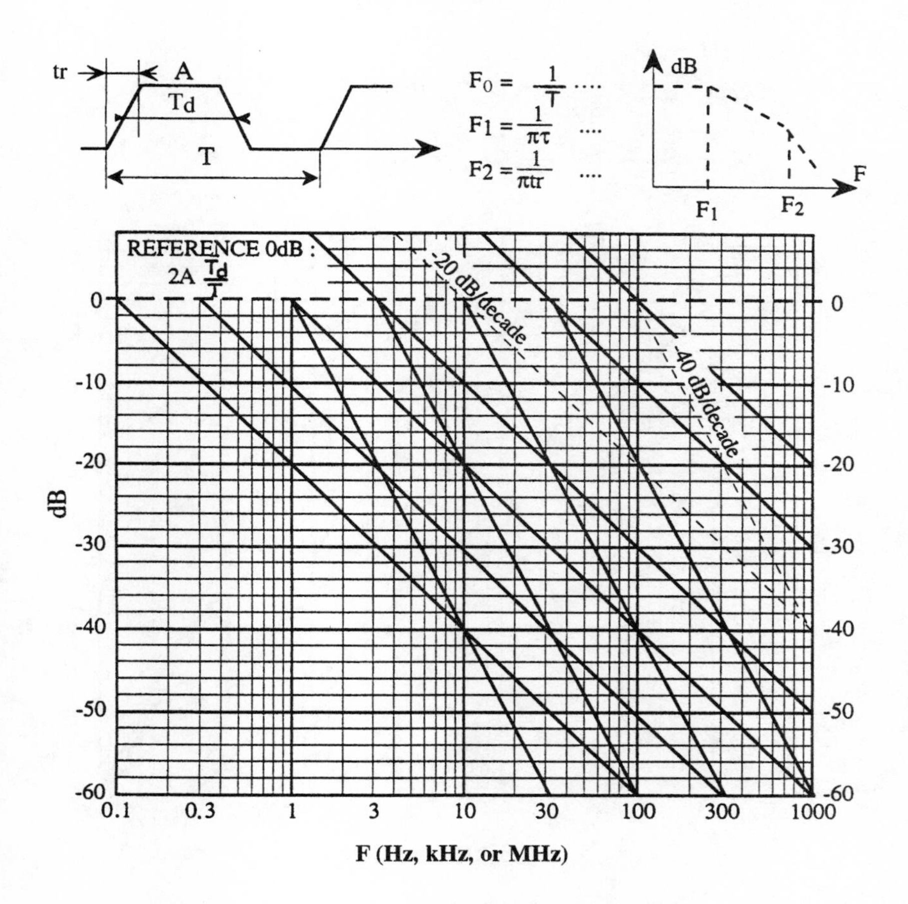

Harmonic #								
Frequency								
Ref. amplitude, dBV or dBA								
Slope decrease, dB								
Amplitude of harmonic #								

Note: If the duty cycle T_d/T is ≥ 32%, the amplitude (peak) of the fundamental is always

$$A(F_0) = A - 4\ \text{dB}$$

Figure 1.1 Nomogram for quick Fourier envelope.

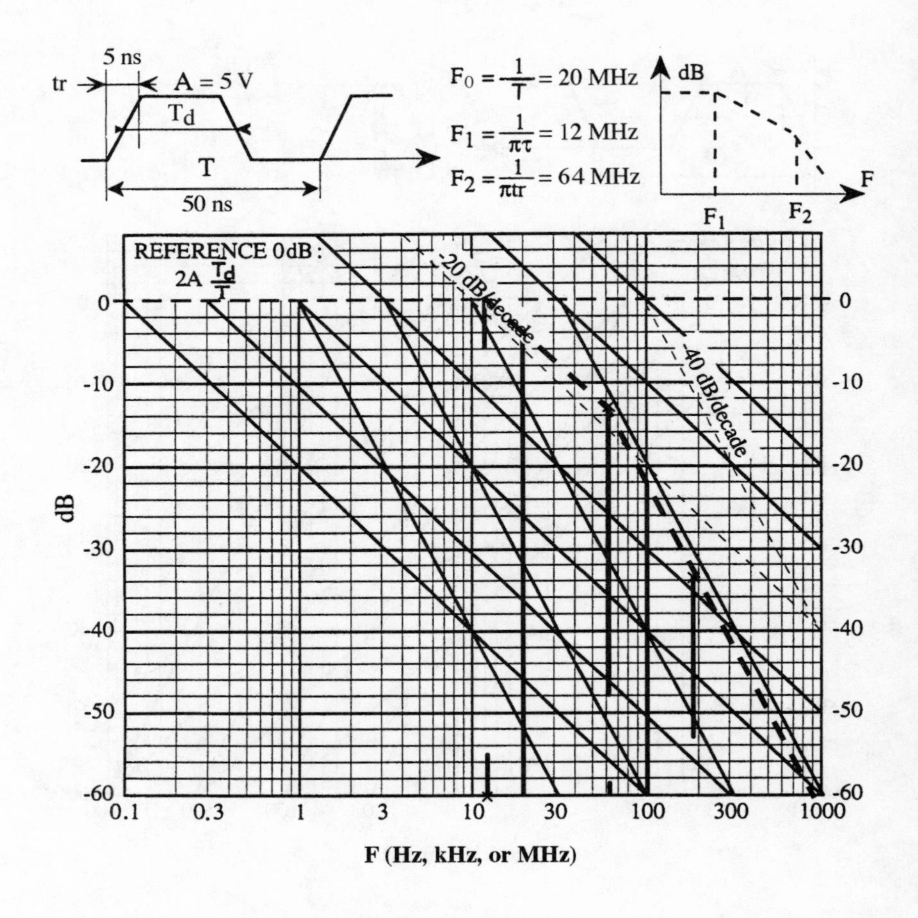

Harmonic #	F_0		3	5	7	9		
Frequency (MHz)	20		60	100		180		
Ref. amplitude, dBV or dBA	14		14	14		14		
Slope decrease, dB	–4		–14	–22		–32		
Amplitude of harmonic #	10		0	–8		–18		
Example: Harmonic #5 (100 MHz) amplitude is –8 dBV, or 0.4 V peak, corresponding to 0.28 Vrms.								

Figure 1.2 Example of quick estimate for frequency spectrum of a 20 MHz clock pulse train.

Then, the starting amplitude of the envelope (the top reference line) is calculated as follows:

$$20\log\left(2 \times A \times \frac{T_d}{T}\right) = 20\log\left(2 \times 5 \times \frac{1}{2}\right) = 14 \text{ dBV}$$

On the template, the frequency F_1 is drawn up to the top 0 dB reference line. From there, a 20 dB/decade slope (corresponding to the 1/F amplitude falloff) is drawn, using the parallel grids, until frequency F_2 (64 MHz) is reached. From this point, a 40 dB/decade slope is drawn. The envelope is now complete and represents the locus for the maximum amplitudes. The amplitude of any significant harmonic, in dBV, now can be found from the computation table below the nomogram. To obtain the amplitude in dBμV, simply add 120 dB to the dBV figure.

A few clarifications may be helpful regarding this simplified envelope process.

1. The envelope shows the locus of the maxima. For pulses train with 50 percent duty cycle, only odd harmonics exist, and they are aligned on this maximum envelope. For smaller duty cycles, only a few harmonics actually reach this maximum—the others drop over series of arches. However, this is the highest harmonic level that generally will create the maximum EMI.
2. The envelope is unipolar, not showing the phase of the harmonics. In fact, there is a 180° phase reversal at each multiple of $1/T_d$. Nevertheless, EMI receivers or spectrum analyzers are insensitive to the phase, displaying the absolute values.
3. When the signal period is relatively long (e.g., in the range of microseconds or more), frequency F_0 is low, and there are several harmonics in a megahertz interval. Two situations may occur:
   - EMC specifications, and the corresponding receivers, use typically resolution BW of at least 10 or 100 kHz. With EMI sources such as digital circuits clocks and switch-mode power supplies, the noise will appear as narrowband, since the frequency interval between any two harmonics is > BW.
   - On the other hand, broadband addition of spectral components should be considered for signals or noise spikes with low repetition rates, e.g., less than a few kilohertz. These include motor brushes, ignition noise, rectifiers noise, printer heads, and so forth. In this case, BB amplitudes are expressed in μV/kHz or μV/MHz.

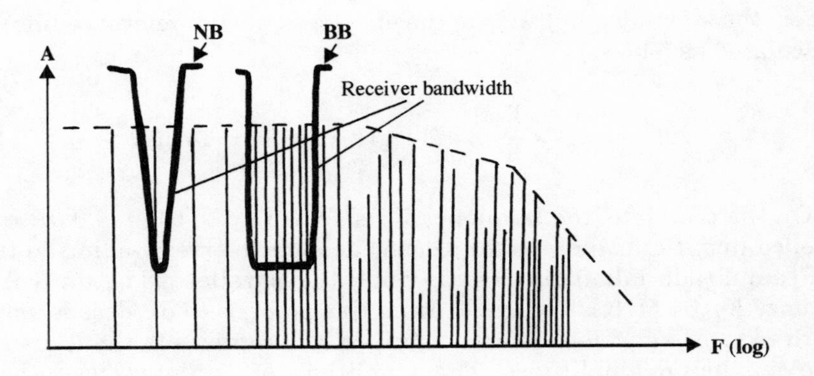

1.5 Sources, Victims, and Coupling Paths

The three elements of an EMI episode are the source, the victim, and the coupling path. Table 1.2 shows the potential ingredients.

Some types of equipment can be sources of EMI via the wanted and unwanted signals they generate. They are characterized by the amplitudes, waveforms, and frequencies of their signatures. Some equipments can be victims of EMI via the unwanted energy they receive. They are characterized by their sensitivity and bandwidth.

EMC solutions can be placed at either end of the coupling path, or in between, to make the sources less emitting or the victim less susceptible. There are engineering constraints and cost aspects in deciding whether to put the suppression burden on the source, the victim, or the coupling path. Common sense dictates that the suppression be made at or near the source, since one single culprit could jam a multitude of nearby victims. This is not always feasible or even desirable. Some sources (e.g., radio or radar transmitters, ESD, lightning, and so forth) are inevitable parts of our environment, and an equipment must be able to function in their presence. Thus, in many instances, fixes will be applied at the victim's side or over the coupling path.

The "source and victim" dichotomy is typical of problems in the field and during the development cycle of a product (self-jamming). In other cases, no EMI problem has manifested yet, but the equipment fails to meet the EMC tests: It is the test instrumentation that plays the role of a potential victim (emission test) or source (immunity test).

1.6 EMI Coupling Paths

In many cases, the actual sources or victims are either out of reach or cannot be modified, and only one possibility remains: Introduce sufficient attenuation somewhere along the coupling path.

Table 1.2 Ingredients of an EMI Situation

Sources of Conduction/Radiation	Typical Freq. Range		
	kHz	MHz	GHz
Radio transmitters			
broadcast, TV			
communications			
radar, navigation			
telemetry			
portable telephones			
antitheft and surveillance			
ECM, high-power microwave weapons			
Receivers local oscillators			
Industrial, scientific, and medical RF devices			
Computers, peripherals, digital circuits			
Switch-mode power supplies and inverters (dc/ac, dc/dc, ac/dc)			
Variable-speed drives			
Light dimmers			
Motors, generators, switches, and relays			
Solenoids, actuators			
Fluorescent and neon lights			
Arc welders			
Electrified fences			
Electric trains (pantograph/catenary)			
Engine ignition			
Natural sources:			
lightning			
ESD			
Possible Coupling Paths			
Conduction			
common impedance (ground, etc.)			
power mains			
interconnecting cable			
Radiation or induction			
antenna-to-antenna			
box leakages			
field to loop, or wire			
loop, or wire to field			
wire to wire (crosstalk)			
Receptors/Victims			
Radio receivers			
public radio and TV			
communications, mobile and portable			
radar, navigation			
Analog sensors and amplifiers			
Position detectors			
Industrial controls			
CRT monitors			
Computers and logic circuits in general			
Ammunition, ordnance, electroexplosive devices			
Human beings (biological hazard)			

Example The H-field from an ac power distribution in a building is disturbing a CRT monitor, causing a fluttering of the picture. Assuming that neither the power wiring or the CRT can be modified, the following options exist to attack EMI along the coupling path:

- Install a thick ferrous conduit or raceway around the ac wiring.
- Install a magnetic foil shield around the CRT.
- Reorient and move the CRT workstation away.

Because of these options, it is important to keep the potential EMI coupling paths constantly in mind. Often, several paths exist in parallel, and all contribute to the problem. Fig. 1.3 shows conceptually the flow of interference from source to victim via the five basic coupling paths. These are illustrated in the following paragraphs. Although this manual is not intended for calculations or math modeling, it is often useful, before or during an EMI firefight, to know at least the order of magnitude of the coupling coefficient to verify the likelihood of some such coupling mechanism. This way, likely candidates can be retained for investigation, and time will not be wasted on false routes.

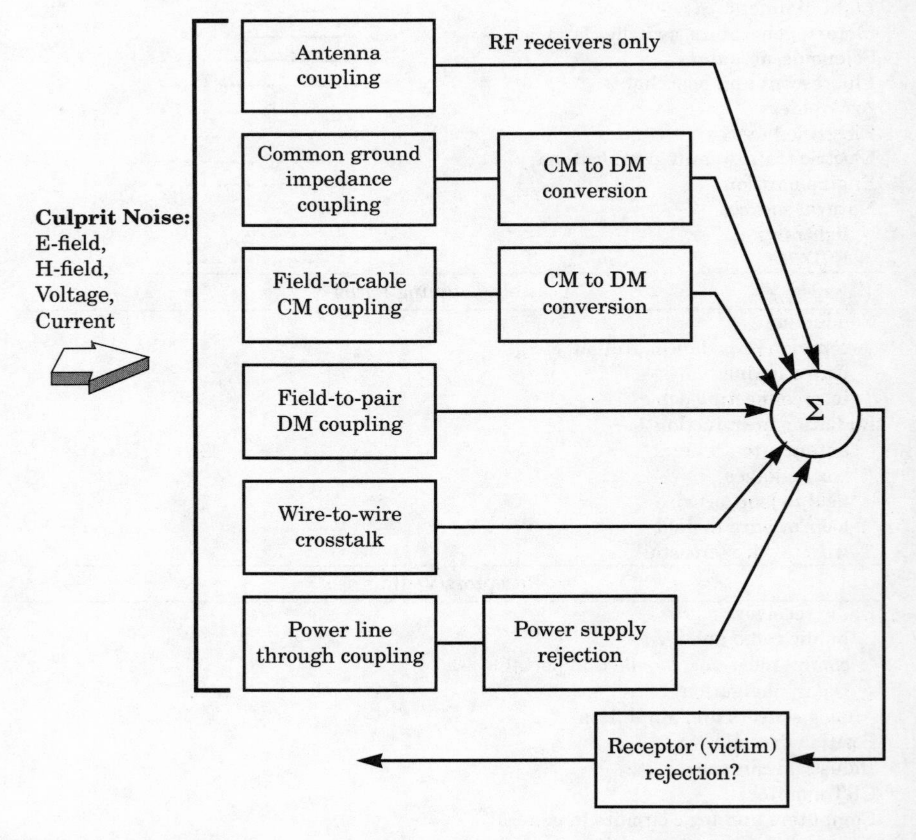

Figure 1.3 Flow diagram of source to coupling path to victim scenario. *Courtesy of EMF-EMI Control.*

1. **Common impedance coupling** is a situation in which source and victim are sharing some common conductor—generally a return path, power bus, ground structure, and so on. The I × R, or I × ωL drop in this path will affect the victim by causing a CM voltage in series in the circuit loop.

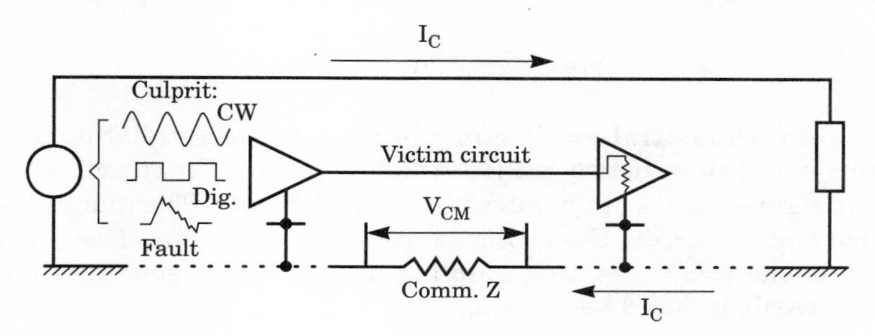

2. **Field common-mode coupling** is a mechanism by which a radiating source induces a CM voltage in the loop formed by victim circuit(s) and a common reference plane. The loop can be visibly materialized (with length L and height h) or invisible, such as a cable far above any ground surface.

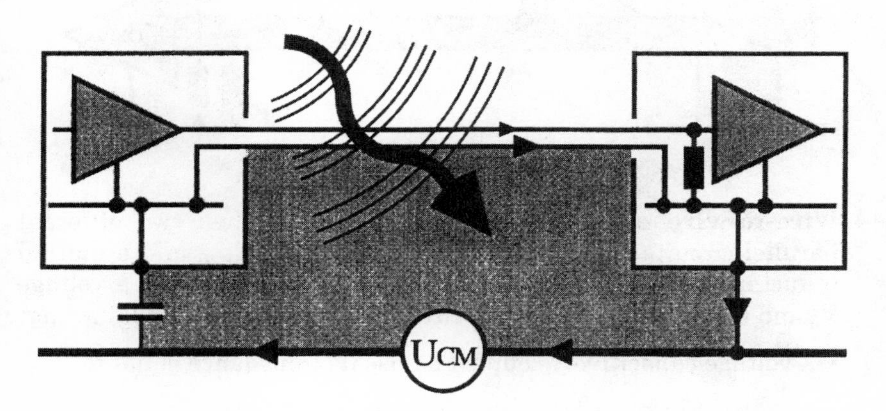

If the field is predominantly magnetic, the CM voltage appears in series in the ground loop, and its value is

$$V_{CM(volts)} = -(\Delta B/\Delta t) \times \text{area}$$

where ΔB = variation of magnetic induction in Tesla. If in air, B in the formula is replaced by $4\pi \times 10^{-7} \times H(A/m)$.

If the field is electromagnetic, known by its E-field value,

$$V_{CM}(\text{volts}) = \frac{L \times h(m^2) \times F(MHz) \times E(V/m)}{48}$$

This equation is applicable up to ≈ L(m) = 150/F(MHz). Beyond this limit, the maximum possible induced voltage is simply

$$V_{CM} = 2 \times h(m) \times E(V/m)$$

3. **Field differential-mode coupling** is the direct radiation pickup in a wire pair or between a trace and its return. The chassis or local earth reference is not involved in such a DM loop. The induced voltage appears across the victim's terminals. Due to the close proximity of the two wires, this coupling mechanism is generally minor and easily reduced by twisting.

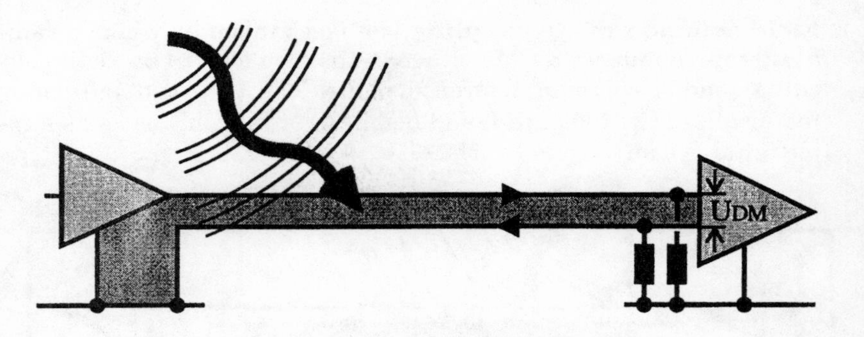

4. **Wire-to-wire coupling (crosstalk)** occurs when two different parallel circuits exhibit a mutual capacitance, C_{1-2}, and a mutual inductance, M_{1-2}. Therefore, if the culprit circuit carries a voltage V_c and a current I_c, the victim circuit will see both of the following:

   - a voltage capacitively coupled across its impedance equal to

   $$V_{cap} = R_V C_{1-2}\, \Delta V_c/\Delta t$$

   where R_V is the parallel combination of victim near-end and far-end resistances

   - a voltage magnetically coupled in series, equal to

   $$V_{mag} = M_{1-2}\, \Delta I_c/\Delta t$$

Such coupling can occur in DM and CM as well, if the culprit line is carrying CM noise.

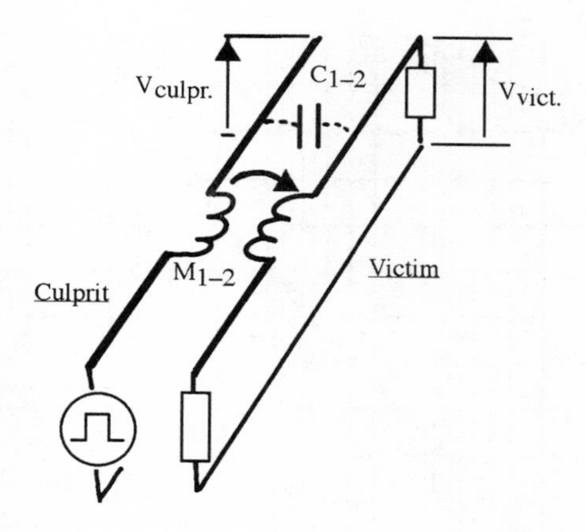

5. **Power Mains coupling** is a phenomenon by which the ac or dc line is disturbed by some EMI source and conveys this disturbance into other users.

6. **CM-to-DM conversion** is an extremely important factor in understanding and reducing EMI coupling. This is the conversion ratio, always ≤ 1, by which:

   - Regarding *susceptibility*, a CM voltage (coming from coupling no. 1 or no. 2), finally appears as DM across the victim's input.
   - Regarding *emissions*, a DM signal sent on a line creates CM currents and voltages along a ground loop via a more or less controlled ground path.

In both cases, we want this ratio to be as small as possible, because it reveals the propensity of the system to be susceptible to (or be an emitter of) EMI. Many of the fixes described in this manual are intended to reduce CM/DM conversion.

A classic illustration of mode conversion is the CM/DM factor shown in Fig. 1.4. With a 10 m link or ordinary wire pairs, if zero-volt signal references are grounded at both ends to the chassis (this one, then, earthed), the DM/CM ratio will be poor (about –6 to –10 dB up to the first cable resonance).

If, instead, the zero-volt reference is lifted from the chassis ground on at least one end (the wire pair providing the signal return), the induced CM voltage (V_i) can push no current into the cable, except through the unavoidable parasitic capacitance, C_p. The figure shows that, with a typical value for C_p in the 100 pF range, the DM/CM ratio that grades the CM immunity of the link can be as small as –80 dB at ≈100 kHz. However, as EMI frequency in-

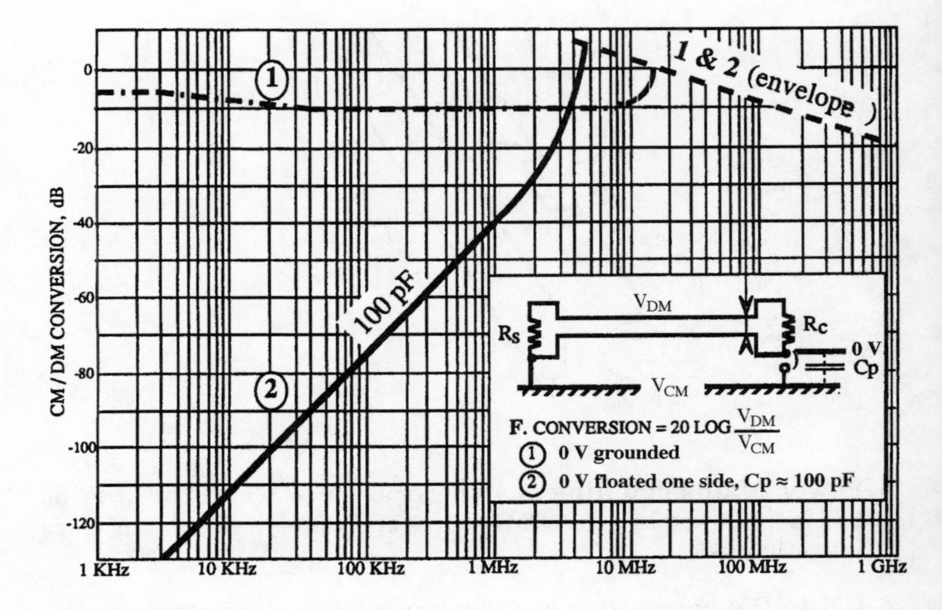

Figure 1.4 CM-DM conversion factor, example for a 10 m unbalanced link, AWG22 pair, $R_s = R_c = 100\ \Omega$.

creases, a point is reached (in the 5 to 7 MHz range for the 10 m cable shown) at which floating is no better. Beyond this peak, the coupling goes through a series of resonances and antiresonances with no clear advantage in floating. This is why, against EMI problems in the megahertz region and above, single-point grounding is often abandoned in favor of multipoint grounding.

1.7 Victim Rejection

Input stage rejection is the last opportunity for the victim circuit to be immune to incoming noise. In this respect, victim circuits can be grossly classified into two categories, based on their amplitude versus frequency:

1. **Tuned circuits,** which exhibit the characteristics of a bandpass filter. They are very sensitive in-band (typically from less than 1 µV to a few millivolts) and rather insensitive out-of-band (typical rejections of 60 to 70 dB if no front-end audio rectification occurs). Common examples include radio/TV/radar receivers, telephones, modems, and some instrumentation amplifiers.

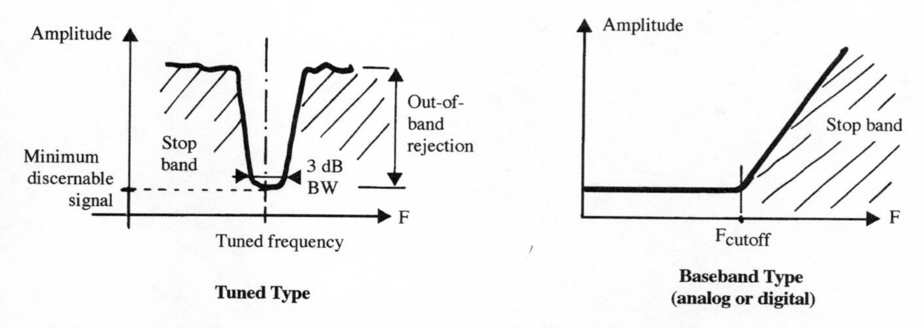

2. **Baseband circuits,** which behave as lowpass filters. They are less sensitive in-band (typically a few millivolts to several volts) but more open to unwanted signals due to their wider bandwidth. Typical examples include digital devices (Fig. 1.5), op amps, and video amplifiers.

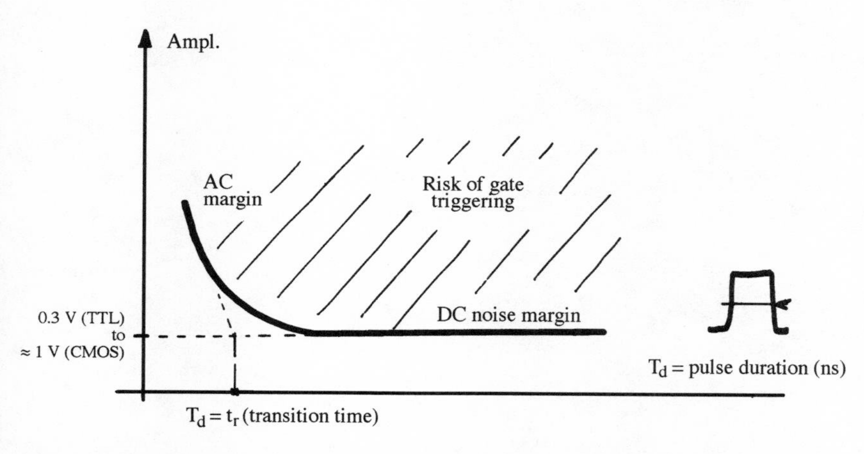

Figure 1.5 Time domain equivalent of the amplitude vs. frequency sensitivity curve, for a digital circuit.

In both cases, if the EMI frequency (or spectrum) is out-of-band, a rejection can be expected. Appendix H shows some typical curves for out-of-band rejection of most common configurations. If the out-of-band EMI level is very high, as for instance a 300 mV, 27 MHz unwanted signal arriving on the input of an op amp whose cutoff frequency is 100 kHz, the rejection will be much less than the theoretical one (which, in this case, should be 48 dB). This is due to an effect called *audio-rectification.* It typically happens when out-of-band HF amplitudes exceed ≈10–30 mV, where nonlinear components in the input, such as a base-emitter junction, will rectify some of the HF carrier.

Chapter

2

Optimal Selection of EMI Fixes

EMI suppression techniques, as explained previously, can be applied at various levels, depending on the location, availability, economics, and other factors specific to each situation. This also is strongly influenced by the time the EMI problem appears in the product life: preproduction, production, retrofit, or field situations.

In the selection guidelines provided hereafter, the potential coupling mechanisms have been sorted out in three major families.

1. **Conduction coupling.** Culprit and victim are interconnected through power, signal, or ground conductors.
2. **Radiation coupling.** Culprit couples to the victim through space propagation.
3. **Crosstalk.** There is no discrete connection between culprit and victim, but the proximity of their respective wirings or traces produces mutual capacitance and inductance).

Table 2.1 shows the appropriate location of fix, depending on the type of coupling. Table 2.2 is a more precise matrix for applying EMI fixes. Some of them require significant changes in the equipment or its installation; others require little or no additional parts but merely a relocation of components or cables.

When a product design is essentially complete, it is no longer feasible to make fundamental changes such as:

- changing from a plastic to metal box
- replacing a noisy logic family with a quieter one
- replacing a single-ended transmission link with a differential one

Table 2.1 Where to Apply the Fix?

	Principal effect of fix*		
Actual location of fix	Conduction coupling	Crosstalk	Radiation coupling
At source level			
1. Reduce generated frequency spectrum to what is strictly necessary:			
▪ Decouple/filter the source circuit output.	✔	✔	✔
▪ Select technologies with slower rise time, lesser dV/dt or lesser di/dt.	✔	✔	✔
2. Reduce loop areas in source circuit.			✔
3. Move noisy components away from box apertures and seams.			✔
4. Shield source components.			✔
5. Use transient suppressors.	✔		✔
6. Shift operating frequencies.	✔	✔	✔
Across coupling paths			
1. Use HF filtering on I/O cables (of either source or victim) along their path, preferably at box or system frontier.	✔	✔	✔
2. Use CM reduction techniques:			
▪ balanced transmission	✔	✔	✔
▪ floating	✔	✔	✔
▪ isolation transformers	✔	✔	✔
▪ ferrites	✔	✔	✔
▪ optoelectronics, etc.	✔	✔	✔
3. Use twisted and/or shielded cables and shielded connectors.		✔	✔
4. Metallic raceway or companion cable or braid		✔	✔
5. Reduce common ground (zero-volt, chassis, or earthing) impedance.	✔		
6. Reduce loop size of interconnect cables.		✔	✔
7. Improve shielding of source or victim equipment cabinet.			✔
8. Use shielded room at source or victim site.			✔
9. Separate cables into families.		✔	
At victim level			
1. Reduce bandwidth to what is strictly necessary.	✔	✔	✔
2. Decouple/filter input ports.	✔	✔	✔
3. Decrease input impedance.	✔	✔	✔
4. Reduce loop area in victim circuits (including use of multilayer boards).			✔
5. Shield victim components.			✔
6. Use transient suppressors.	✔		✔
7. Shift operating frequency.	✔	✔	✔

* Called "principal" because no EMI is exclusively conducted or radiated.
Source: courtesy of EMF-EMI Control

Table 2.2 Fixes Application Matrix

Fix	Class	Coupling paths: Common-ground impedance	Power line to box, box to power line	Field to cable, cable to field (CM)	Field to pair (loop), pair (loop) to field (DM)	Crosstalk	Field to box, box to field	Field to room
Floor impedance reduction	A✧	✔						
Grounding straps and spacers		✔						
Transient plate	A✧	✔		✔				
Power line filtering, CM, DM	C✧		✔					
Signal line filtering	C✧			✔	✔	✔	✔	
Ferrites	A✧		✔	✔		✔	✔	
Isolation transformer, power	D✦		✔					
Line conditioner/UPS	D✦		✔					
Transient protector (surges only)			✔	✔	✔			
Isolation transformer, signal	D✦	✔		✔				
CM transformer (balun)	B✧	✔		✔				
Metallic raceway, companion braid	B✧	✔		✔		✔		
Aluminum foil	B✧					✔	✔	
Copper tape		✔		✔		✔	✔	
Shielding mesh/band, zipper	A✧			✔	✔	✔		
Twisted wires	B✦				✔	✔		
Shielded cables and hardware	C-D✧			✔	✔	✔		

Classes:
✧ = efficiency is null or mediocre at LF, improves as frequency increases
✦ = best efficiency is at LF, degrades as frequency increases

Expected Attenuation Range
Class A, 0–20 dB
Class B, 20–40 dB
Class C, 40–60 dB
Class D, >60 dB

Table 2.2 Fixes Application Matrix ***(Continued)***

		Coupling paths						
Fix	Class	Common-ground impedance	Power line to box, box to power line	Field to cable, cable to field (CM)	Field to pair (loop), pair (loop) to field (DM)	Crosstalk	Field to box, box to field	Field to room
EMC gaskets	C-D✧						✔	
Conductive paint and fabric	C✧						✔	✔

Classes:
✧ = efficiency is null or mediocre at LF, improves as frequency increases
✦ = best efficiency is at LF, degrades as frequency increases

Expected Attenuation Range
Class A, 0–20 dB
Class B, 20–40 dB
Class C, 40–60 dB
Class D, >60 dB

Instead, you have to deal with what exists and try to improve it. Sometimes EMC resources are there, but not fully employed. Often, they are nonexistent. For instance, returning to the previous list as an example:

- **Box level.** If plastic, the enclosure can be metallized using copper or nickel paint spray. If metallic but leaky, seams, cable entries, receptacles, etc., can be made RF tight by gasketing.
- **Noisy logic.** Overly sharp transitions can be smoothed with RC or ferrite filtering.
- **Transmission links.** Single-ended types can be turned into differential without changing the driver/receiver pairs by adding balancing transformers and so forth.

Options for fixes are even more limited when the EMI problems occur in the field, which often precludes any serious change inside the product. Only "downstream" changes are feasible, such as external filters, cable protection, or site modifications.

In Chapters 4 through 12, which describe various practical fixes, these two basic situations—final product development or field operation—have been addressed. For easy retrieval and quick interpretation, each solution is depicted at the end of its relevant chapter or section by a sort of "identity card" format. This is written with minimal text and organized in a format similar to the customer information printed on medicine packages, as follows:

Indications

When should one use this fix? It is used to cure what? It is employed following which symptoms? This is an "or" list. Each indication is a valid case for using this fix.

Prescriptions

How do we use this fix? How are the components arranged, and of what value are they? How should they be installed?

Contraindications/Warnings

When is this fix useless? What are its limitations? What are the possible side effects?

When trying a fix on a given problem, remember that EMI often uses several coupling paths or entry/escape ports in parallel. As an example, assume that, from the same source, two coupling paths, A and B (e.g., ground noise and crosstalk) are causing, respectively, a 10 dB and 6 dB contribution to a total interference. Applying a 10 dB reduction to mechanism A will not decrease the problem by 10 dB, but by ≈4 dB. If, instead, one applies a 10 dB fix to path B, because the change on A was disappointing, the global improvement will be only 3 dB. Only when both coupling paths are treated will the overall interference be reduced sufficiently.

For each family of fixes, a list is provided that shows some commercially available components, with manufacturers' names and references. This does not imply a specific preference, but simply that these are sources with which the author is familiar.

Chapter

3

Diagnostics, Troubleshooting Techniques, and Instrumentation

3.1 EMI Problems during Prototype and Qualification Phases

Assuming that a certain degree of EMC engineering has been incorporated into the original design of an equipment, it is strongly recommended to validate and optimize these engineering choices.

During prototype or qualification phases, you can be confronted with the following problems to solve.

3.1.1 Real EMI Problems

- *Self-jamming.* The product under development does not perform correctly, because of internal EMI problems (intrasystem EMI).
- *Lab environment.* If the system is not installed (and it seldom is) in a shielded room during normal, non-EMC, activities, some noise sources in the lab or plant, or their vicinity, could disturb the prototype's operation.

The approach for such situations is basically the same as the one described in Sec. 3.4, for on-site field problems. There is one major advantage, however: The product is still somewhat "flexible" for some internal modifications (PCB, packaging, and so on), and the local ambient sources can be isolated more readily.

3.1.2 Specification Compliance Problems

- Excessive emission levels (conducted or radiated)
- Poor immunity levels (conducted or radiated)

In this case, no real EMI problem exists yet, and may never exist. But specifications are made to guarantee system EMC in most conceivable site configurations, with a high degree of confidence. Noncompliance problems can appear at several stages (the earlier, the better):

1. Engineering EMC evaluation on the lab bench, using simplified diagnostic methods and instrumentation, usually performed on subassemblies or early prototypes, sometimes without the convenience of a shielded room
2. Prequalification EMC test performed on preproduction samples
3. Formal EMC test

The degrees of freedom for making cost-effective changes are reduced as the situation evolves from stage 1 to 3. The next sections will describe proper methods for coping with any one of those situations.

3.1.3 Optimizing the Sequence of Development/Prequalification Tests

At this stage, you are exploring and validating the actual EMC behavior of an equipment that has never been formally tested. Therefore, it is important to use your time and resources efficiently, on a prototype that is probably in high demand for all the other items of a typical (and usually time-squeezed) development program. The following suggestions should be considered (Fig. 3.1).

A. Emissions tests can be performed more quickly. It is not necessary to watch for possible malfunctions, as will be the case for susceptibility. The equipment just needs to be turned on and run. The improvements you will eventually apply as a result of these tests are practically always beneficial to immunity.

1. Among emissions tests, conducted emissions are the fastest and easiest ones, because
   - The test setup is relatively easy to install.
   - It is rather easy to track culprit sources.

   This includes conducted emissions (CE) on power cables from 150 kHz up to 30–50 MHz, and eventually down to 10 kHz using a line impedance stabilization network (LISN) or current probes.
2. Next come radiated emissions (RE) tests via:
   - Conducted substitution method, using clamp-on probes
   - Near-field probes

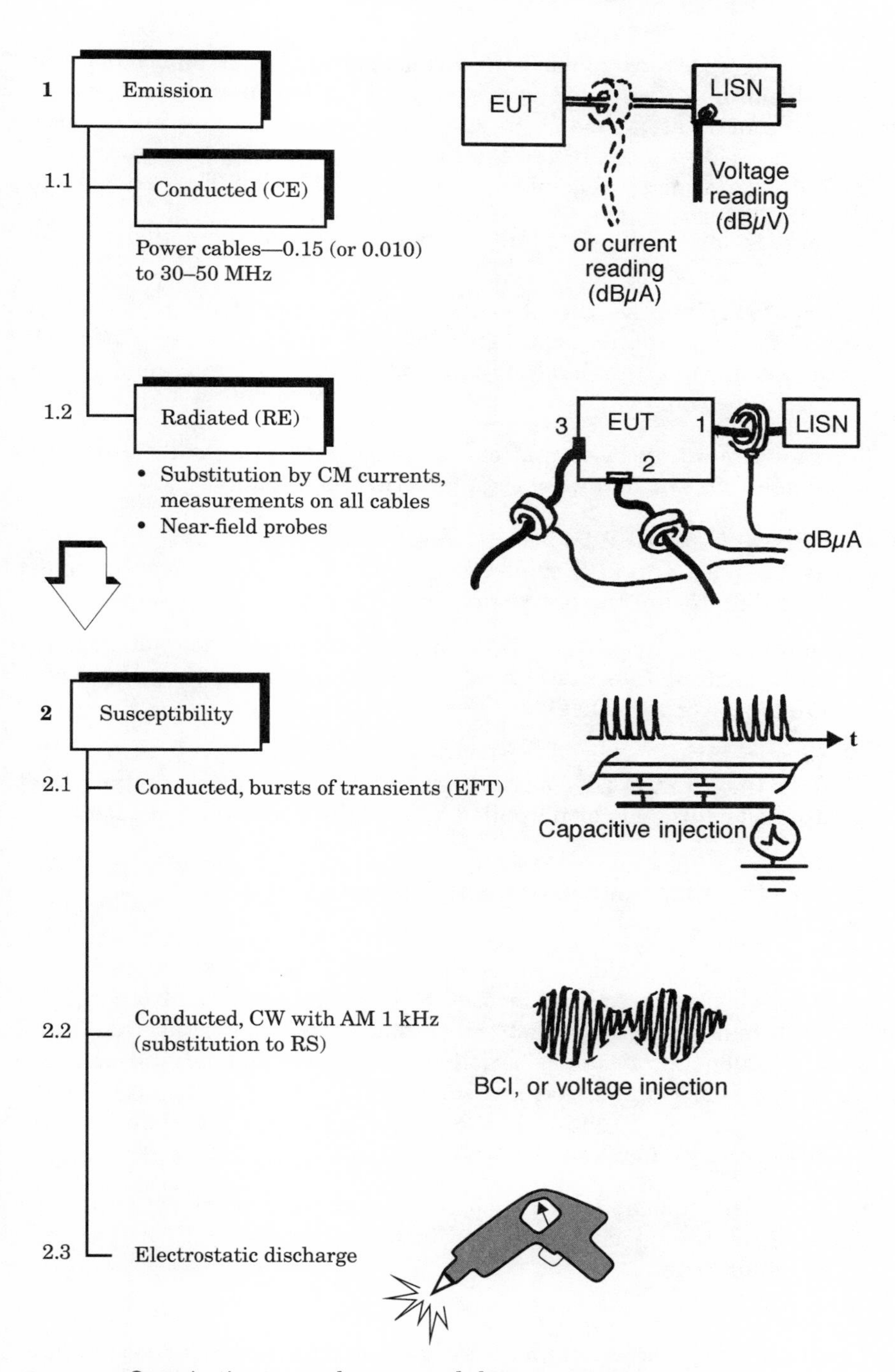

Figure 3.1 Organization tree and recommended test sequence.

B. Immunity tests rank third in performance ease. As for emissions, conducted immunity tests are easier to perform, because they do not require the mandatory use of an anechoic room. In fact, in many cases, immunity tests via conducted methods, using CW or pulse stimuli, are the only ones that are feasible outside a real EMC lab. They allow us to:

- Track, isolate, and eventually harden the victim components or circuits.
- Validate or improve the filtering, shielding, and other solutions applied to the coupling paths. This includes also *software solutions* to mask transient problems: errors detection + self-recovery, watchdogs, and so on.

Of the immunity tests that can be performed practically in the development lab, the following sequence is recommended:

1. Bursts of electrical fast transients, or EFT, per IEC 801-4 (or 1000-4.4) on power and I/O cables. Certainly, this is the easiest and most revealing test of the entire program.
2. Immunity to RF, CW fields, by a substitution method such as BCI (bulk current injection) or voltage injection using coupling/decoupling networks (CDNs).
3. Electrostatic discharge (ESD) testing. This is relatively easy to perform (next to the EFT in that respect) but can be destructive if not applied cautiously or if applied to a totally nonhardened product.*

3.2 Checking for Compliance with Emissions Specifications

Because military (and, later, civilian) EMC specifications have been progressively enforced, this facet of EMC compliance has been a growing nightmare, due largely to the increasing speed of digital circuits and clock frequencies. It takes serious precautions to hold the spurious emissions below limits that are set in hundreds of microvolts (for the conducted part) and 10 to 100 µV/m limits at 3 m (for the radiated part).

This investigation is to be carried in two circumstances:

- During the development phase, to check in advance that the product will meet the specifications limits for conducted (CE) or radiated (RE) emissions

* Testing for lightning surge or other energetic line transients is not recommended during development testing because of the possible damage to the EUT.

- After an unsuccessful EMC qualification test, to analyze the reasons for the excessive emissions and to optimize the solution while the prototype is in the engineering premises

3.2.1 Minimum Requirements for the Test Site

If the EUT is not installed in a Faraday cage, as it is often the case during prototype activities, it at least should be in a location with the following characteristics:

1. It should possess *a quiet RF ambient,* not in close proximity (3 m minimum separation) to powerful noise sources such as fluorescent tubes, air conditioning compressors, elevators, power converters, welding equipment, or machine tools. Upper floor locations, near windows, should be avoided; a preferred zone would be a ground level or basement room, recessed from the building façade.

2. It should be provided with *noise-free power mains,* with the EUT and instrumentation preferably being on a clean ac branch that does not feed other potential noise sources. Since this is not always easy to guarantee, a good precaution is to install an isolation transformer plus an EMI filter at the room power distribution panel. This will isolate, from an RF standpoint, the EUT and test setup from the conducted EMI ambient. In addition, the isolation transformer is practically mandatory to prevent the ground fault detectors from being triggered by the LISN (artificial mains) leakage current. If the EUT is a dc-powered equipment (as is commonly the case with vehicle, airborne, and aerospace electronics), it is recommended to use a set of 12 or 24 V batteries to ensure perfect isolation from power mains ambient noise.

3. It should include a *test ground plane.* Depending on its size, the EUT may be standing on a workbench, as is often the case with prototypes, or floor standing. In either configuration, it is absolutely necessary that the entire EUT, associated cables, and EMC instrumentation be resting above a ground plane (in fact, a conductive reference plane) extending sufficiently beyond the perimeter of the whole setup. This is to replace the noisy, undefined, unreachable earth reference of the ac wall outlet. (How does one get, without significant inductance, connected to the real Mother Earth?)

   This plane can be any solid metal sheet such as aluminum, copper, or galvanized steel. The thickness is not important, so whatever can be grabbed from the local model shop, factory stock, or nearby hardware store is adequate. By default, heavy-duty household aluminum foil (Reynolds® wrap, for instance) can be used, with a double layer recommended for better resistance to tearing and scratching.

The metal plane will be the local ground reference for all the test gear:

- Chassis ground for the spectrum analyzer, oscilloscope, etc.
- Chassis ground for the LISN and any accessory that needs to be equipotential

It will also allow for stable, well-defined EUT cables heights above ground, necessary for good test repeatability. For safety reasons (not so much for EMC considerations), the ground plane should be connected to the local ac earth ground.

3.2.2 Instrumentation

The minimum instrumentation requirements for conducted and radiated emissions analysis are as follows:

1. *Spectrum analyzer* (or *EMI receiver),* 10 kHz to ≥1 GHz with sufficient sensitivity (e.g., –100 dBm at 10 kHz BW) and selectable IF bandwidths (10 kHz, 100 kHz, preferably with an "N" connector input). If EUT with clock speeds above 100 MHz are expected, a 1.5 to 2.7 GHz capability is needed.

2. *Low-noise, wideband preamplifier.* This is to make up for the mediocre sensitivity of most spectrum analyzers, which does not allow us to check against certain stringent emission limits. Minimum characteristics are:

   - Gain ≥ 14 dB, flat within ±1 dB, up to at least 300 MHz
   - Noise figure ≤ 4 dB

   A "N"-style connector input is also desirable.

3. *EMI current probe,* calibrated up to 300 MHz.

4. *Line impedance stabilization network* (LISN), also called an *artificial mains network.* It is typically the 50 μH/50 Ω type, which is the most widely prescribed in modern EMC specifications (see Figs. 3.2 and 3.3). If the type of products to be tested are always single-phase ac, a two-section LISN is adequate. If the EUTs can come in a wide variety of supply types (single-phase, three-phase, dc, etc.), a multiphase LISN is preferable. (Remember the possible need for an isolation transformer, as mentioned in Sec. 3.2.1.)

5. *Recorder.* This is not absolutely mandatory, but it is very convenient to compare the results of various iterations. Many modern analyzers incorporate an IEEE bus or RS232 output, allowing di-

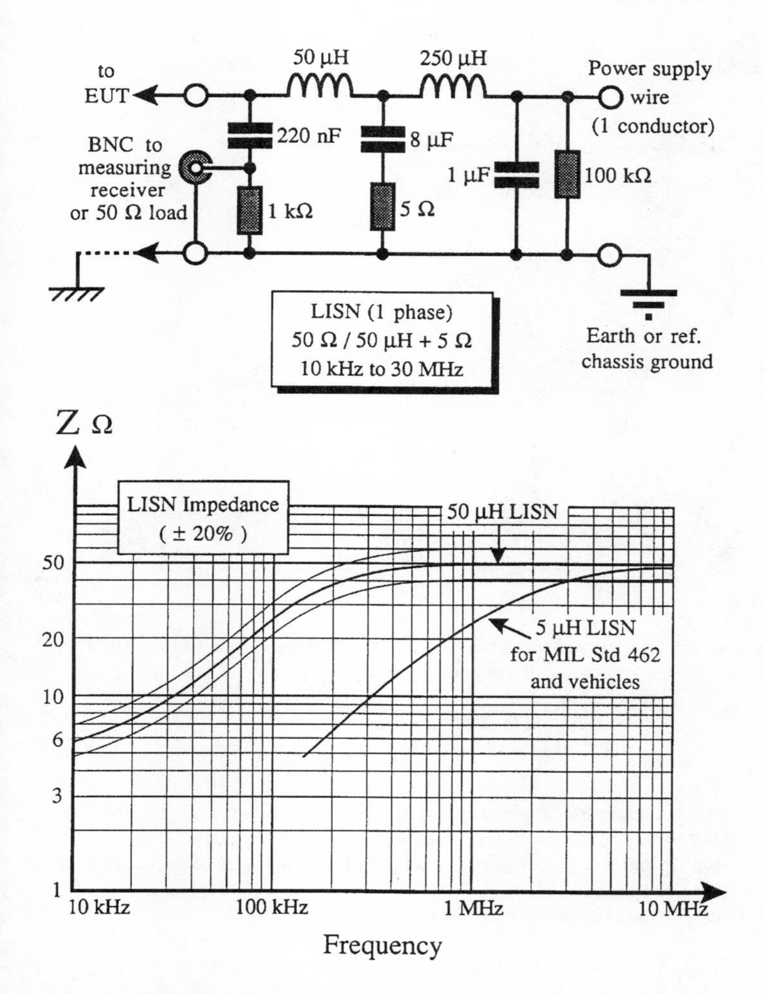

Figure 3.2 Line impedance stabilization network.

rect storage and printout of the screen display. Older, or more primitive, equipment has an X-Y output that can be wired to a strip-chart pen recorder.

6. *Good quality coaxial cables.* Although this may sound like a trivial detail, we have seen many occasions where a significant amount of time and effort was wasted in chasing odd results, nonreproducible data, and other glitches caused by mediocre quality, worn-out cables and coaxial connector gear. Not all coaxial cables are created equal. The good quality, dense coverage of the braid and the perfect, leakproof, circumferential contact of the connector backshells and mating parts are paramount to dependable test results. A

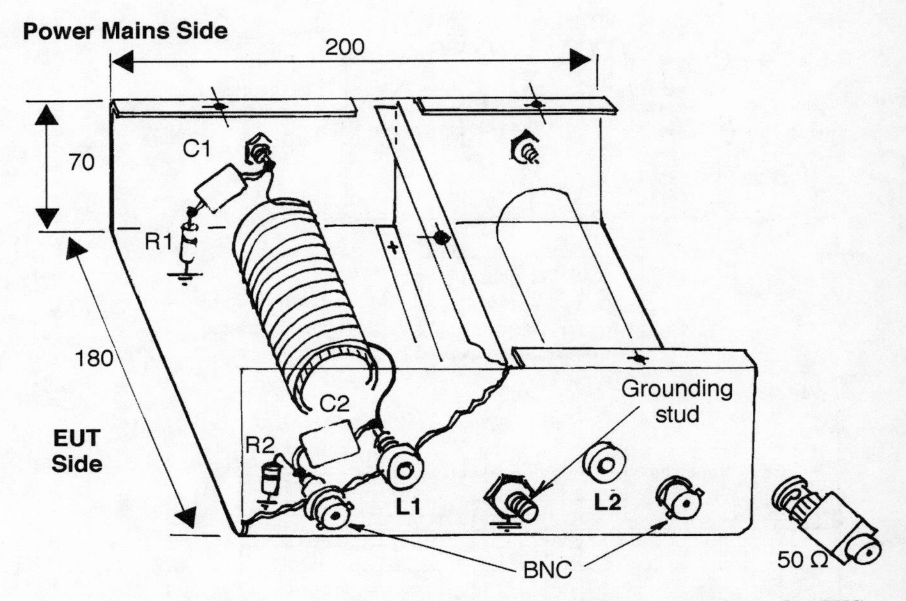

Inductors (2), 50 μH:	56 turns of enamel AWG #16 or #18 wire on a 40 mm dia. PVC or phenolic mandrel.
C1, C2 (2):	0.22 μF, 500 V, ±5%
R1 (2):	5 Ω, 10%, 1/2 W carbon
R2 (2):	1 kΩ, 10%, 1/4 W carbon

- Leads of C2 to BNC and to L1 socket must be *very* short.
- The vertical wall is necessary to prevent lateral HF coupling between L1 and L2.
- Banana sockets for EUT cable can be replaced by a regular ac outlet.
- Dimensions, in millimeters, are illustrative only. The box is made of galvanized or zinc-plated steel, 15/10 mm thick.
- Coaxial 50 Ω load to be put on the unused BNC port.

Figure 3.3 Homemade LISN, for investigations only, usable in the 0.1–50 MHz range. It can also be used for EMI injection when testing the susceptibility of ac input. The RF port can handle 30 Vrms (CW) or 500 V peak (short pulses < 50 μs). The low value of C1 allows use of the LISN without an isolation transformer (no triggering of ground-fault detectors).

brand-new set of RG58 coaxial cables and BNC connectors can do a fair job. An old set of dubious-origin coaxial cables, fitted with BNC connectors showing respectable mileage, can ruin a series of otherwise correct test runs.

Whenever possible, we recommend at least the use of recent, well-maintained 50 Ω/RG58 cables (check the markings on the cable jacket) and BNC connectors. A much preferable choice is to use double-braid coax (RG214 or RG55) with N connectors, although this makes your cable gear more rigid and heavy. There are three reasons for this:

- Double-braid coax cables have a better immunity (typically 30 dB better shielding factor) to ambient noise than single-braid ones. This will prevent your measurements from being corrupted by non-EUT emissions.
- Double-braid coax has lower HF loss. On short runs, and below 100 MHz, this is not critical. Beyond 3 or 10 m of cable, and for frequencies above 300 MHz, the few decibels that the double-braid coax will save in transmission loss can be very precious in a domain where sufficient dynamic range is hard to achieve.
- "N" connectors, because of their threaded fittings, have superior shielding effectiveness and are less prone to side traction abuse than the bayonet locking of BNC. An excellent substitute, although less widely used, is the TNC (threaded version of BNC), which has all the advantages of the "N" style but the same small size and weight as the BNC.

A trade-off between the single braid and the heavy, bulky double braid is the ferrite-coated RG58, especially developed for EMC measurements (see Sec. 6.3).

7. *Antennas.* This is a complex issue when it comes to troubleshooting or investigation prior to (or instead of) actual radiated emissions EMC testing. If the EUT is not in a shielded anechoic room or in an open test site with sufficient free area, the use of regular EMC antennas should be discouraged. Many factors in the surroundings (e.g., proximity to conductive objects, insufficient clearance from the ceiling, and so on) make field measurements in nonstandard conditions extremely questionable and not even repeatable. This is why, in the debugging and analysis techniques described hereafter, we try to stay away from standard antennas. Substitution methods such as cable current or near-field proximity probes are recommended instead.

3.2.3 EUT/Prototype Setup

A typical, temporary location setup is shown Fig. 3.4. All test instruments and accessories are grounded to the reference plane using short straps. It may be convenient, in advance, to install few soldered flat-head screws or soldered nuts for these grounding points, which must be sufficiently far (≥15 cm) from the edges. Conductive copper tape can be used instead, with a large contact area. Notice the location of the EUT and instrumentation cables—not too close from the edges. The EUT is grounded via its normal power cord safety conductor, unless it is a normal practice for its installation (e.g., in vehicle or aerospace applications) to have a direct ground connection.

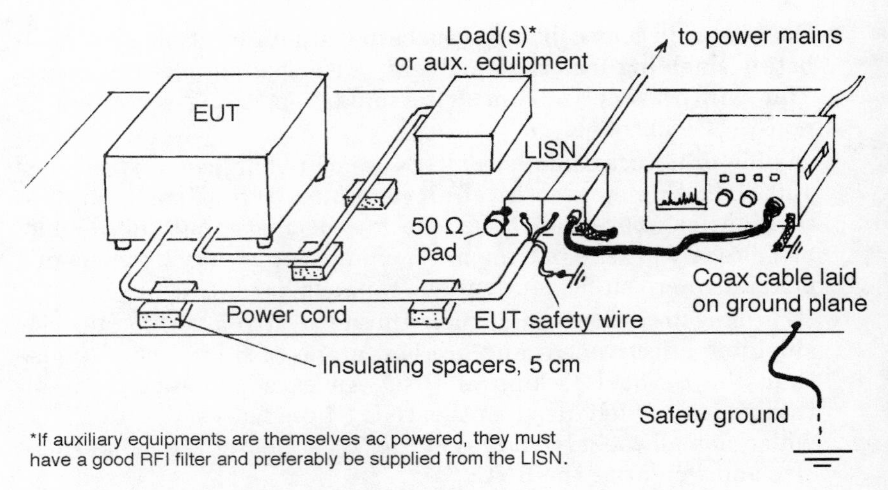

Figure 3.4 Test setup for conducted emissions test.

3.2.4 Conducted Emissions (CE) Compliance

Test Preparation

1. Familiarize yourself with, and keep a record of, the EUT characteristics relevant to its potential EMI signatures. For instance,
   - Are there any switch-mode power supplies?
   - What are the switching frequencies?
   - Note the voltage settings. Many EUT power supplies accept several ac input voltage settings such as 120 and 240 V. Remember, you will need to use both and retain the worst envelope of the two EMI spectra. The low-voltage setting means twice as much primary current, hence more DM noise. The higher setting means the highest dV/dt, hence more CM noise.
   - Are there any broadband sources (e.g., thyristor-controlled 50/60 Hz chopper, brush-type motors, gas-discharge tubes, etc.)?
   - What are the lowest and highest clock frequencies used by the logic? Usually, many of the Fourier series terms (especially the fundamental of clocks and odd harmonics) are detected on power and I/O leads, generally above 10 MHz.

   All the concepts above are useful to prepare a quick list of potential HF sources. To some extent, an intelligent test program will try to anticipate which type of *repetitive* or *random* noise could be present on the EUT cables. This will facilitate the identification of the BB versus NB nature of some measured spectra.

Consider, for instance, the nominal bandwidth for CE measurements from 150 kHz to 30 MHz being 9 kHz (or the closest value on a spectrum analyzer, i.e., 10 kHz). If the EUT contains a power supply with a 30 kHz switching rate, you do not have to worry about BB spectrum and quasipeak reading. All your readings will be NB, corresponding to dBμV rms. On the other hand, using a 100 kHz (or 300 kHz) bandwidth for the same test would give pessimistic results, which *should not be corrected* by using a bandwidth correction factor of –20 log(100 kHz/9 kHz); this is because, the source repetition rate being 30 kHz, the only acceptable correction factor should be –20 log(100 kHz/30 kHz), i.e., –10 dB.

2. Anticipate and prepare a few (ideally, a single) operating modes for the EUT that will exercise the maximum of its internal functions, such as retrieve stored data, interrogate peripheral equipments, print or display, etc.

3. Since you are probably not in a Faraday cage, check out the ambient noise. Although the LISN acts as a filter vs. conducted RF ambient on power mains, the portion of the cabling that is on the EUT side, plus all other cables of the test set, can act as pickup antennas for broadcast and other stable ambient RF fields. If the EUT operates with auxiliary equipments (simulators, active loads, peripherals, etc.), make sure that:
   - These ancillary devices, if supplied from their individual power cable, do comply with an EMC specification level at least as severe as the one you are looking for, or
   - They are equipped with an efficient line filter (even as a temporary add-on).

   The dressing of all cables at 5 cm (2 in) above a ground plane, even for nonmilitary equipments, will limit their ability to pick up RF ambients as long as this height is smaller than a tenth of the wavelength. For instance, compared to a cable hanging loose at 1 m or more, the proximity of a ground plane underneath will reduce by 30 dB the possible pickup of a 10 MHz short wave ambient.

   The final check for ambient background is to run several sweeps of the CE test with the sensors (LISN or probe) in place, EUT switched off, and the analyzer or receiver in "peak-hold" mode. The background level must be at least 6 dB below the CE specification. If this criterion is not met, the CE measurement is, in theory, not feasible. However if the criterion is only violated at a few specific frequencies, a few more steps are worth trying:
   - Record the few frequencies where the maximum background criterion is exceeded.

- If violations are NB, check whether they may be caused by a local culprit within or near your test premises. If this is so, try to switch off this spurious source or to reduce its contribution to your RF ambient. Reorient the culprit equipment, relocate its cables, or equip the cables with large ferrite toroids (two or three turns) near the source (see Chap. 6).
- If violations are BB, try all the above, but if it fails, retain the option of switching to a smaller bandwidth, such as 1 or 3 kHz, when you check the EUT at this specific frequency step. If your EUT signature is NB, its level will not change, while the BB background will decrease by 10 to 20 dB (depending on whether it is random or stable repetition rate).
- If the strong ambient is caused by local radio stations, there is not much you can do. Such ambient levels may easily exceed 60 dBµV as seen at the LISN port, and you cannot, of course, filter the EUT cables. The only chance left is that the EUT emissions be not co-channel with these ambient signatures, which is easily verified, since you will have a record of them.

A word of caution: this check of the background is necessary as opposed to relying on the EUT *on* and *off* check while testing. This is because, in certain cases, an EMI ambient picked up in the test area, appears to be *carried* and even amplified by the EUT wiring. Turning the EUT off and noticing a decrease in dBµV readout could make us believe that this is EUT noise.

Test Routine. Since dynamic range is seldom a problem with CE testing, leave at least 10 dB of attenuation at receiver input. If available, insert a surge limiter, which often incorporates a 6 or 10 dB attenuator. Verify that the spectrum analyzer is not saturated. A prudent habit is to check, especially with very noisy EUTs, that a 10 dB attenuation added on the input settings brings exactly a 10 dB decrease in readout.

Sweep the prescribed frequency range, using the "peak, max-hold" mode if available, overlaying 5 or 10 sweeps.

Repeat for each one of the power input leads, making sure that the LISN ports on the untested leads are fitted with a 50 Ω coaxial load.

Interpreting the Test, Diagnosis, and Fix

1. *Where specification violations appear* (Δ dB), return to these specific frequencies and try identifying the culprit source(s) by turning off, or removing selectively, the following:
   - Each one of the switch-mode regulators, dc-dc converters, etc., if there are several

- processor cards with fast logics
- motors (brush-type)
- repetitive spark sources or discharge tubes
- thyristors

2. *NB or BB?* Most specs allow for some relaxation when EMI is BB (low repetition rate, compared to the receiver bandwidth). Our deliberate choice of a peak detection was made to speed up the measurement, but it may penalize BB emissions. A second chance exists if the violation is due to BB noise. Focusing on the only frequency of concern, switch to the quasipeak (QP) detection mode, or vary your BW to determine the NB or BB nature. If BB, apply the proper relaxation factor or BW correction.

3. *CM or DM?* This is important to know for selecting the appropriate fixes. DM emission (i.e., phase to neutral, or + to – wires) typically dominates below few hundred kilohertz, as it is generally caused by the first harmonics of the switch-mode regulator current through the front-end capacitor ESR and ESL. It *increases* with full load or when the primary input voltage is set at minimum (Fig. 3.5).

   This can be emphasized by one of the methods of Fig. 3.6, using a current probe or a special DM–CM splitter made by AEMC Corp. (See the manufacturer references later in this chapter.) CM emission (any wire-to-ground plane) typically dominates above a few hundred kilohertz. It usually increases when the input voltage is at maximum and keeps the same value at minimum load.

4. Is the CM current, for out-of-spec frequencies, also leaking on I/O signal cables? This reveals a lack of CM decoupling on the secondary side of the EUT power supply, and a flow of CM current across zero-volt references and structures.

   Caution: *Never turn off the EUT while the spectrum analyzer is connected to the LISN. Unless you have inserted a surge limiter in the RF input, the overvoltage caused by the current switched off in the 50 μH inductance can damage the analyzer.*

5. While monitoring with the current probe and spectrum analyzer, try to reduce all excessive spectral lines by at least a Δ dB + 6 dB margin:

   - Improve filtering at subassembly level (less costly than increasing main filter).
   - Check for deficiencies in main filter mounting.
   - Check the main filter schematic (from manufacturer's data). Is there enough inductance and capacitance for common or differen-

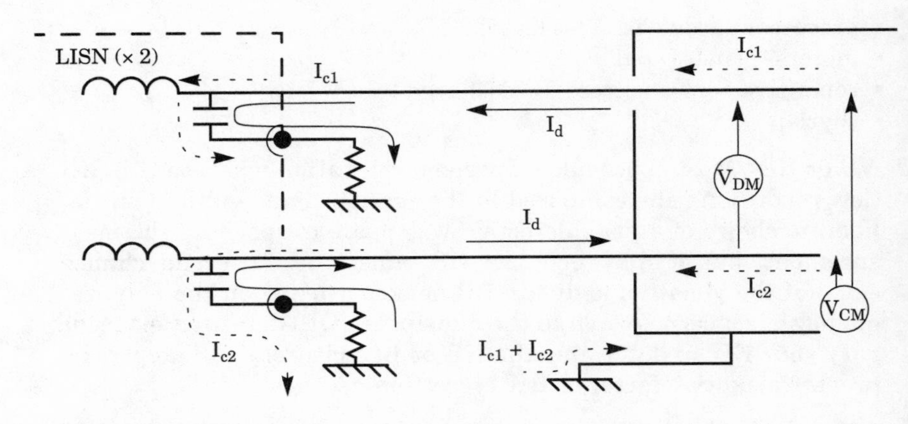

Equivalent circuit:

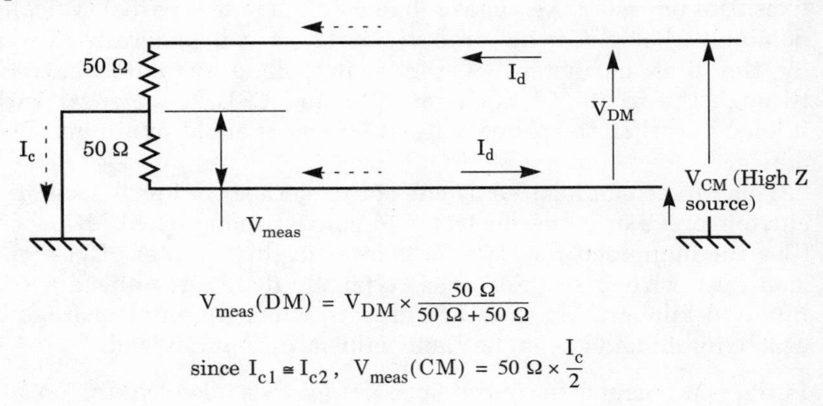

$$V_{meas}(DM) = V_{DM} \times \frac{50\ \Omega}{50\ \Omega + 50\ \Omega}$$

$$\text{since } I_{c1} \cong I_{c2},\ V_{meas}(CM) = 50\ \Omega \times \frac{I_c}{2}$$

Figure 3.5 Understanding DM and CM current paths through the LISNs. The EMI voltage reading is a combination of DM and DM.

tial mode attenuation? *Beware of H-field reradiation of the EUT on its own power cord (one of many EMI traps).*

6. Validate step 4 by a formal retest. Check that all doors and panels are closed.

3.2.5 What to Do If No LISN Can Be Inserted

Sometimes, an engineering prototype does not adapt easily to a LISN connection. For instance, the power input may be dc, with heavy-gauge conductors and special power plugs that cannot be matched to the LISN output. Also, typical LISNs are limited to 16 or 30 A. 100 A LISNs are available but are quite expensive and not easy to rent. In

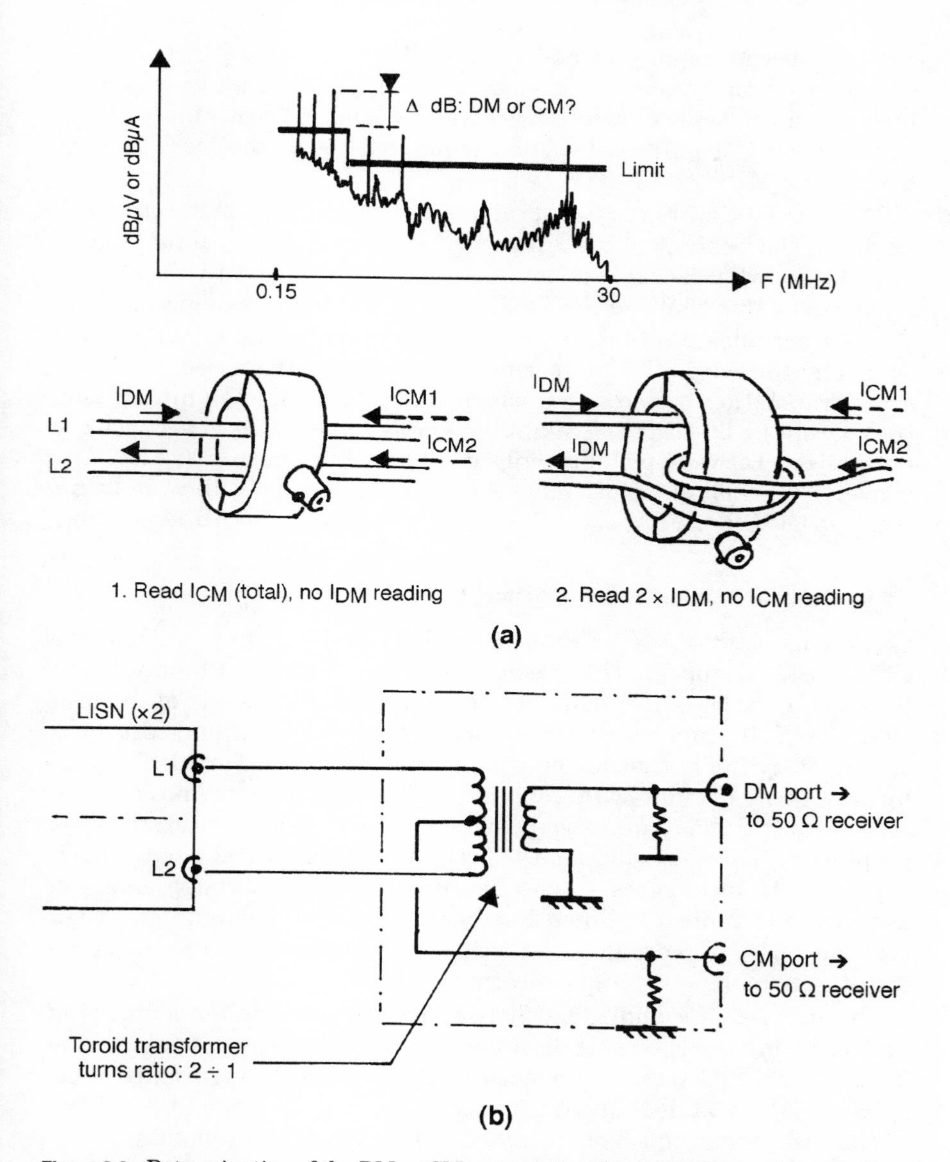

Figure 3.6 Determination of the DM or CM nature of conducted EMI (a) using a current probe and (b) using a DM–CM splitter.

this case, the following setup can be used instead, and the limit criteria be modified accordingly. The power input wires should be individually separated (taken out of their overall jacket, if any) over a minimum 1.50 m length. At least 1.30 m of this length will be pressed against the ground plane using adhesive copper tape. This will create

≈200 pF of CM bypass. In addition, a minimum three turns of these conductors will be passed through a large ferrite ring, providing at least 2 μH/turn (see Sec. 6). This makes a kind of "poor-man's LISN," inviting the CM current to flow predominantly in the local ground, above ≈ 1,5 MHz.

The assumption here is that, above few megahertz, this setup will force the CM current to see a lower impedance than the actual mains impedance (whatever this latter is) such that the current will take the maximum value that the EUT's CM source will ever be able to inject.

If the specification limit is expressed in current, the interpretation is straightforward. If the specification limit is expressed in dBμV, some translation is necessary, which is not as simple as multiplying the current by 50 Ω, as one might be tempted to do. This is because the normalized LISN impedance only reaches 50 Ω above 500 kHz. The conversion graph of Fig. 3.7 allows us to translate a CE limit in dBμV (such as CISPR or FCC, class B or A) into an equivalent dBμA reading.

3.2.6 Radiated Emission Compliance, Substitute Method

Complying with radiated emissions (RE) limits [whether they are FCC, CSA, European (EC) regulations, or Military Standard 461 (MIL-STD-461) specifications (let alone TEMPEST)] is one of the considerable EMC challenges during product development and testing. A typical storyline is that of the development engineer "woodshedding" his best blend of rules-of-thumb and company's home-grown recipes in an attempt to get a prototype cleaned up and free of internal noise problems. Then, the equipment is brought to an EMI test site to "see if it passes." In most cases, it does not, unless a very thorough emission analysis and control regimen has been completed beforehand.[1] Then comes a series of redesign and costly test iterations to bring the emission spectrum below the specification limit.

The methods recommended herein are time-savers for identifying and reducing out-of-spec radiations without the need of taking the EUT to an EMC test lab for each trial. We already mentioned (Sec. 3.2.2) that a good deal of RE prognosis can be done without resorting to EMC antennas, anechoic rooms, and 1 or 3 m measurements.

The basis of the investigation method is that any external cable, by the CM current it carries, can easily radiate more than the boxes themselves—up to 200 to 300 MHz. This is because cables' mere geometric dimensions, usually exceeding 1 m, turn them into efficient antennas, while the PCB traces and small internal wiring represent dipoles or loops with sizes one or two magnitudes smaller. Above 300 MHz, where wavelengths become shorter than 1 m (i.e., quarter wavelength < 0.25 m or 10 in), the cable-as-antenna contribution progres-

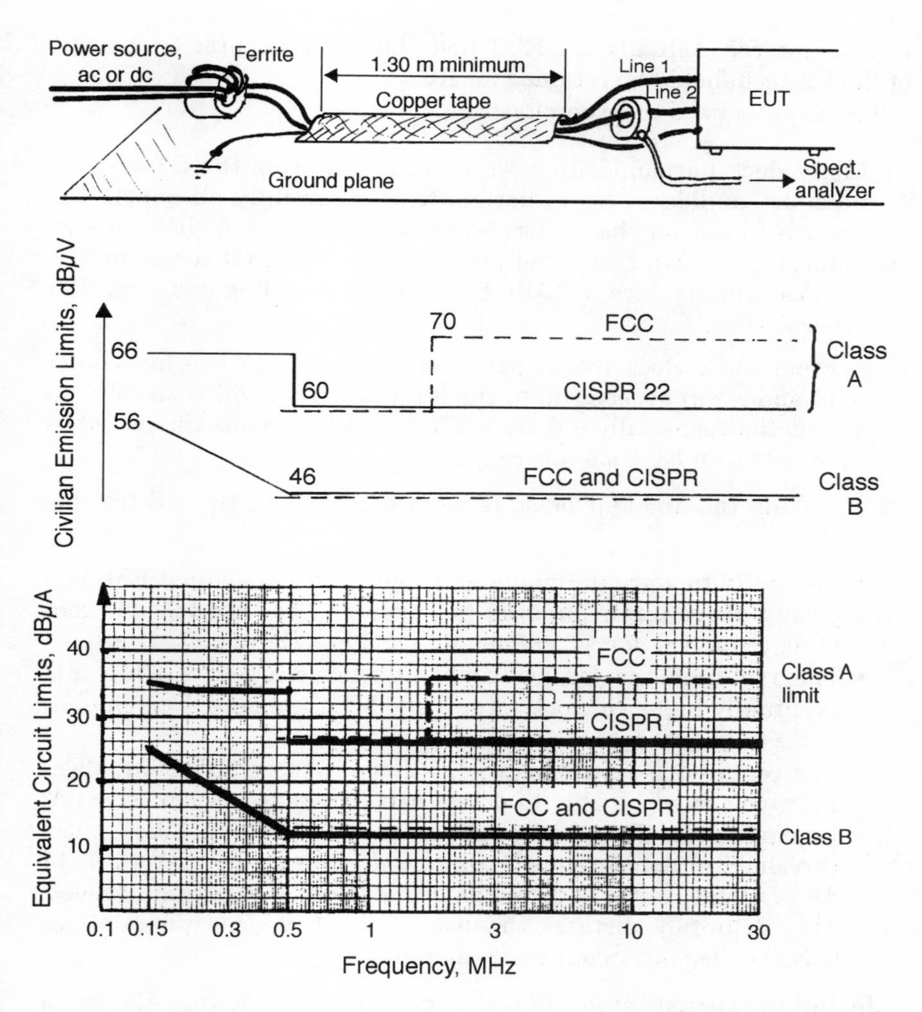

Figure 3.7 Conducted emission limits on mains wires when a LISN cannot be installed. IdBμA translated from the voltage limits by IdBμA = V_{limit} (dBμV) − Z_{LISN} (dBΩ). Translated limits apply to the current on one wire, whether DM or DM.

sively "shrinks," while the box internal radiators compete with, and eventually override, cable radiation.

Therefore, since we often deal with these two families of contributing antennas, we will try to find out whether the cables, the box, or both are likely to violate the RE limit.

Test Preparation. The preparation is essentially the same as for CE (Sec. 3.2) in terms of anticipating the EMI frequencies that likely will be encountered. Actually, the EUT installation and operating mode(s) of the CE test should be retained for RE.

Two aspects need to be emphasized:

1. Many clock harmonics that were not seen during the CE test are possible candidates for radiation. Record carefully all quartz frequencies used on the PCBs. Make sure you record their precise value, e.g. "14.275 MHz," not just "14 MHz." When it comes to harmonics ranking 15th or 25th, the error in rounding down reaches several MHz.

   From these clock frequencies, print a list of *all* the harmonics (odd and even) at least up to the 30th one. This will ease your investigation, especially if there are some CW ambients that could be misread as an EUT signature.

2. Checking the ambient noise is more crucial with the RE test because:

   - Proximity to a ground plane no longer protects against EM field pickup above a few hundred megahertz. (The field-to-cable coupling reduction, for a 5 cm height, vanishes above 600 MHz.)
   - The requirements for low ambient expand up to 1 GHz, and eventually higher if clock rates ≥ 200 MHz are used. *This applies to the auxiliary equipments as well.*
   - The required measurement sensitivity is more stringent. Basically, you must be able to detect cable currents as low as 1 μA. Depending on the type of current probe, this translates into a threshold of detectable signals ranging from 2 to 5 μV (6 to 14 dBμV). This, in turn, demands a low level of background noise. This definitely dictates the use of the low-noise preamplifier (LNA) called out in our instrumentation list.

Install the current probe on the longest I/O cable (within ≤30 cm of the corresponding EUT connector zone) for measuring the ambient noise, which is seen as a spectrum of CM current. Keep a record of this spectrum or store it in the spectrum analyzer's memory. Preferably, this record of the ambient noise level will use the same spectrum analyzer settings (see the next section) as the actual RE measurements.

The final step is to prepare a data form or spreadsheet for the CM current readout. (See the blank form shown in Appendix J) The conversion of a given E-field in dBμV/m down to the equivalent dipole current is briefly described in Appendix G. Thus, for each E-field limit, we can derive a corresponding CM current criteria, I_{CM} max (Table 3.1a).

The values in Table 3.1a correspond to a no-margin situation. When measurement uncertainty is accounted for, a practical criteria result is

Table 3.1a Common-Mode Current Limits for Radiated Emission Verification

At some frequencies, FCC/CSA and CISPR differ by ≈2 dB. The most severe requirement of the two has been applied.

FCC, CSA, or CISPR (11 or 22)

I_{max} (dBμA)	F →	30–230 MHz	230–400 MHz	400–1000 MHz
	Class A	20	26	N/A
	Class B	10	16	N/A

Table 3.1b Practical Pass-Fail Criteria for CM Current Check

	F →	30–230 MHz	230–400 MHz	400–1000 MHz
If:	Class A Class B	I < 18 dBμA I < 8 dBμA	I < 24 dBμA I < 14 dBμA	OK. It will pass the RE test.
If:	Class A Class B	I > 30 dBμA I > 20 dBμA	Do not waste time bringing the EUT to a test site. It certainly will exceed the limit.	

Table 3.1c MIL-STD-461(C)* RE02 Limit, Aircraft Category

F, MHz →	50	100	200	300	400
NB limit					
E_{spec} (dBμV/m)	25	30	34	37	38
Translated I_{CM} limit (dBμA)	10	8.5	8	7	6
BB limit					
E_{spec} (dBμV/m/MHz)	72	68	65	69	72
Translated I_{CM} limit (dBμA/MHz)	57	47	38	39	40

* For MIL-STD-461D, the values differ by a few decibels.

shown in Table 3.1b. Between the two criteria, there is some latitude for trying your luck. This depends on the following:

- How costly is your engineering effort vs. EMC lab cost?
- How critical is your $/dB investment, i.e., the cost penalty of a slight overdesign with mass-produced equipment?

For an EUT that must satisfy MIL-STD-461C specifications, Table 3.1c has been prepared. It takes into account the reduction in radiation brought about by the MIL-STD type of setup, with cables at 5 cm above the ground plane. The approximate correlation for MIL-STD-461, up to ≈400 MHz, is

$$I_{dB\mu A} = E_{spec} + 19.5 - 20\log F_{MHz}$$

This current, as seen by our clamp-on probe, will appear as a probe voltage:

$$V_{out}\ \text{probe (dB}\mu\text{V)} = I_{CM}\ \text{(dB}\mu\text{A)} + Z_t\ \text{(dB}\Omega\text{)}$$

where Z_t is probe transfer impedance, from the manufacturer's calibration curve.

From this, the entire instrumentation response is

$$\text{dB}\mu\text{V (true probe output)} = \text{dB}\mu\text{V (displayed)} - \text{LNA gain} + \text{coax cable loss}$$

Example Assume we are planning an RE test against FCC or CISPR 22 limits, focusing first on the 30 to 230 MHz band. For a specific type of current probe, this part of the measurement spreadsheet will appear as follows:

F, MHz	30	50	100	150	200
1. Probe Z_t (dBΩ)	14	14	14	14	12
2. LNA Gain (dB)	20	20	20	20	18
3. 2 m RG 58 + BNC connector loss (dB)	0.3	0.3	0.5	0.5	0.6
4. I limit for FCC/CISPR class B (dBμA)	10	10	10	10	10
Limit, translated into dBμV as displayed, 4 + 1 + 2 − 3	43.7	43.7	43.5	43.5	39.4

This of course, is the "no-margin" limit.

Test Routine (Fig 3.8). Set the spectrum analyzer (or EMI receiver) to the proper bandwidth and detection for RE measurements. For instance, for FCC- or CISPR-type limits,

- RBW: 100 or 120 kHz (can be reduced to 30 or 10 kHz when focusing on certain specific NB signatures).
- Detection mode: peak, sweep in "max hold."
- Video BW: ≥300 kHz.
- Span: Although, for the final curve plotting, you can use the entire limit span (e.g., 30 MHz to 1000 MHz), it is better, during the investigation phase, to use a finer span such as 5 or 10 MHz/div., giving a better resolution to identify culprit frequencies.

Thereafter, the recommended procedure is shown in the flow diagram of Fig. 3.8.

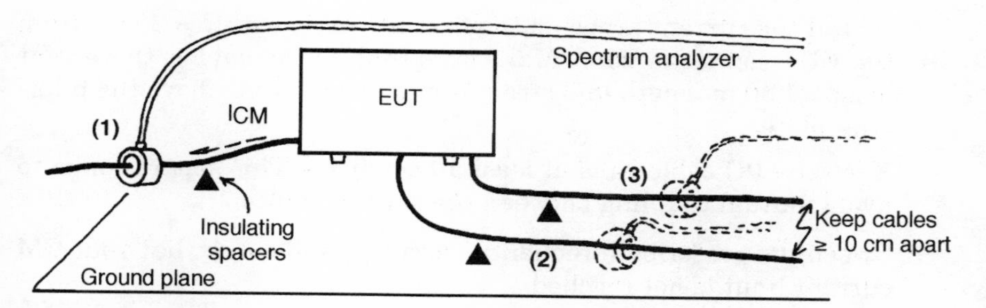

Ⓐ Set the current probe approx. 20 cm from the EUT face. Scan distance for max. reading.

Ⓑ Verify that none of the readings from cables 1, 2, or 3 above exceeds the conducted EMI substitution criteria for RE (Table 3.1).

Ⓒ

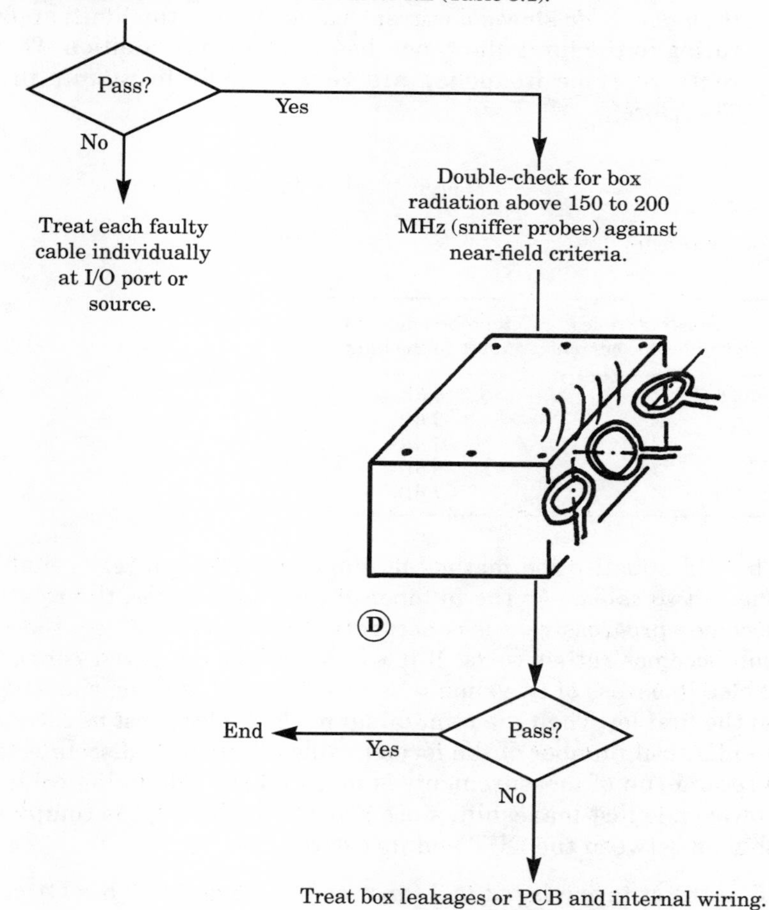

Figure 3.8 Test routine for RE evaluation, substitution method, during development.

1. Install the current probe on the first I/O cable within ≈ 30 cm from the EUT connector zone. Make sure that, for all cables, there is at least a 1.50 m length laid straight and evenly 5 cm above the reference plate.

   Keep the I/O cable runs at least 10 cm apart. This separation is to avoid mutual coupling between the several cables.

2. Sweep the prescribed frequency spectrum and check that your CM current limit is not reached.

3. Repeat for each cable or harness, including the power cord.

4. Verify that none of the CM currents spectrum readings on cables C1, C2, C3, et al., reaches the limit. If, at some frequency, *more than one cable* shows a current value close to the limit, apply a derating to the limit allowance, based on an rms addition. These currents, at same frequency, are very unlikely to *radiate* in phase. Therefore,

$$I_{TOT} = \sqrt{I_1^2 + I_3^2 + I_3^2}$$

For example,

Number of cables showing same current	Reinforcement to apply to the limit
1	0 dB
2	–3 dB
3	–5 dB
4	–6 dB
5	–7 dB

This RE substitution method by the CM current is very reliable with one or two cables. As the number of cables increases, the relationship becomes progressively less accurate. Beyond five cables, the correlation becomes rather weak. If $n > 5$ (including the power cord), the I/O cables in excess of five should be disconnected and the checking made on the first lot. Then the remaining $(n - 5)$ cables must be reconnected, an identical number of the former cable lot must be disconnected, and a second run of measurements is made on the remaining cables. This, however, is just makeshift, since you are modifying the complex interactions between the EUT and its cable.

5. If the criterion is not met (answer "No" in decision box C), you have to treat individually each cable where the CM current is too high. Since the PCBs probably cannot be significantly redesigned, this

means that you have to seal the escape ports or the cable itself. This includes:

- Filtering the I/O ports using filtered connectors, feedthrough filters, ferrites, and so on
- Checking the zero-volt to chassis equipotential around I/O ports
- Inspecting the internal wiring of the EUT to spot possible crosstalk couplings
- Using shielded cables or, if already shielded,* improving their shield-to-chassis connection

You must continue hardening until the current probe readings show that the CM current limit (or the amount of Δ dB reduction from the unmodified EUT) is satisfied on all cables. Watch to be sure that your fix on one cable has not just shifted the problem to another frequency (another clock harmonic, for instance) or another cable, since you have modified some CM impedance.

6. If the cable check is OK (answer "Yes" in decision box C), it is prudent to check also for box radiation above 150 to 200 MHz. There is no magic frontier line in this competition between radiating cables and radiating boxes. It is simply a matter of conductor dimensions and conductor heights above a metal surface (CM loop areas, in a sense). Typically, with EUT box dimensions less than ≈50 cm, say a 19-inch rack size, the lengths of PCB traces and inner wiring, ribbon cables, and so on do not exceed 20 to 30 cm. This is the wavelength for 1 to 1.5 GHz, and these elements are not efficient radiators until they approach a quarter wavelength, whereas below 150 to 200 MHz, external cables (whose length typically exceeds 1 m) are very efficient radiators. Of course, with very large EUTs such as a full-size cabinet, internal wiring between racks, backplanes, and so on, enhanced by door seam leakages, can cause the box to compete with cable radiations at lower frequencies.

Checking for box radiation (box D in our flow diagram) requires the use of a near-field probe or *sniffer.* Such probes are commercially available or can be homemade (Fig. 3.9).

Although a wide variety of probe types (E-field, H-field) shapes and sizes exist, a good trade-off for "sniffing" around a box, cabinet, or other radiator is an H-field loop with a diameter in the range of 2.5 to 7.5 cm (1 to 3 in) (Fig. 3.10). This will provide a sufficient sensitivity when located at 5 to 10 cm from the EUT sides. Smaller probes (e.g.,

* The substitution method is perfectly applicable to shielded cables, since a CM current over the shield reveals a certain amount of shield leakage.

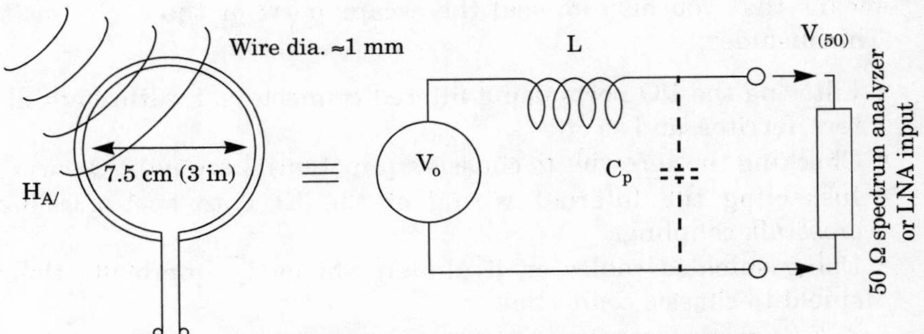

For the dimensions given, L = 250 nH, C_p = 0.3 pF.
First cutoff frequency is reached when $L_\omega = 50\ \Omega$, i.e.,
$F_{C1} \approx 30$ MHz.
Below F_{C1}, $V_{(50)}$ increases with F.
Above F_{C1}, $V_{(50)}$ is independent of F.

In a sine wave H-field, $V_o = \omega \bullet \mu_o \bullet S \bullet H$.

$$V_o = 2\pi F \bullet 4\pi \bullet 10^{-7} \bullet \pi r^2 \bullet H_{A/m}$$

For the loop shown (r = 0.0375 m),

$$V_o\ (\mu V) = 0.35 \bullet F_{MHz} \bullet H\mu A/m$$

So, at 30 MHz and above (up to at least 500 MHz),

$$V_{50\Omega}\ (\mu V) = H_{(\mu A/M)}$$

In the far field, 0 dBµA/m ≡ 51.5 dBµV/m, since E/H = 377 Ω. Therefore, the probe factor @ F ≥ 30 MHz is

$$HdB\mu A/m = VdB\mu V_{(50\Omega)}$$

or

$$EdB\mu V/m = VdB\mu V_{(50\Omega)} + 51.5$$

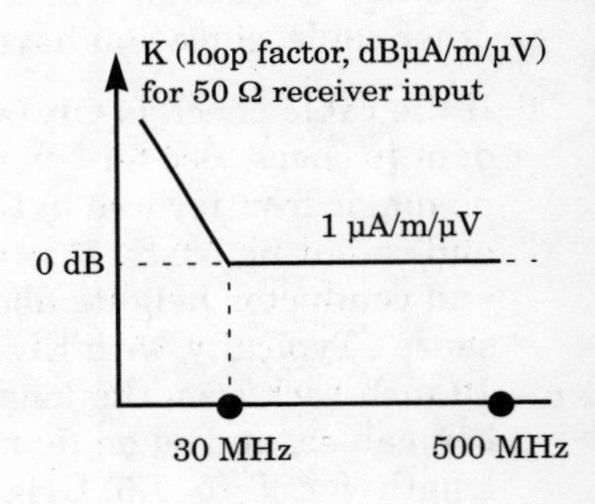

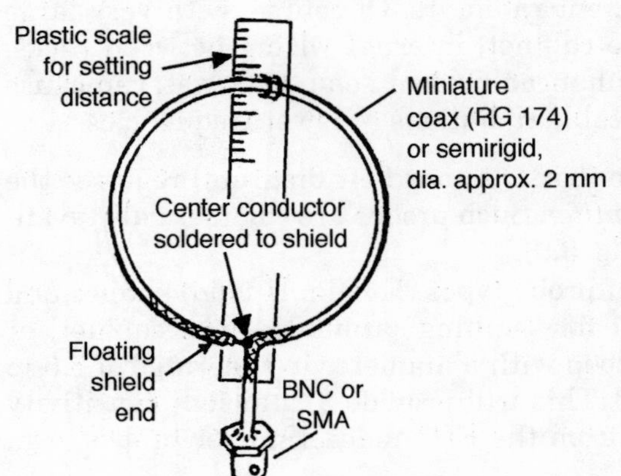

Practical realization of the 7.5 cm dia. loop with an electrostatic shield to prevent near-field capacitive coupling.

To avoid mismatch above ≈100 MHz:

- Keep the cable to receiver very short (<30 cm), or
- Sacrifice 6 dB of sensitivity and install a 50 Ω resistor at the probe output, or, better still,
- Install the LNA right at the probe output.

Figure 3.9 Example of a homemade, calibrated, near-field probe for close-up emission measurements. To avoid large errors at periodic line resonances, consider the matching precautions.

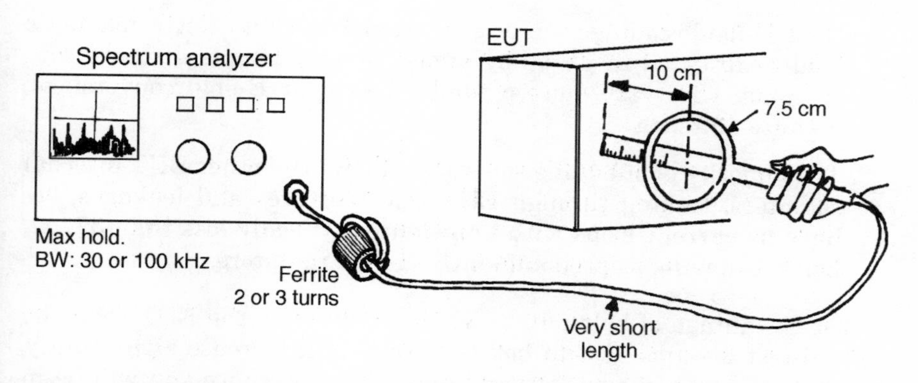

F_{MHz}	30	100	150	200	300	500
Class B limit @ 3 m (dBμV/m)	40	40	40	40	46	46
Nearness correction, with proper near-field/far-field factors	53.5	43	39.5	37	33.5	29.5 → (far field)
Transposed limit @ 0.1 m (dBμV/m)	93.5	83	79.5	77	79.5	75.5 →
Limit, expressed as probe output (dBμV)	42	31.5	28	25.5	28	24
Readout for 10 dB margin or dynamic range	(32)	(21.5)	(18)	(15.5)	(18)	(14)

Notes:

- At all frequencies, compliance of box emissions vs. class B can be checked with a 10 dB margin, even without the need of an LNA.
- The nearness correction was calculated using a conservative assumption of $1/D^2$ for near-field. This leads to the following formula for adjusting $H_{(3\,m)} \rightarrow H_{(0.10\,m)}$

$$\text{Corr.} = \frac{D_{NF-FF}}{3\text{ m}} \times \left(\frac{0.10}{D_{NF-FF}}\right)^2$$

with $D_{NF-FF} = \frac{\lambda}{2\pi} \approx \frac{48}{F_{MHz}}$; hence Corr. (dB) = $-83 + 20 \log F_{MHz}$ up to ≈500 MHz.

- The probe coaxial cable should be of good quality, with some ferrite loading to prevent parasitic field pickup by the coaxial shield.

Figure 3.10 Example of 3 m RE limits (class B CISPR or FCC) transposed to 0.10 m close-in measurements using the 7.5 cm sniffer.

1 cm dia.) can be used when searching hot spots on a "bare bones" PCB, because the searching distance will be proportionally closer. The advantages of an H-field probe are as follows:

1. In near-field conditions, the H-field is less affected than E-field by the proximity of other objects, such as your body, instrumentation, etc.

2. Near H-field readings can be converted easily into electromagnetic fields, expressed in μV/m, by applying some simple distance conversions. These are more credible than near E-field conversions, because of item 1.

3. The majority of offending sources on PCBs and other EUT internal radiators, leaking through EUT box apertures and leakages, behave as current loops with impedances typically less than 377 Ω, hence behaving as predominantly H-field radiators.

The advantage of being 10 to 30 times closer to the EUT than the formal test distance is that box radiations will increase significantly, whereas the ambient background noise will not. Since you will be in near-field conditions for a some part of the spectrum (at 10 cm from the EUT, one must be above 500 MHz to reach the far-field conditions), it is important to *keep a very constant distance* from the EUT faces. In the proximal region, the field roll-off such as $1/D^2$ or $1/D^3$, and a small shift in source-to-antenna distance, will cause a significant readout variation.

Depending whether they are calibrated, close-in field probes can give the true value of the H-field in dBμA/m, or simply a relative indication, to grade the improvements realized. Note that true H-field reading is necessary if you have no baseline of a formal RE test to begin with.

To explore the EUT faces and find the radiating hot spots, a good practice is to draw a scanning perimeter around the EUT at a fixed distance of 5 to 10 cm. The probe should be fitted with a distance spacer (a sacrificed piece of a plastic ruler, for instance). Rather than keeping the probe at hand, a small fixture such as a miniature vice or "two-hands" PCB holder with ball joints will allow us to explore the perimeter at constant increments (e.g., 10 to 15 cm). (The closer the antenna, the smaller the increments, since—in theory—step size should not exceed ≈1.5 times the source-to-antenna distance.)

For each point, the antenna should be oriented successively in the X, Y, Z axes, focusing on the highest reading.

A substantial amount of time can be saved by anticipating where the hot spots might be. Plain metal faces do not radiate, but discontinuities do, especially if there is a current path nearby. Look for zones such as

- Cover seams and butt joints.
- Ventilations and display openings.
- I/O connector plates (the I/O port flanges of typical PC packaging are frequently involved).

- Cable penetrations, including filter areas (filters can be highly radiating devices if not perfectly bonded), and eventually fiber-optic entries.

Depending on the EUT's stage in the development/production cycle, this investigation will result in some hardening of the equipment itself, which may be addressing:

- *The PCBs:* loop area reduction, clock-trace guarding, IC decoupling, and eventually individual shields on top of large ICs (MCMs, PGAs, microprocessors)
- *The internal packaging:* daughter-board-to-motherboard loop reduction, crosstalk avoidance, wire-to-chassis loop reduction.
- *The box shielding:* aperture control, gaskets, tight grounding of I/O receptacles, etc. This is the "brute force" approach, but sometimes it is the only option left when the calendar is the driving element.

This method may appear to be slow and fastidious. However, it is the author's experience that, considering the cost of actual RE testing (typically $1,200 to $2,000 per day of pure EMC lab cost), benchtop investigations are indeed cost-efficient. In fact, some EMC practitioners have refined this investigation technique as a quasi-official test policy within their companies.

Investigation after Failing a Formal RE Test. The method also can be used after a first, unsatisfactory RE test in a calibrated open-field site or shielded anechoic room. The difference with the former development routine is that, this time, there is one additional clue, and a precious one: the actual RE curve that will be your reference to grade further progress. Since you know exactly by how many decibels the specification limit was violated, every decibel you will scratch on the CM current reading will be a decibel deduction of radiated field. The procedure, after the first RE test, will start as follows (Fig. 3.11):

1. *While the equipment is still on the test range,* disconnect *all* external cables except, of course, the power cord.

2. Rerun the RE test to see if the EUT passes.

3. If the answer is "Yes" (see decision box in flowchart), this means that the cables are the radiation vectors.

4. If answer is "No," it can be one of two things:

   (a) The only cable left—the power cord—is radiating.
   (b) The box itself is also radiating.

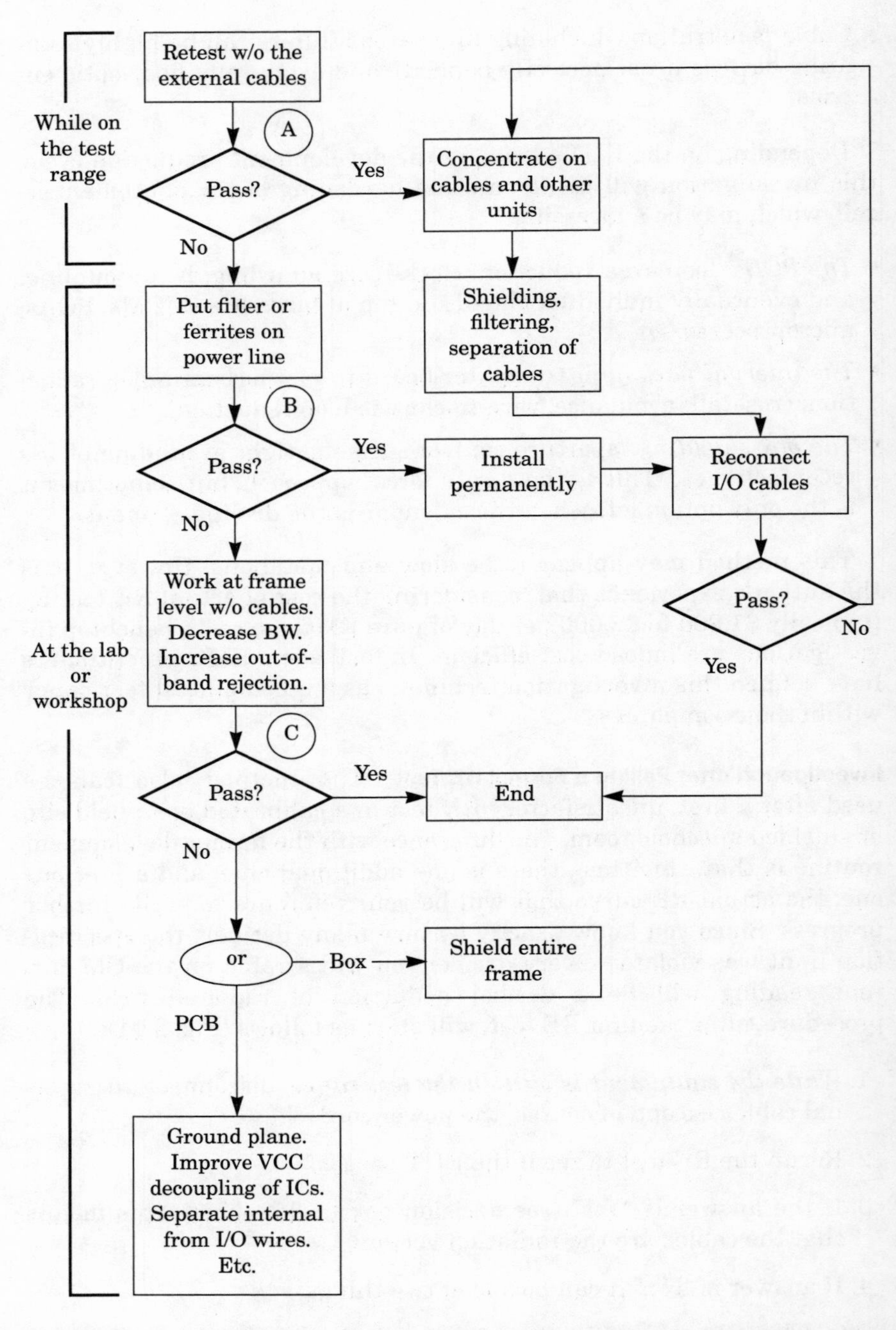

Figure 3.11 Strategy to use when EUT is failing RE test at the standard EMC test site. *Courtesy of EMF-EMI Control.*

For checking (a), insert one or two large CM ferrites on the power cord, with at least two (three are preferable) turns for each, very close to the EUT power cord entry. Press firmly the whole power cord against the ground plane.

Rerun the RE test. If the EUT now passes, then the power cord was the last antenna [item (a)]. Otherwise, item (b) is true: The box also radiates.

5. One last check is needed if item (a) was true. It is prudent to reconnect the I/O cables and verify RE while the power cord radiation is still inhibited. This is because the power cable emission, although the dominant one, may have masked some lesser, but still out-of-spec, radiation from other cables. If this is true, some rework will be needed is this area as well (as in item 3, "Yes").

At this juncture, you can take the EUT back to the engineering lab and proceed to the reduction of emissions by the method seen before; for each frequency violation, you will have to reduce the CM currents by the amount of decibels (plus some margin) by which you were exceeding the limit. Bear in mind that if the box alone was also exceeding the limit [item (b)], you must *fix the box first.*

Extension of the RE Substitution Method to PCBs. The method of CM currents and near-field probes can be extended to radiated emissions analysis of PCB prototypes. This is especially practical for products where the majority of functions and user's controls are gathered on a single board, with little or no additional subassemblies.

The PCB should be placed over the reference plane, isolated by spacers of a few centimeters (Fig. 3.12). Ideally, the spacers should recreate the actual distance of the PCB above its normal resting surface. The placement of I/O cables and other items is similar to what is shown in Secs. 3.2.3 and 3.5.1. The test routine for measuring CM currents is identical. For near-field probing, the dimensions of the H-field probe have to be smaller than for box shielding survey, e.g., 1 to 2 cm dia. Since hot spots near ICs or critical traces need to be detected with good accuracy, the probe distance is smaller too, e.g., 1 to 2 cm. Due to the $1/D^2$ or $1/D^3$ dependency, it is even more important to keep a constant distance from the target.

When searching for PCB emissions, a direct probe of voltage or current can lead to the culprit component by focusing on the pins or traces that exhibit the highest spectrum, especially above hundreds of megahertz.

The voltage spectrum can be directly picked up on the IC pins or traces using a highpass probe tip (Fig. 3.12). The problem is one of reference, since we are chasing CM voltages, not DM. This can be controlled

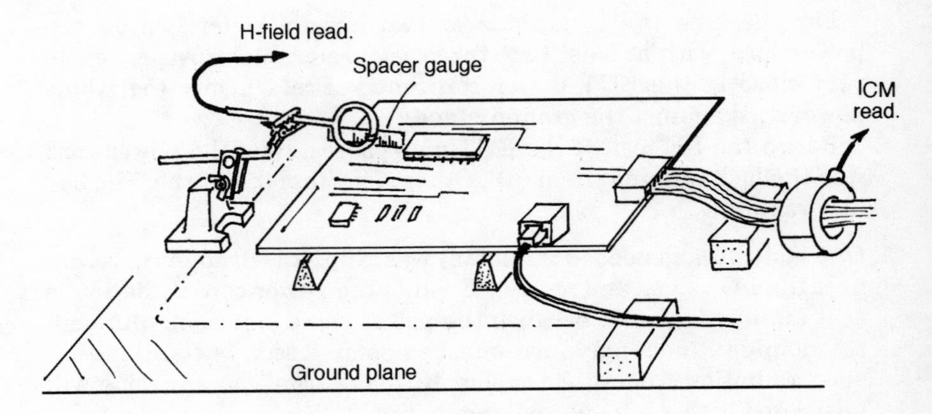

(a)
General layout for RE investigation

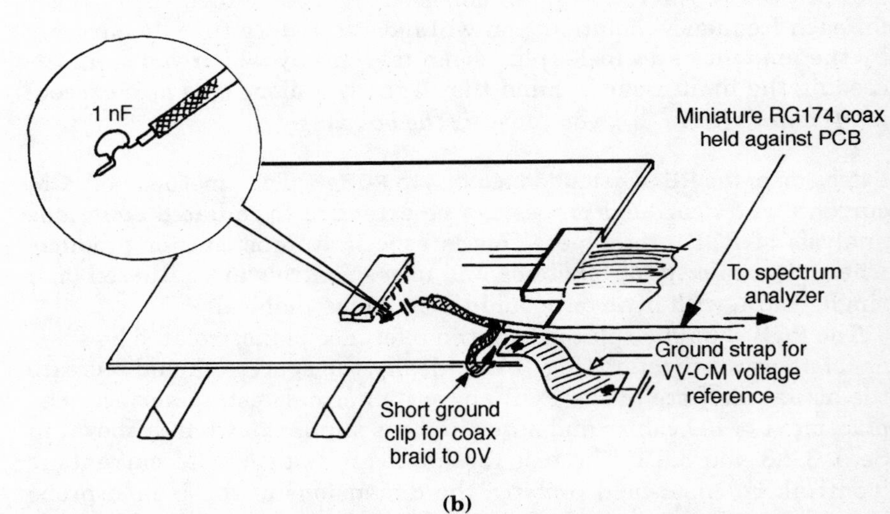

(b)
Setup for measuring noise voltage spectrum on specific IC pins or traces. This allows identifying hot leads with high spectral contents above approx. 50 MHz. With respect to FCC or CISPR class B, any harmonic exceeding approx. 1 mV (60 dBμV) of CM voltage is a possible suspect.

Figure 3.12 Example of PCB setup for RE investigation.

by grounding the zero-volt reference of the PCB to the reference plane, at the connector side. The spectrum analyzer probe is referenced to this zero-volt point by a short strap, and voltage is measured from there. The coaxial cable should be a miniature one and should be laid on the PCB surface to minimize the potential pickup loop.

The current spectrum can be tracked with a miniature surface-current probe like the ones developed by B. Danker, from Philips, Netherlands (Fig. 3.13). For instance, residues of clock harmonics, when found on traces that do not carry clock signals, reveal crosstalk coupling that could turn low-speed signal traces into clock radiating antennas (Figs. 3.14 and 3.15).

3.3 Checking for Compliance with Immunity Specifications

Immunity test instrumentation and procedures are amply described in relevant standard documents and literature (see the references at the end of this chapter), and it is not the purpose of this handbook to replicate them. Instead, we will insist on simplified testing approaches, using basically the same instrumentation but in a more expeditious way, to evaluate the susceptibility level of a prototype, or some of its subassemblies, against strong CW or pulsed disturbances. These disturbances can be caused by:

- Other parts of the system (intrasystem EMI) that are not yet fully integrated
- Certain types of environmental threats

Only a limited set of tests will be carried out on the prototype, as explained in Sec. 3.1. Preferably, they will be done only after the emis-

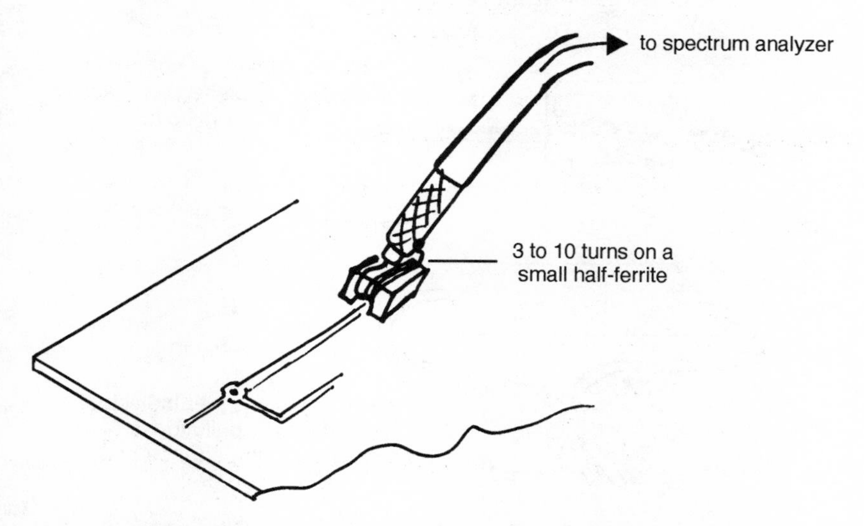

Figure 3.13 Miniature surface current probe. This probe can be used to spot excessively noisy traces or leads. Output is typically 1 to 3 μV/μA from 30 MHz up.

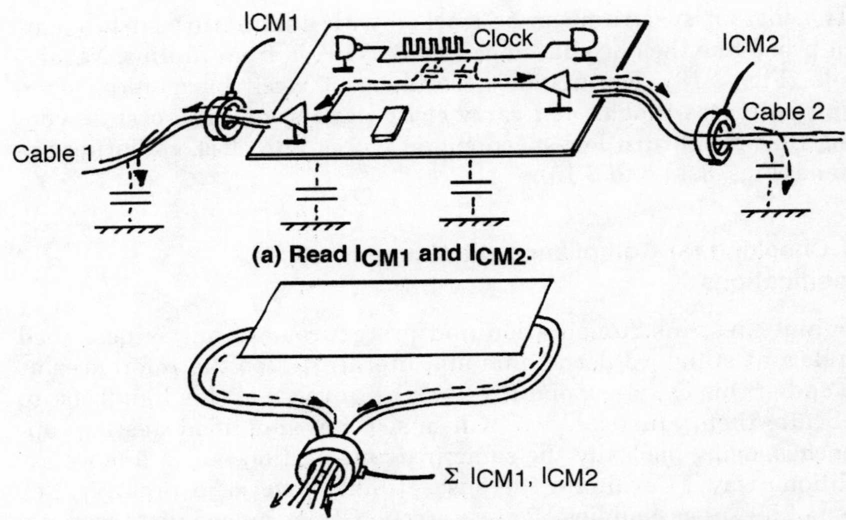

Figure 3.14 How to detect I/O cable pollution by internal crosstalk on a PCB.

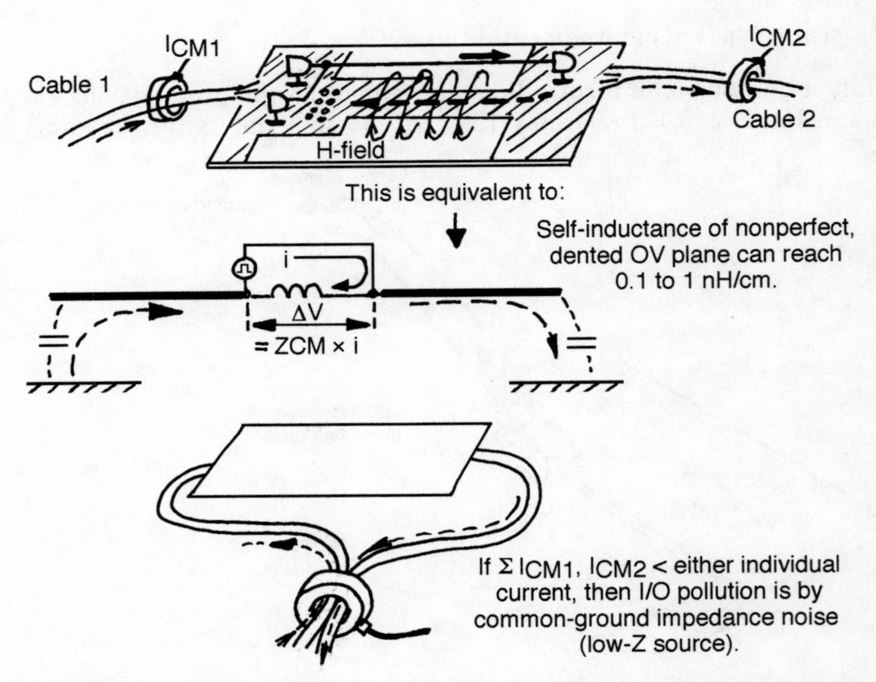

Figure 3.15 How to detect I/O cable pollution by a common-ground impedance problem within a PCB.

sion tests have been carried out and, eventually, their noncompliance has been investigated.

Notice that, in these tests, we will always try to reach, within a reasonable span, the *actual threshold of susceptibility* for the EUT in order to have a figure of our *compatibility margin,* as compared to the immunity requirements.

3.3.1 Minimum Requirements for the Test Site

The requirements for an informal susceptibility test are basically the same as for emissions (Sec. 3.2.1) except that the quiet RF ambient requirement is exactly the opposite:

RF Ambient Considerations. A quiet RF ambient is not necessary. Quite the contrary, susceptibility tests will generate strong HF or pulsed interference, which may disturb other equipments in the vicinity. Therefore, it is prudent to ban, or temporarily discontinue, the use of any sensitive electronic equipment such as instrumentation (other than EMC), computers and office equipment, oscilloscopes, etc., in immediate proximity—say, within less than 5 m from the test setup.

Filtered and Isolated Power Mains. For the same reciprocal reasons as above, it is recommended to filter, and if possible isolate, the ac power branch circuit that supplies the EUT. Since a LISN was used for emission tests, it can be used as well with CS testing, along with the isolation transformer.

Test Ground Plane. This is similar to emission requirements (except for ESD testing which may require, for indirect ESD, a different set of metallic planes).

3.3.2 Conducted Susceptibility Test Preparation

For all susceptibility/immunity tests, it is mandatory that the EUT can be set in an automatic (preferably self-looping) mode, such that it performs its normal functions continuously. This can be done by employing an auto-test mode or specific emulation software that

1. constantly activates the EUT circuits, emulating as closely as possible a real application without the need of operator intervention
2. can indicate clearly a malfunction or disability by
   - displaying an error or wrong data
   - printing a report
   - reset or lockup

- activating an LED, a buzzer, etc.

without the need of any external measuring instrument (e.g., oscilloscope, data logger) unless it is connected via an optical fiber link.

This last point is very important. Ideally, no external monitoring equipment should be used to check a fault condition, because the additional probes and cables, and the monitoring device itself, can cause the EUT to fail at lower levels, giving incorrect test results. In addition, the oscilloscope, data logger, remote keyboards, ATE, audio metering equipment, and so on could be disturbed by the test, giving misleading information.

With large systems, it is sound policy to prepare application software that exercises every specific operating mode, corresponding to specific hardware areas. This will improve the efficiency of the hardware test program itself, improving the probability to hit the worst sensitivity windows, without running an excessively long test. This also facilitates error diagnosis.

During this development phase test, it is preferable to limit the auxiliary equipments (AEs) to a minimum and to use, everywhere possible, passive loads at the end of I/O cables. A typical example is a Tx/Rx link in a self-looping mode. No other cable than those normally used with the EUT should be installed, except optical fiber. As for emission testing, the AE(s) must have a same grade of immunity as the one we are testing the EUT for, or an efficient EMI filtering scheme.

3.3.3 Electrical Fast Transients

Of all EMC tests, the electrical fast transient (EFT) burst test is certainly the simplest to install, easiest to perform, and one of the most meaningful. Owing to its wide broadband spectrum, covering kilohertz up to at least 60 MHz, it will track down most of the EMC weaknesses of digital as well as analog equipments. The envelope of the bursts can be mistaken by logic circuits as a valid series of bits, while individual short pulses can trigger edge-sensitive gate inputs. With analog circuits, the envelope of the bursts can be detected by sensitive amplifiers in a similar manner as pulse-modulated CW interference (see Sec. 2.7).

Because it is meant to simulate the coupling of switching transients (contact repetitive arcing and so on) from nearby power devices, the test is applied not only via the mains connections but also to any signal/control cables, according to IEC 801-4 or 1000-4-4.

Instrumentation and Setup

- The EFT generator must conform to IEC 801-4 or equivalent (Figs. 3.16*a* and *b*).

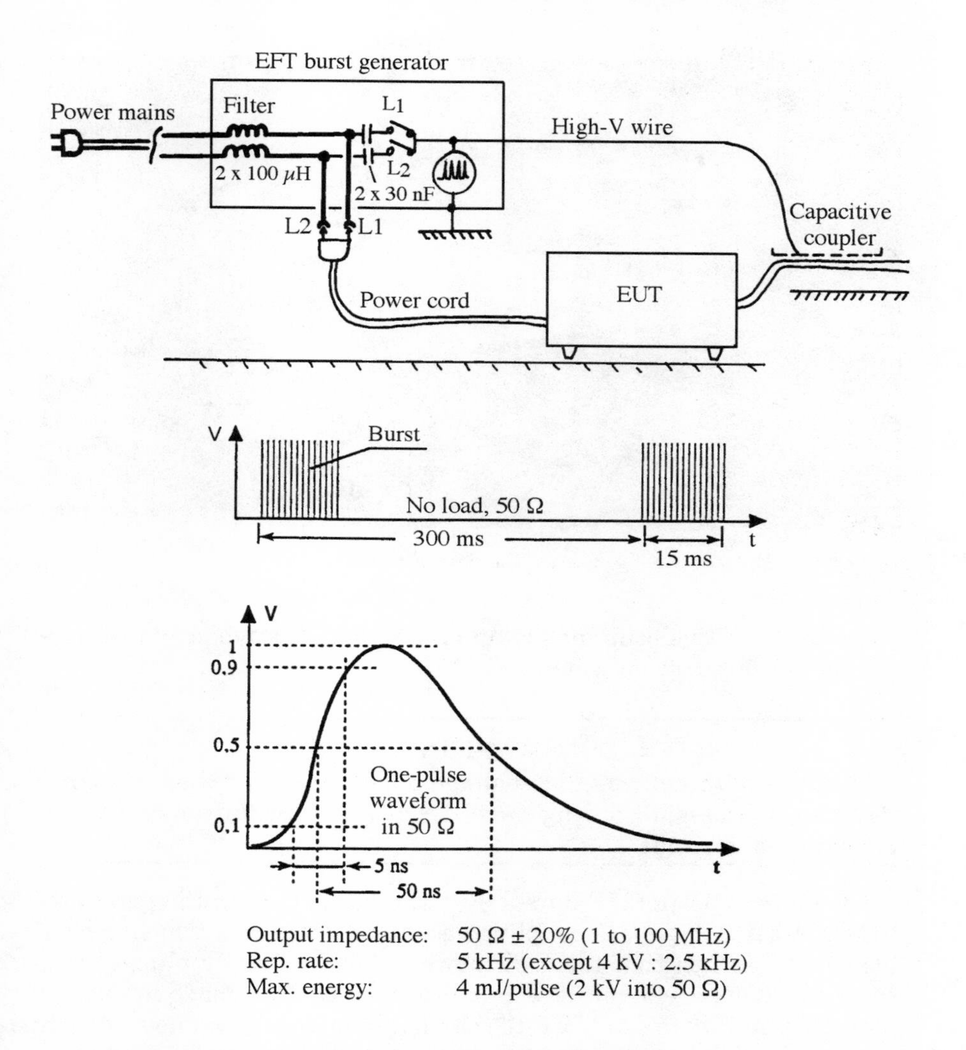

	Test voltage ±10%	
Severity level	Power mains (kV)	I/O lines (kV)
1	0.5	0.25
2	1.0	0.50
3	2.0	1.00
4	4.0	2.00

Figure 3.16a Fast transient generator.

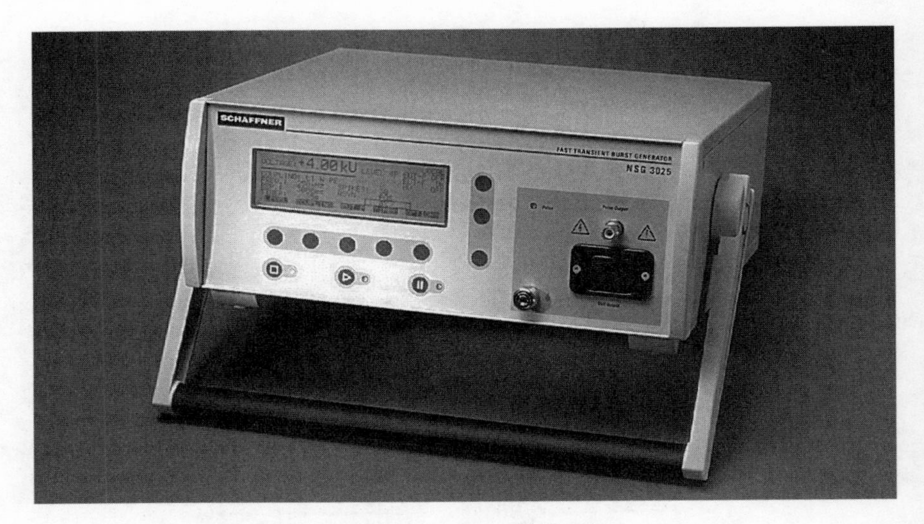

Figure 3.16b Fast transient generator. *Courtesy of Schaffner.*

- Use a capacitive coupling clamp or, by default, kitchen-type aluminum foil, 30 to 50 cm wide.

WARNING

The generator delivers high-voltage pulses, up to 4 kV, on live parts that can be accessible during the test. Operation by untrained personnel can be dangerous.

For superimposing HF bursts over the power mains, the generator incorporates a highpass coupling/decoupling network, which resembles a LISN, such that the EUT power cable is directly plugged into the generator's power outlet. The injection of EFTs is made internally, with a possibility to select L1, L2, earth (green or green/yellow), or any combination of the three. In any case, the injection is a CM type, line-to-ground plane. If the generator does not have such suitable internal ac coupling network, or if the EUT power cable does not fit into the generator outlet (e.g., because of conductor's size), you can use, by default, the capacitive clamp injection. In this case, since a LISN probably cannot be inserted either, it is prudent to be isolated, in HF terms, from the ac mains upstream by inserting CM ferrite rings (at least two turns) at the far end of the power cable, near the lab power outlet.

Capacitive injection to I/O cables is applied via a special capacitive clamp, a 1 m sort of fold-over channel, which is driven by a high-voltage output cable directly off the generator. This clamp is the standard coupler for qualification test. However, its size (not to mention its cost) and weight makes it impractical for a breadboard prototype EUT. In this case, a tight wrap of aluminum foil, about 30 to 50 cm wide, will provide the 60 to 100 pF of distributed capacitance needed for the CM injection simultaneously on all the wires of the tested cable. This wrapped sleeve, being the hot plate of the test, must be kept isolated at ≈5 cm height above the test ground plane, like all other EUT cables (Fig. 3.17). The HV injection cable should be ≤ 1 m long.

The generator box must be resting on the ground plane and grounded with a very short braid strap. This is a critical detail, because too much inductance in the return path would affect the test results. The generator should not be too close from the EUT: Although it is a decently closed metal box, a 4 kV pulse with a 5 ns rise time is a hefty transient that radiates a strong field nearby. Close exposure of the EUT to this field could degrade the test results.

The AE(s), if any, must be capable of enduring this type of test, as explained in Sec. 3.3.2, or equipped with several turns of CM ferrite toroids, *close to the AE box.*

EFT: Test Routine. *For all injections,* start with the lower level first, for at least 30 s, for each polarity. Do not switch abruptly from (+) to (–); always step through an "off" mode in between. Do not touch the EUT while testing. Do not let the HV cable hang loose close to the EUT.

1. Start with power mains first, at a low level. The IEC 801-4 severity levels can be used, if nothing else has been specified. Severity level 3 is generally satisfactory. However, since it is a development test, it is advisable to try one level above to know your EMC margin.

2. Increase the level step by step, addressing all combinations of (+), (–), L1, L2, L1 + L2, and so on. Record run/fail levels when reached.

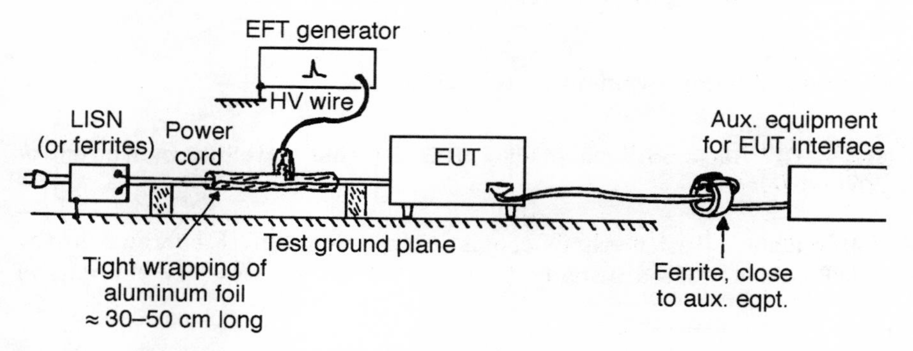

Figure 3.17 Development stage test setup, when EUT cannot be fed through the EFT generator.

3. Switch off the L1, L2, earth selectors.
4. Inject onto each I/O cable, keeping the capacitive sleeve within ≈1 m from the EUT. If there are shielded cables, the sleeve is put over the shield, which normally has some sort of insulation jacket. While testing the I/O cables, keep the EUT power cord plugged into the generator, since it provides an HF isolation from the power mains.

A suggested test log format is shown in Appendix J.

EFT: Diagnosis and Fixes. If the test has failed, reduce generator output down to a "no-fail" level, as a reference, on all cable ports. Check for

- missing or incorrect cable shields connections (pigtails, etc.)
- ungrounded metallic connectors
- incorrect bonding of connectors plate to main chassis
- lack of, or improperly mounted, CM filtering capacitors (long leads)
- I/O wires running too far inside the equipment before they are filtered
- incorrect mounting of power-line filter

Improve immunity by

- correcting any of the above
- replacing plastic connectors by metallic ones (for shielded cables)
- insertion of filtered connectors (e.g., male/female adaptors) on critical ports
- CM ferrites—split cores (two or three turns) very close to cable entries

Validate the improvement by retesting.*

3.3.4 CW Radiated Susceptibility (RS): Substitute Method by Current or Voltage Injection

Introduction. Test methods consisting of injecting HF currents in the EUT cables existed already, to a limited extent, in early versions of

* Note: If a fix does not seem to work, do not remove it. Add another one. EMI is a multipath mechanism.

Military EMC specifications such as MIL-STD-461A, also called CS01, CS02. The British Air Force improved the method greatly circa 1985 under the name bulk current injection (BCI). The general idea is that the generation of strong, quasi-uniform RF fields over large test objects is always a difficult and expensive chore, especially below ≈50 MHz, where emissions antennas that could fit inside a Faraday cage have a mediocre radiation efficiency in terms of volts per meter produced per watt of RF drive.

In a manner reciprocal to the substitute emission method (Sec. 3.2.5), the assumption is that, up to ≈150 to 300 MHz, depending on EUT size, the external cables are the most efficient receiving antennas. Thus, if one injects onto these cables *the same RF current as would be seen if they were actually illuminated by the prescribed field,* this will replicate the EMI scenario without the expense of bulky antennas and amplifiers, anechoic room, and the like.

This is the basic concept, and it usually gets enthusiastic reactions in symposium lectures. As usual, reality is not as idyllic, for two reasons.

1. The coupler for injecting the CM current must be a *noninvasive*, i.e., preferably a current clamp. To avoid loading the tested cable by the mirror impedance of the RF generator, it must be a *weak-coupling* current transformer. For instance, with a 4:1 turn ratio, the generator impedance seen in series on the tested cable is $50\ \Omega/(1/4)^2 = 3\ \Omega$, which is acceptable. The price we pay is that the coupler efficiency is not optimum, due to mismatch and coupler losses. A typical value for a decent probe efficiency is approximately –10 dB; i.e., it takes 10 W of power to inject 1 W into a typical EUT cable. This roughly corresponds to simulating a 10 V/m rms, plus 20 percent AM, incident field over a cable located ≈ 0.50 m above ground.
2. When EUT cable lengths are reaching quarter-wavelength resonances, they in turn become radiating antennas; i.e., most of the injected power is wasted in radiating an external field instead of getting into the EUT. Hence, the very problem that was meant to be avoided resurfaces: The whole setup is exciting room, or site, resonances. In other words, this injection test needs to be performed inside an anechoic room.

So what is left of the benefits of the BCI method? Significant advantages do remain.

1. Quarter-wavelength resonances excepted, you do not dispose all of the wasted power into the air, generating strong radiated fields.

2. With a controlled height of the EUT cables, 3 to 5 cm above the ground plane, a reduction effect exists in the cable-to-field coupling, below 600 MHz. Since the BCI method is not to be used above approximately 300 MHz (civilian specs define 230 MHz as the upper range for this method; MIL-STD-461D prescribes it up to 400 MHz), this 3 to 5 cm height results in a >6 dB reduction of this side-effect radiation.

3. The influence of human body and foreign objects around the test bench is not critical, as it would be with an actual radiated susceptibility test setup. Therefore, the EMC investigations are easier to carry out, in comparison to having to constantly come in and out of an anechoic room.

These three items are favorable for development testing, particularly in a lab environment that is not an EMC test chamber. However, some precautions are needed regarding your neighbors; otherwise, you may interfere with AM, FM, TV, and other services, as far as few hundred meters away, in a free-space type propagation.

Precautions Regarding Ambient, for BCI Test

- Do not practice such a test in the upper levels of a multistory building, or near the windows. Try to use a room located at or below ground level and in the center core of the building. (These are already general guidelines for test sites, as described in Sec. 3.2.1.) The best situation, of course, would be to use an anechoic room.
- This is a swept test. Do not halt too long at each discrete frequency step, especially in the AM, FM, TV bands.

Instrumentation and Setup. The RS test, even simplified by the BCI method, is the heaviest and most cost-intensive of the whole test program. In addition to the instrument cost, it is mandatory that the test be performed by an experienced EMC engineer.

The following instrumentation is needed (Fig. 3.18):

1. *Sine wave generator,* 0.1 to 300 MHz minimum, 1 kHz AM capability, preferably internal. The tracking generator of a spectrum analyzer could be adequate, but it should accept amplitude modulation and be programmable to sweep by increments and halt-on-step.

2. *Wideband power amplifier,* 0.1 to 300 MHz, with protection against output mismatch. Specify an "N" connector output. The amplifier power depends on (a) the efficiency of the coupler you will be using and (b) the E-field level you have to simulate for RS compliance.

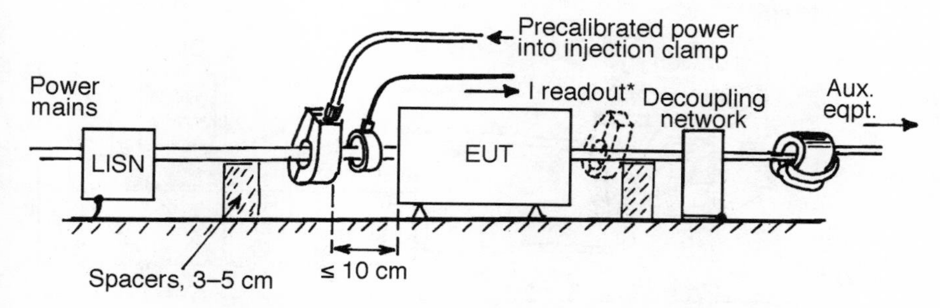

Figure 3.18 BCI test setup, shown for injection on power leads. Clamp is moved to I/O cable next.

Based on average coupler + mismatch losses of 10 dB, the following table gives the required power including modulation margin, plus 6 dB margin, to explore actual susceptibility thresholds.

Power Requirements for BCI, Based on Typical Injection Clamp

E-field requirements	AM modulation	Open voltage* to apply to EUT cable	Amp. power for 150 Ω CM impedance
1 V/m	50 – 80%	1 V	0.16–1 W
3 V/m	50 – 80%	3 V	5–10 W
10 V/m	50 – 80%	10 V	60–100 W

*The open voltage is based on average cable height in actual installations of ≈0.50 m. Different heights will result in different open voltage—but different CM impedance as well.

3. *Injection clamp,* with a sufficient window size to accommodate the largest EUT cables (2.5 to 4 cm recommended). Homemade devices are not encouraged for this clamp. It is a critical accessory, and it is better to let the professionals make it. An N connector is preferred.

4. *Current reading probe* for injected current monitoring (same as for emission tests).

5. *Calibration jig,* for precalibration power adjustment (Fig. 3.19).

6. *50 Ω coaxial load,* with power rating in accordance with the preceding table.

7. *Directional coupler*—not absolutely mandatory, but strongly recommended for forward power monitoring.

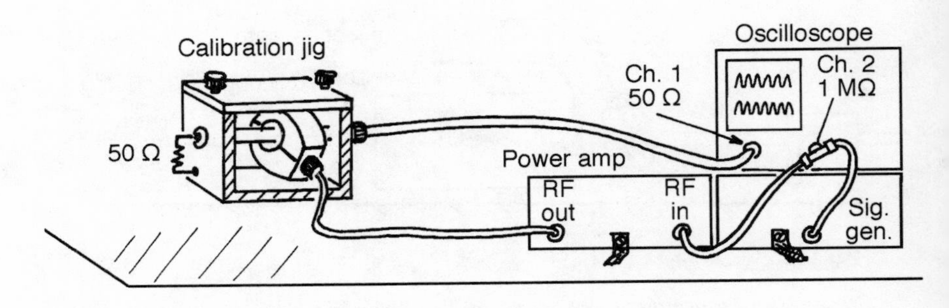

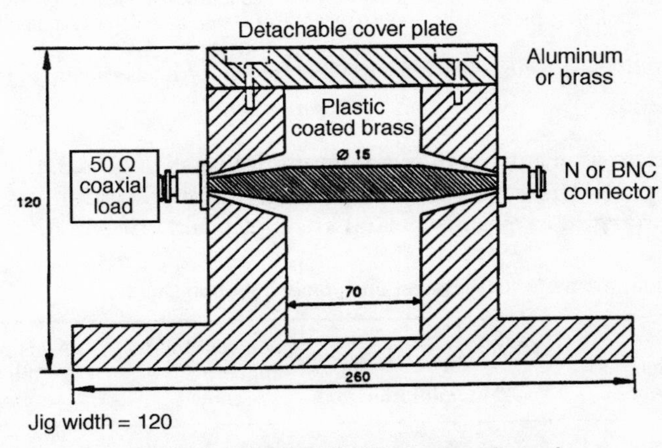

Detail of calibration jig (dimensions in mm)

Figure 3.19 Calibration setup for BCI test, using simply an oscilloscope for monitoring (assuming the power amplifier gain is well characterized). More accurate monitoring of the injection drive can be made at the RF power output using a 40 dB directional coupler and replacing one of the oscilloscope channels with a spectrum analyzer.

8. *Spectrum analyzer* or *EMI receiver* for monitoring power (or current). Current reading can also be made with an oscilloscope, since sensitivity is not a problem for BCI. A 50 Ω coaxial two-way switch can be used to avoid two instruments.

9. *LISN,* for power leads.

10. *Set of decoupling networks.* Investigations can be made without such devices, but with less accuracy, because of mismatched line ends. The decoupling networks provide a constant CM impedance of 150 Ω above 100 kHz, which helps in reducing mismatch oscillations and high-voltage standing wave ratio (VSWR). It is preferable that decoupling networks be provided with connectors that match the type of I/O cables of the EUT (Fig. 3.20).

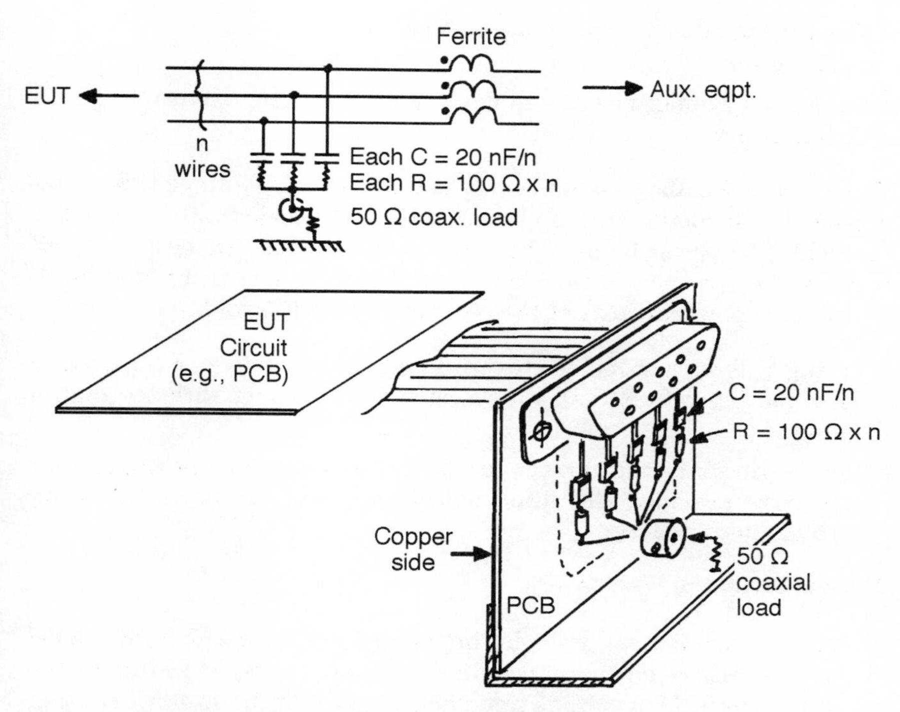

Figure 3.20 Example of a homemade, 150 Ω decoupling network, using a mating connector for easy EUT interface. The BNC socket is for the 50 Ω load, but it can also be used as an injection port with a 50 Ω generator instead of the injection clamp.

Current Adjustment. A great deal of writing and sweat have been expended over the control of BCI current. Should we monitor a constant current (i.e., the power being constantly adjusted)? Or should we monitor a constant power (i.e., regardless of the actual current, the applied power is the one determined during the precalibration phase with the 50 Ω + 50 Ω jig)? This all boils down to few simple facts:

- The harness under test is a severely mismatched transmission line, with respect to its CM characteristic impedance.
- At some frequencies, resonances and antiresonances will cause huge variations in the loop impedance seen by the injection clamp, varying from zero to infinity.

Given that we have three variables (V, I, P), any two of which will dictate the third, it is obvious that:

- One cannot pump a prescribed current into an infinite impedance, even with infinite power.

- One cannot develop a prescribed voltage across 0 Ω, even with an infinite power.
- Even a very small power will develop an infinite current into a null impedance.

Considering the above, what counts is that the open voltage that is impressed to the loop is the one it would see if it was actually exposed to the field. The power being set by the precalibration, the best trade-off to generate this open voltage is to keep this same forward power level, whatever the actual current that results. This will avoid:

- Trying to force a specified current in an open loop (e.g., EUT circuits with floating references), which causes overtesting with too high a voltage.
- Trying to reduce the current to a specified value when the designer may have provided CM filter capacitors, resulting in undertesting with too low a voltage.

BCI Test Setup and Routine

1. Install the injection probe in the test jig. Adjust the generator level for the prescribed current, indicated by the voltage reading in the 50 Ω output. This voltage must be exactly half the prescribed open voltage.
2. Record the forward power across the frequency spectrum for the RS testing (default = 0.1–300 MHz), using an *unmodulated* CW signal.
3. Install the injection probe and monitoring probe on the first EUT cable, as shown in the figure, within ≤15 cm from EUT I/O port.
4. Set the required modulation (e.g., 50 or 80 percent AM).
5. Scan the entire frequency range using the predetermined frequency steps (see next section). At each frequency step, check the forward power against the prerecorded calibration levels, without modulation. Readjust if necessary.
6. Whenever an EUT malfunction is noted, reduce the power and determine the threshold at which the malfunction appears. Note both the forward power and current.
7. Repeat for each EUT cable.

Frequency Increments. Frequency sweeping by increments is preferable to continuous, analog-type sweeping. Use proportional increments rather than fixed steps. For instance, with increment sizes such as $0.01 \times F_0$ (Ref. MIL-STD-462D, Appendix A), a scan from 0.1 to 100 MHz requires approximately 700 steps. This is a trade-off between the monumental time it would take to explore all the susceptibility windows and the risk of hit-or-miss by using excessively large increments. The 1 percent increment rule is based on the assumption that some EUT susceptibility windows (unknown to the operator until the test is finished) can be very sharply tunable, with Q factors > 10 (Fig. 3.21).

Since only a few susceptibility windows will exhibit such behavior, one can significantly reduce the test duration by setting a test level 3 dB higher and using steps of 0.02 or 0.05 F_0, then returning only to the weak spots for a finer search.

At each step, the dwell time depends on the EUT time to respond and, furthermore, to display the malfunction. Outputs that activate a mechanical function (meter movement, servo motor, etc.) are slower to react than video displays or digital functions. By default, a value of 1 second per step can be selected, but this time needs to be augmented by the settling time of the signal generator sweeps, plus eventually the forward power adjustment.

Diagnosis and Fixes. This is similar to EFT testing.

Other Methods of Injection: CDNs and Electromagnetic Clamp. Two other methods are possible for RF injection. One, called out in IEC1000-4-6, consists of using complete sets of coupling/decoupling networks (CDNs) to inject a prescribed CM voltage into each wire of the tested I/O cable. The CDN is a highpass filter (sort of a LISN, used as an injector) such that the injected current goes into the EUT only, and not into the auxil-

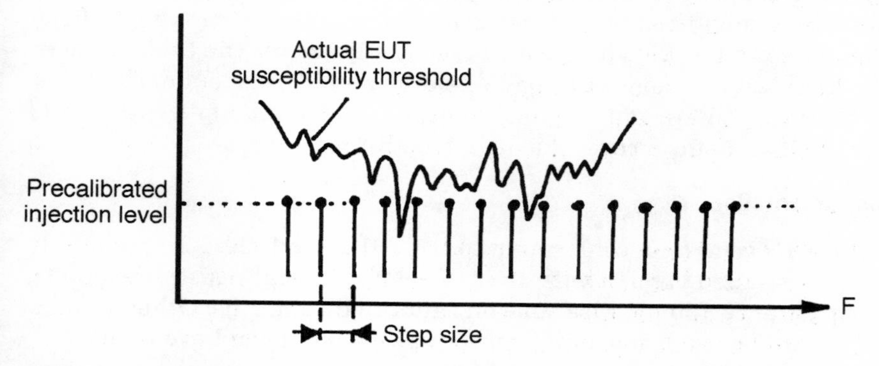

Figure 3.21 Illustration of the hit-or-miss risk, using frequency increments, when the EUT exhibits sharp resonances in its susceptibility profile.

iary equipment (AE). This method is less wasteful of power than the BCI, but it requires a huge number of different CDNs, according to the number of different interfaces that can be tested. For cables with more than two or four wire pairs, the method is very impractical, compared to BCI.

The electromagnetic clamp combines the magnetic injection of BCI with a capacitive injection via a slotted metal sleeve. Therefore the cable sees simultaneously an electric and magnetic coupling, with a directional effect: the resulting power is sent to the EUT side, practically no power being sent toward the AE. This is a very attractive method, with an excellent efficiency, requiring less power for the amplifier. Unfortunately, as of 1999, the electromagnetic clamp is rather large and bulky, and not easy to use on a development bench. If progress can be made to ameliorate that drawback, this certainly will become the perfect injection method of the future.

RS Test Methods Other Than Injection. Not considering formal radiated susceptibility tests with antennas, equivalent RF illuminations can be performed in a stripline or TEM cell. Both are well suited for testing small EUTs and easily can generate high-voltage fields (≥100 V/m) with a moderate power expense. However, the cable illumination is limited to vertical E-field, and the method is more a susceptibility test for the box alone or for a prototype PCB. The stripline is easy and inexpensive to construct, but this is an open-sided structure, and it radiates significantly outside.

3.3.5 ESD Test

The ESD test is a powerful trigger of potential EUT failures. Not only does it cover an extremely wide spectrum in a glance (the 1 ns rise time corresponds to an occupied spectrum of 320 MHz), but, when applied in a conducted way, it generates an extremely strong EM field. Compared to the EFT test, which essentially excites the cable aspects, the ESD test also addresses box shielding and PCB layout deficiencies. In a sense, you are zapping, in 100 ns, a broad spectrum that it would take half an hour to cover during a true RF test.

Instrumentation

- *An ESD generator,* conforming to IEC801-2 (or 1000-4-2), preferably the post-1990 version with the HV relay. Internal resistance: 330 Ω, capacitance 150 pF. Rise time on calibrated load, 1 ns. Other simulators can be used, depending on the specification you have to meet.
- *A field coupling plate,* if the EUT has to be tested for indirect electrostatic discharge (IESD).

EUT Setup. The EUT preparation is the same as for all other immunity tests (Fig. 3.22*a*), with a special emphasis on the proper decoupling and immunity levels of the AE, if any. The location of the simulator ground lead is a sensitive parameter: It should be attached to the ground plane on the same side of the EUT as the face being tested.

Test Routine

- Decide on the indirect (IESD) or direct (DESD) discharge procedure and discharge type on EUT areas: contact or air discharge (Fig. 3.22*b*).
- Identify and mark the discharge points either on the EUT faces, using a water-based pen or on a sketch of the EUT.
- Start with a low level (e.g., 2 kV) in a repetitive mode (5 or 20 pulses per second) for a rough survey of all the discharge areas.
- Increase the level progressively by 2 kV steps until you reach the ESD criterion plus a 1 or 2 kV margin.
- If failures are detected, return to the failed area with a single-shot mode to allow for EUT software recovery (if any). Run a minimum of ≈50 discharges per point (this is because the ESD pulse is a very short-duration, random phenomenon, and only applying a few discharges would make this a hit-or-miss process).
- Record the run/fail levels (an example of data reporting form is provided in Appendix J).
- Document clearly the type of error/malfunction.

ESD Test: Diagnosis and Fixes. If the test fails (e.g., EUT does not meet severity level 4, 8 kV contact discharge and 15 kV in air):

- Record and review carefully:

  (a) the run/fail levels at all discharge points
  (b) the type of error/malfunction associated

- When trying fixes such as:

  – shielding (or shielding improvement) of I/O cables
  – shielded connectors
  – filter connectors, ferrites, decoupling capacitors
  – better bonding of loose panels, doors, etc.
  – better bonding of accessible switches, keys, etc.
  – reduction or gasketing of slots and seams

  always grade your progress via a new run/fail map.

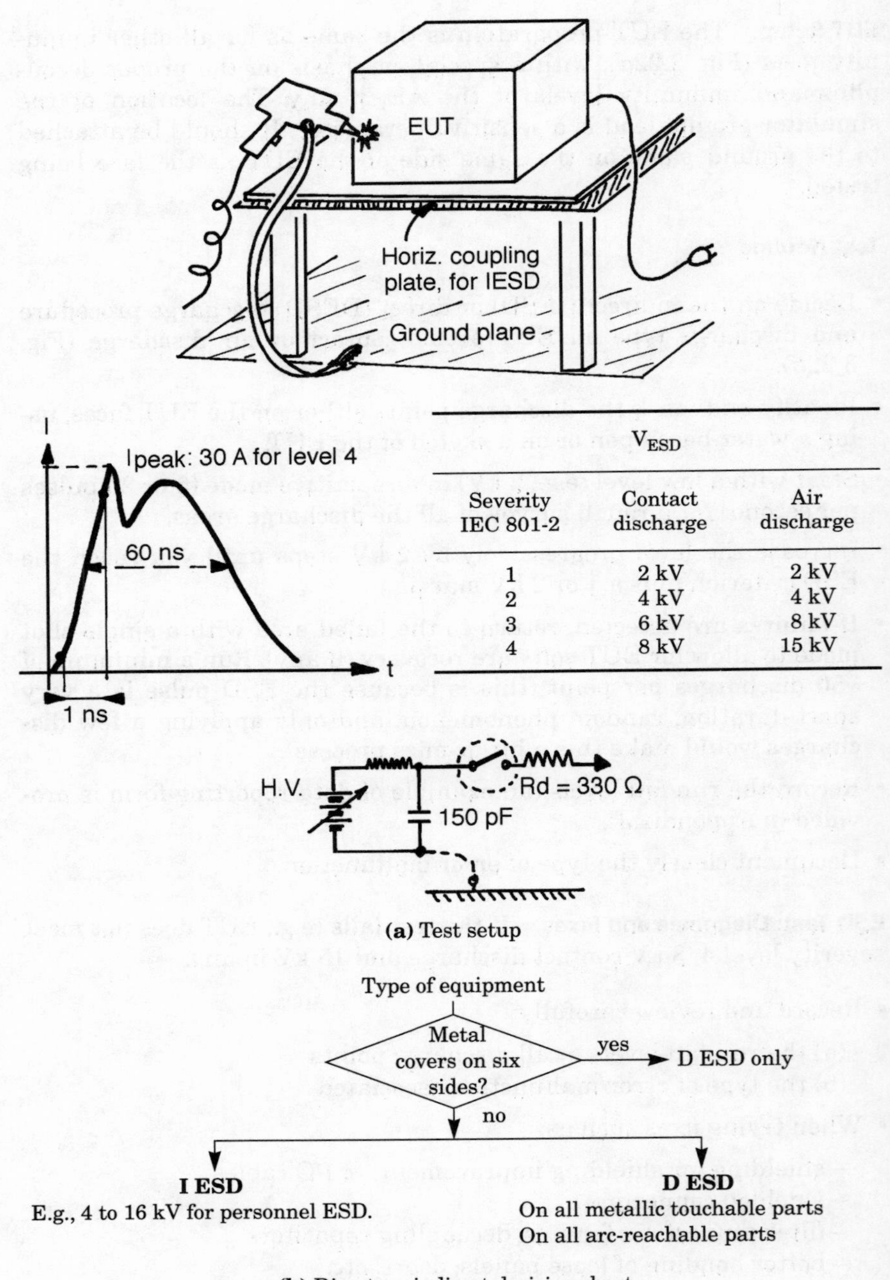

	V_{ESD}	
Severity IEC 801-2	Contact discharge	Air discharge
1	2 kV	2 kV
2	4 kV	4 kV
3	6 kV	8 kV
4	8 kV	15 kV

Figure 3.22 Typical setup and schematic of ESD simulator for human body-type ESD, per IEC 801-2.

- Try to visualize the ESD current paths and how they could be affected by fixes; this often gives a clue to the validity of a fix.
- Never remove a fix that seems not to work. Accumulate them until you achieve success. Then, remove fixes sequentially to find out which were useless.

3.4 EMI Problems in the Field (What to Do When Equipment Fails)

EMI problems occurring at customer's site often present a difficult challenge for the engineer or technician. The equipment is hardly modifiable internally, so EMC has to be upgraded, usually by external additions or by modifying the physical installation. It could be argued that, at the time this handbook was written, most electrical/electronic equipment sold on the market has to meet reasonably severe EMC specifications such as FCC or CSA (for the USA and Canada), European Norms, VCC (for Japan), and so on. This is even more true for the military and aeronautical industries where stringent norms have been in force since at least the mid 1970s. Therefore, it may seem that susceptibility and emission incompatibility problems simply should not happen anymore.

Although things have significantly improved since the legal (and contractual) enforcement of EMC standards, the ideal situation where "nobody disturbs anybody" is still far away, if ever within reach. There are many reasons for this, including

- The presence of older equipments that do not comply with any EMC specification
- Inadequate cabling and grounding practices by unaware contractors
- Specific EMI environments that are more demanding than usual
- Certain sensitive equipments that by nature are impossible to harden (e.g., e-beam microscopes, magnetic sensors, and sonar detectors)
- Certain equipments which, by nature, generate strong RF emissions (e.g., industrial, scientific, and medical equipment)

EMI troubleshooting on an installed system, generally at customer's site, can become very frustrating at times, particularly when you, and the people around you, think you have tried absolutely *everything*. If this happens, step back mentally, get out of the bushes to see the trees, and reflect again on the simple facts we related at the beginning of this manual. Ask yourself:

1. What and where is the source?
2. What is its frequency range?
3. What and where is the victim?
4. Which coupling path(s) can the EMI possibly take?

Don't take anything for granted. EMI hide-and-seek is a very deceptive game. Bonding points that appear to be good conductors may not be. Grounding connections that people swear have been installed may be missing. Parts that seem to be floating often are not. Capacitors become inductors at HF, while inductors become capacitors. The six people asked about the facility's earthing scheme may give you, with absolute conviction, six different versions. Good common sense recipes fail, not because common sense does not work but because you have overlooked one or more ingredients. If nothing else, tracking down EMI teaches us modesty, even if, when things finally become obvious, we say, "I knew it!"

The strategies recommended here are ones that have been field tested over many years. It may occasionally happen that, with a little bit of luck, someone using black magic and blindfold recipes will get faster results. But in the long view, we have found that methodical approach always produces documentable and reproducible success.

3.4.1 Before Moving to the Problem Site

In many instances, we have found that hurrying to a problem site in response to a panic call may lead the specialist to commit two errors:

1. He does not collect enough data to prepare one or more plans of attack.
2. He brings either too many or too few pieces of equipment, and either way will leave behind the one precious accessory that does not exist at the site.

It is somewhat as if a surgeon were cutting through a patient's skin without any preconceived idea of what he is after. Forewarned is forearmed; collect detailed information in advance.

1. Interrogate the users for possible correlations.
   - Is the problem intermittent or continuous? If intermittent, is it predictable (can it be re-created on request)? This may affect the instrumentation needs.

- Does the failure correlate with a specific time of day? With certain loads on/off the power line? With the operation of local or portable transmitters?

2. What instrumentation is available on the site?

   - Oscilloscope? Bandwidth? Is it a memory type?
   - Spectrum analyzer? Frequency range?
   - Antennas?
   - Transient recorder for intermittent problems?

   Make sure you bring whatever is necessary to complement the available equipment.

3. If there are signs that a nearby radio transmitter could be the problem, try to obtain transmitter data from the operator, the FCC, or other sources, including

   - Transmitter power and antenna gain
   - Frequency
   - Transmitter/antenna distance and direction from the site

4. In many cases, some rough prediction can be done in advance. For instance, in case of a possible CW ambient field, a quick estimate of the field (see Appendix C) at the site may indicate this as a likely cause. A calculated field strength of more than 1 V/m, especially at VHF and above, indicates a very possible cause. Calculated fields of less than 0.1 V/m eliminate this candidate.

5. Based on (1) through (4) above, make some hypothesis about what the source and coupling paths could be.

6. Finally, plan some advance strategies for diagnosis and fix. Do not stick to one plan. Prior to engaging in combat, a general has one or several backup plans in case the enemy does not do what is expected. Foresee some "what-if" actions.

3.4.2 Upon Arriving at the Site

Visual Inspection of the Victim Equipment

1. Does the victim have power-line EMI filters? Do they address both CM and DM?

2. Are the filters mounted correctly through a metallic bulkhead or on a metal wall?

3. Examine the grounding scheme. Multiple ground loops usually exist.

4. Examine interconnecting control and signal cables. Are they shielded? How/where are the shields grounded? Are they running close against other power cables carrying large currents?
5. Are the signal or power cables run in metallic raceways?

Visual Inspection of the Site

1. Are there other heavy-load users on the victim's power branch circuit?
2. Do people use portable transmitters? What power and frequencies?
3. Are there any nearby radars? FM/TV transmitters?
4. Any there nearby air conditioners? RF arc welders? Neon signs? Power converters or ac motors variable-speed drives? Dielectric heaters, and so on?

Continuous or Quasi-continuous Problems. By "continuous" or "quasi-continuous," we mean that the problem is either *always* there or occurs at several times within a reasonable observation period so it can be probed and demonstrated over and over. Assuming that someone has made certain that the equipment itself is in good condition (no self-jamming or marginal conditions), a continuous problem is a blessing for the troubleshooter because it allows him to find the source faster and evaluate the fixes more easily.

The procedure to follow is shown on the upper-right branch of the flow diagram in Fig. 3.23 after answering "yes" at the first question (question box A). This will lead you to the power or I/O cables as a possible coupling media to explore, which will be addressed in the next section.

Intermittent Problems. Unfortunately, intermittent problems constitute a large share of field calls, and this implies a longer preliminary routine to locate the cause. This is indicated under the first vertical branch of the block diagram in Fig. 3.23, once you have answered "no" to the first question (box A). It implies a patient search and interrogations to find evidence of a correlation between the failure and an intermittent operation (turn-on, turn-off, or changes in loading) of some equipment in the surroundings. "Surroundings" may be quite vague, and this could refer to an elevator in the hallway, or an arc welder in a garage a block away, or even a CB transmitter or portable telephone. In any case, if correlations are found and confirmed by reproducing the problem (answer "yes" at box B), we have progressed to the situation where the source is identified, and we can concentrate on the coupling path in next section.

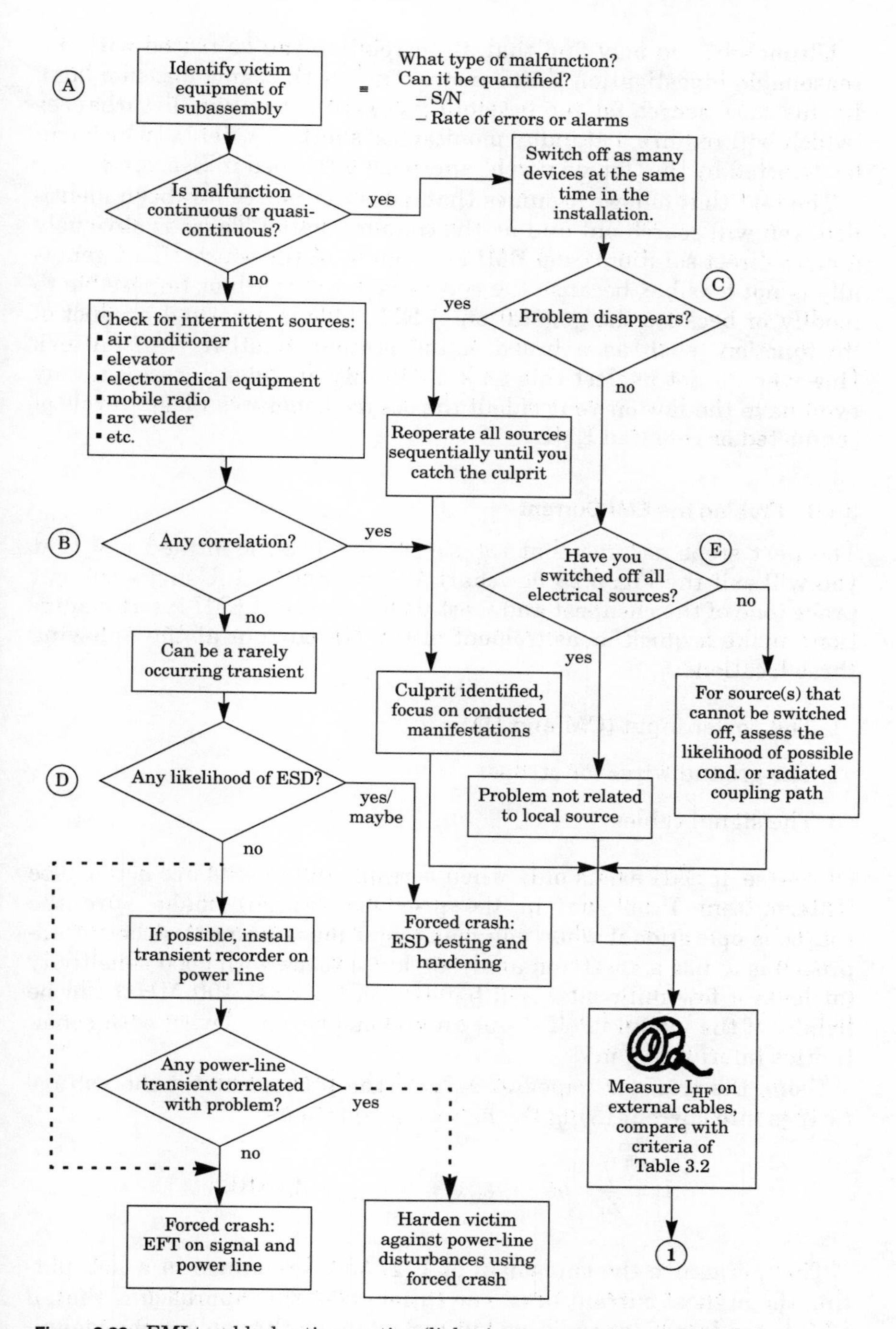

Figure 3.23 EMI troubleshooting routine (lightning events not considered).

Ultimately, you may find that no correlation can be traced within a reasonable investigation time (answer "no" at the same decision box). In this case, search for the possibility of rare power-line disturbances (which will require a standby monitor) or short transients (which can be detected by the "forced crash" approach described in Sec. 3.4.4).

The text that follows assumes that, once the source has been identified, you will search out and fix the coupling paths. There is obviously a more direct solution: Stop EMI generation at the source! This generally is not feasible, because the source is out of reach or impossible to modify, or because the generation of EM fields is a normal product of its function (such as a broadcasting station or other transmitter). However, do not neglect this as a possibility. In some cases, you may even have the law on your side if the source generates illegal levels of conducted or radiated EMI.

3.4.3 Probing the EMI Current

The next steps assume that the source has been identified and that you will exit the Fig. 3.23 flowchart at termination 1. Using a current probe (one of the cheapest and most useful pieces of EMI instrumentation), make a quick measurement of the HF current at the following three locations:

1. The power input (CM and DM)
2. The ground wires (or straps)
3. The signal cables

Of course, if EMI exists only when certain equipments are active (see "Intermittent Problems" in the previous section), make sure the source is operational when you make your measurement. A better approach is to use a spectrum analyzer, but a scope with good sensitivity (at least a few millivolts) and bandwidth (at least 100 MHz) can be helpful if the victim itself is not an analog or radio device with sensitivities inferior to 1 mV.

Using the transfer impedance, Z_t, of the probe, convert the voltage ratings into current using the following equations:

$$I = \frac{V}{Z_t} \quad \text{or} \quad I_{dB\mu A} = V_{dB\mu V} - Z_t(dB\Omega)$$

Then, organize the categories of 1, 2, and 3 readings in a list, putting the highest current first. The rationale of this approach is that, if EMI is conducted (or radiated but picked up by the cables), the magnitude of the RF current will be a clue to the coupling path(s) and its (their) relative importance.

At this juncture, a quick estimate already can be made of the likelihood of any of these cables being an EMI path. Take the highest figure (in milliamps) of these currents and multiply it by the input impedance of your victim's circuit (e.g., a few hundred ohms for digital, a few tens of kilohms for op amps, and 50 or 75 Ω for radio communication receivers).

Compare the result with the intrinsic sensitivity of the victim's front end. For instance, 0.1 mA flowing on a cable terminated by 120 Ω produces 12 mV of noise in the worst possible case, which is well below the sensitivity of any digital device. On the other hand, the same noise would be a very possible threat for radio communication equipment, a modem, a low-level process control amplifier, or other sensitive devices.

Exiting the flowchart via termination 1 brings you to the next chart, Fig. 3.24, which allows you to individually fix or eliminate the cables as possible coupling paths. Finally, an issue exists at decision box F in the figure, whereas the cables are not the (or not the only) coupling path. This indicates that the EMI radiates through the box leakages, and you must conduct the corresponding investigation.

3.4.4 Noncorrelatable Failures: "Forced Crash" Technique

Returning to the first part of the routine (see Fig. 3.23), if the answer at box B was "no," i.e., no correlation has been found, an assumption has to be made about the likelihood of either power-line disturbances, fast transients, or ESD.

The first kind of EMI generally relates to either some regularly timed occurrences, stormy days, or other malfunctions noticed in the same building.

The second type, since it has not been identified by turning off all potential sources (box B), may be caused by one of the many contactors, relays, or solenoids in the installation—generally within no more than a few tens of meters from the victim.

The third type, ESD, often has a seasonal relationship, typically increasing during cold and dry weather. It also depends on human activity and the type of floor covering.

"Forced crash" is a technique by which you decide that you cannot wait for a random, hard-to-catch problem to show up, and you deliberately inject onto the equipment a transient pulse (or train of pulses) that approximates the suspected EMI threat. Starting with low levels, the injected transient is increased until the equipment exhibits malfunctions, degraded performance, or other symptoms. If this threshold

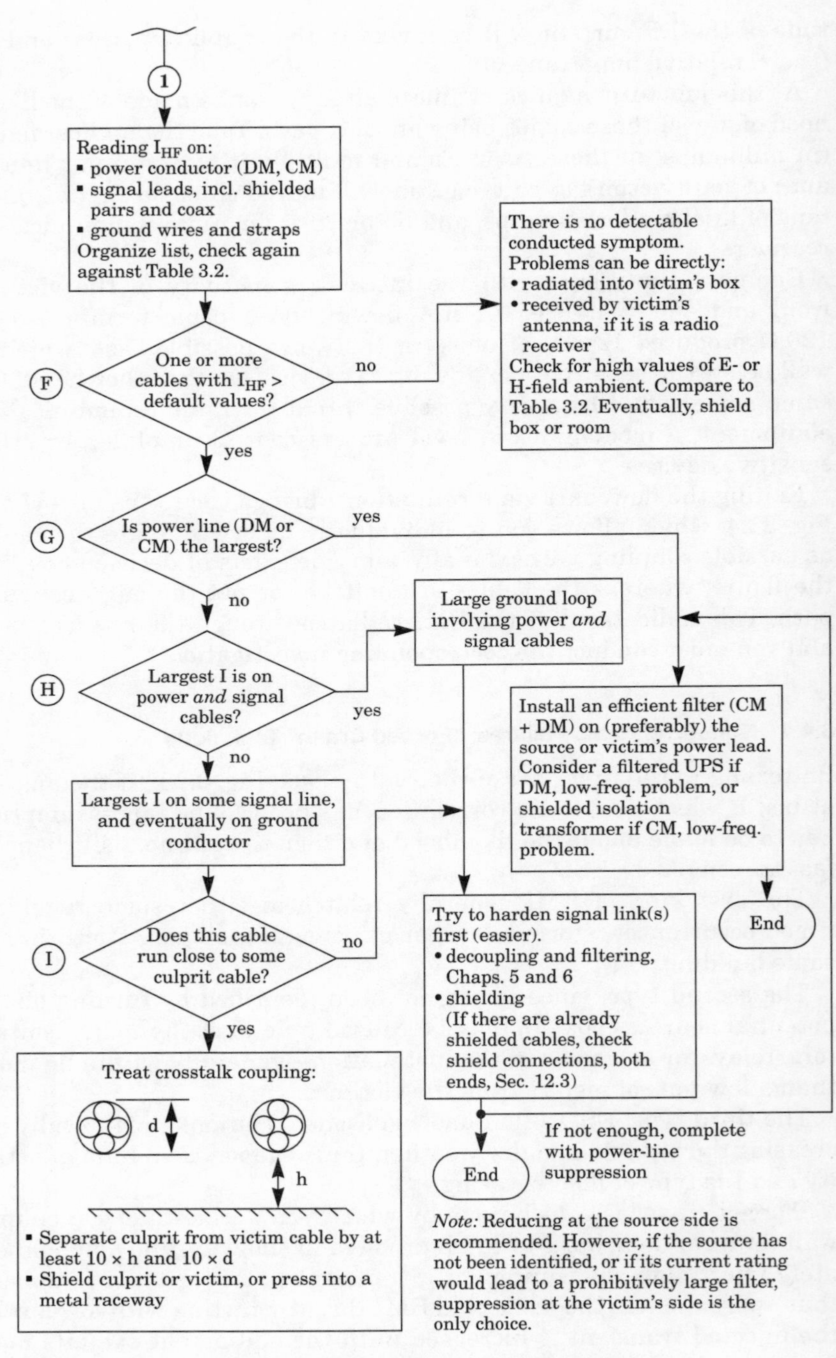

Figure 3.24 EMI troubleshooting routine (continued from Fig. 3.23).

is lower than, or in the range of, expected disturbances for this kind of site (Table. 3.2), you will apply appropriate fixes until the equipment meets or exceeds the immunity level typical of its category. Then, after an observation period of days or weeks, you should notice that the problems have disappeared or that their number is significantly reduced.

Table 3.2 Orders of Magnitude for Sure Immunity and Sure Susceptibility When Tracking Field Problems *(courtesy of AEMC, France)*

Interference	No problem	Problems ...
HF E-field,* 10 to 1000 MHz	0.3 V/m @ worst freq.	10 V/m @ worst freq.
H-field, mains freq.	0.3 A/m (CRT, microscope, etc.)	1000 A/m, "standard" equipment
CM HF current* on mains cable	10 mA @ worst freq.	300 mA @ worst freq.
CM HF current* on signal cables	0.3 mA analog, 3 mA digital	100 mA, 1 A shielded cable
RMS voltage* on mains	0.4 V, bad regulator	10 V @ worst freq.
CM HF voltage* at electronic input	3 mV low level, 300 mV digital	10 V without opto, 300 V with opto
Leakage current** on green wire	1 A, but ac protection	100 A, video + H-field

Note: These are default values for all types of equipments (except radio receivers, which are very sensitive in band). "No problem" values are the ones below which *no equipment* can be susceptible. "Problems" are values for which *any equipment,* even fairly hardened ones, will be disturbed.
*Multiply the permanent RMS value by about 100 (typically from 30 to 100) for a peak impulsive "equivalent" value.
**Allow 1 A rms/MVA for electronic equipments.

The assumption is, *"If it withstands the standard immunity level on site, the equipment is vaccinated against any type of short, fast-rising transients, even if the actual cause of the problem is never to be found."*

The forced crash method is quite appropriate to use with EFT and ESD, as explained next, and can be used for power mains disturbances as well.

3.4.5 Forced-Crash Techniques with EFT

The EFT test setup and routine are basically the same as for development test (Sec. 3.3.3). The differences are as follows:

1. You probably do not have a reference ground plane. If the victim equipment is in a frame or located near a metallic structure, this will be the reference plane for the EFT generator. Otherwise, you need to install a temporary one, such as a layer of aluminum foil, just underneath the generator (Figs. 3.25a and b). Since EFT addresses cable-induced noise, do not connect the artificial ground plane to the EUT chassis, since this would misrepresent the actual CM return path.
2. You are not testing a development prototype with certain diagnosis commodities, but an equipment actually in service. Be on the lookout for symptoms that have been already detected by the users. Make sure that your forced crash does not lead to severe material damage or even personnel hazard. Try to inhibit temporarily the peripherals that could create such risk.
3. If the EUT power cord does not fit the generator ac outlet, use the foil wrap like for the signal cables (see Sec. 3.3.3). If certain cables are inaccessible or cannot be taken out of their raceways over a sufficient length for the foil wrap, you can, at the very least, inject the EFT on the metallic raceway itself.

Test all cables, starting from a low level. The EUT should be able to reach "run" (no-fail) levels at least equal to the IEC 801-4 recommended severities shown in the following table:

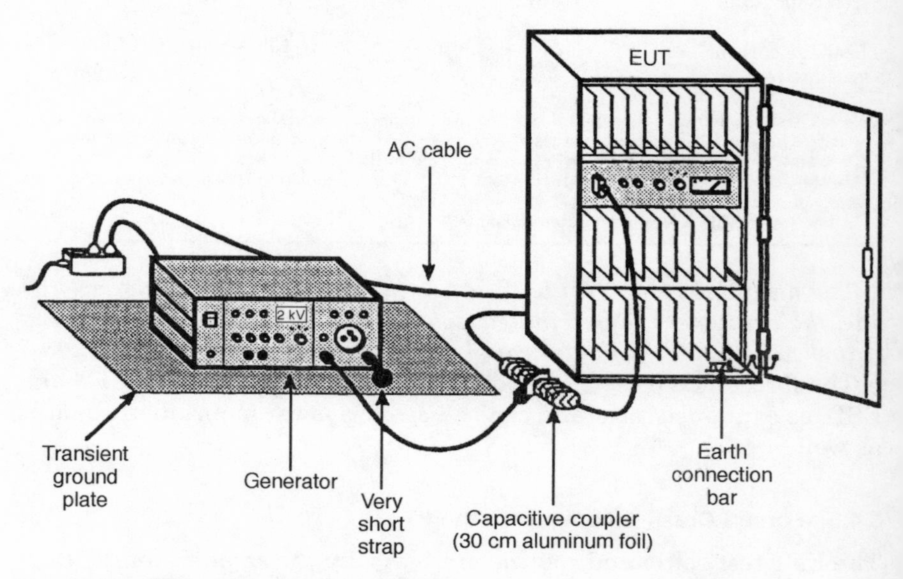

Figure 3.25a EFT on-site injection method.

Figure 3.25b In-situ "forced-crash" investigation with electrical fast transients (EFTs). The simulator is resting on, and grounded to, the metallic structure of the automated belt conveyor. The cable under test is slightly removed from its raceway and wrapped in aluminum, serving as the hot electrode for EFT injection. The HV cable is kept short, approximately 1 m maximum. The generator is an early, but extremely dependable and robust, model (NSG 225, by Schaffner).

	EFT on power mains	EFT on signal lines
Residential/commercial	2 kV	1 kV
Industrial	4 kV	2 kV

If these immunity levels are not met, see the fixing hints in Sec. 3.3. The author has seen many situations in which an installed system had "fail" levels as low as 250 V.

3.4.6 Forced-Crash Technique for ESD

This is to be used when there is indication that the problem could be an ESD (decision box D), or in any case if the EFT test cannot be made.

Forcing an EMI failure with an ESD is a powerful diagnostic tool and a highly localized stimulus. It precisely reveals weak spots in the equipment. Since the pulse is calibrated, progress can be quantified and a susceptibility map can be drawn.

The procedure is similar to the in-lab test (Sec. 3.2.6).

1. Install a layer of aluminum foil underneath the EUT. This ground plane is to stabilize the RF reference during the test and improve repeatability such that the results can be compared to typical immunity values. Do not try to ground the EUT to this plane, since it is normally earthed via its power cord.
2. Mark or identify the discharge areas for direct (metallic equipment) or indirect ESD (plastic), following the guidelines of Sec. 3.3.5.
3. Connect the ground lead of the zapper to the test ground plane (Fig. 3.26).
4. Start with low value, with a minimum of 30 discharges per point.
5. Increase the level by 1 kV steps until you reach a "fail" level or the ESD criterion, whichever comes first.

For an installed equipment, with an IEC 801-2 type of gun, the following "no-fail" levels must be aimed for, depending on the site:

	Contact ESD (direct or indirect)	Air discharge
1. Well-controlled relative humidity, concrete or metal floor, or antistatic carpet	4 kV	4 kV
2. Ordinary office environment, no antistatic precautions	6 kV	8 kV
3. Very dry atmosphere, high level of human activity	8 kV	15 kV

Warnings

- Due to the high voltages that the ESD zapper can deliver, this method could be dangerous for an equipment whose vulnerability is unknown, or even for personnel (especially bearers of cardiac peacemakers).
- This is not a damage-check test: do not discharge directly on parts that are directly wired to sensitive electronics such as connector pins, membrane switches, LEDs, etc.

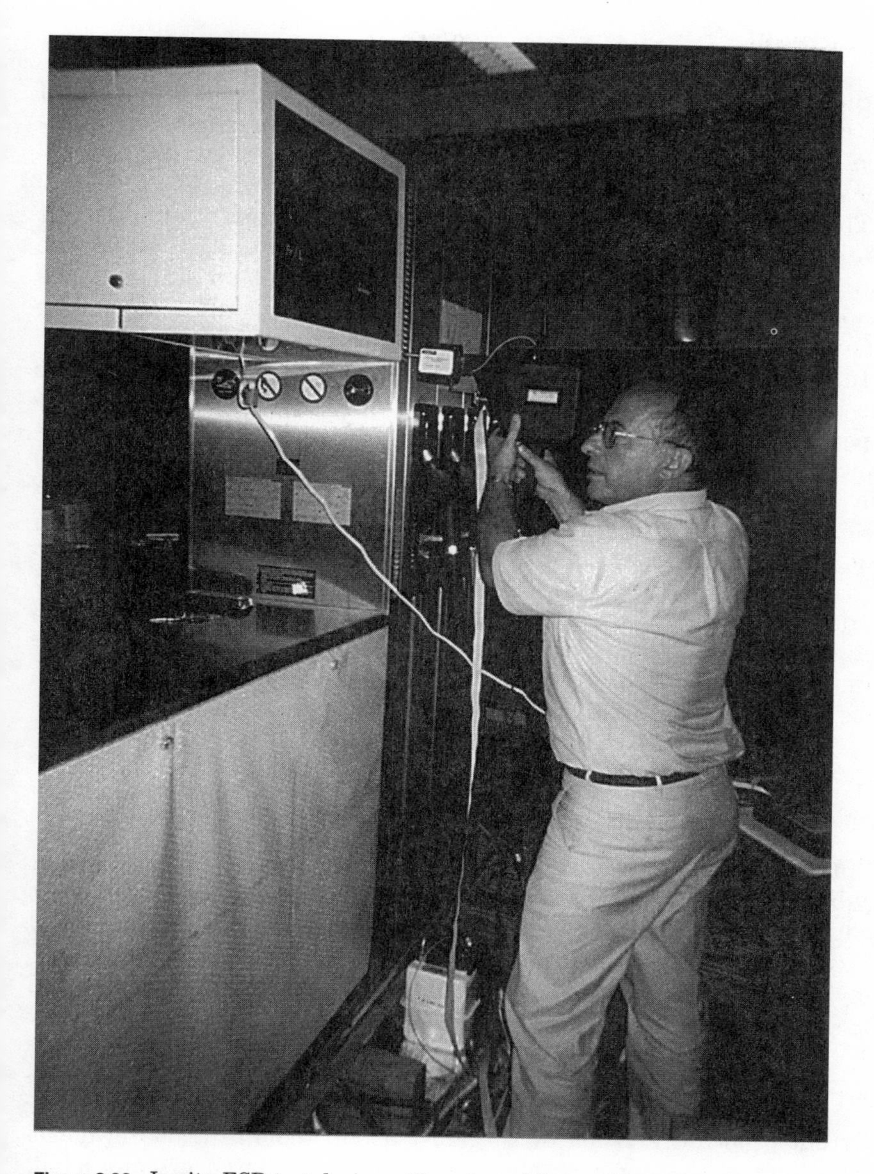

Figure 3.26 In situ ESD test during a "forced-crash" procedure. Here, the author is applying a direct electrostatic discharge (DESD) with the HV relay near a front panel seam, a typical weakness zone. The simulator grounding strap drops vertically to reach the temporary ground plane (aluminum foil) beneath the EUT.

- Because of the strong radiated field caused by ESD, check that no sensitive equipment is being used within less than 3 m.

3.4.7 Power-Line Monitoring

Although it is related to power-line transients, there is a domain of disturbances that the EFT forced-crash test does not cover: long-duration power mains disturbances such as energetic surges with durations in excess of 0.1 ms, short dropouts, and undervoltages.

These, too, could be approached by a forced-crash test, but this would require inserting a power-line disturbance simulator in the equipment power leads. This is a difficult test to do on site, so, if we suspect power-line disturbances, a solution is to install a power-line monitor such as those built by Dranetz, BMI, and others.

Installing such a "spy" device will be for naught if you do not define the types of disturbances that you consider to be tolerable and which types are intolerable. Try to get a figure of the actual power disturbance immunity of the victim equipment. It normally should appear in the manufacturer's technical specifications. Otherwise, use the CBEMA template (Fig. 3.27) as a default, with some safety margin (e.g., 0.9 times the indicated tolerance values) to account for installation variances. If, after an observation period, the monitor has recorded disturbances exceeding the preset threshold, and you have noticed time-correlated equipment failures, you have to fix the problem by using one of the power conditioning or surge suppression fixes described in Chaps. 9 and 10.

3.5 How to Evaluate Fix Results: Current and Field Probes

While you progress through your EMI diagnostics and fix iterations, it is absolutely necessary to have an estimate, even a coarse one, of the results you are getting. Current and field probes are traditionally used for calibrated measurements during standard EMI testing. However, they are also valuable tools to grade the improvements as you step through your troubleshooting and fixes.

3.5.1 Current Probes

Current probes are probably the most useful tools in tracking noise. Regardless of the noise-coupling mechanism (radiated or conducted, CM or DM) *it will finally show up somewhere as a high-frequency current*. When such a probe is clamped on a whole cable harness (embracing incoming and returning signals), it will read only the CM current (normal DM currents cancel each other's flux).

A large proportion of conducted problems (especially above a few megahertz) and practically all EMI radiated problems are strongly related with the presence of CM current in I/O cables. Therefore, any de-

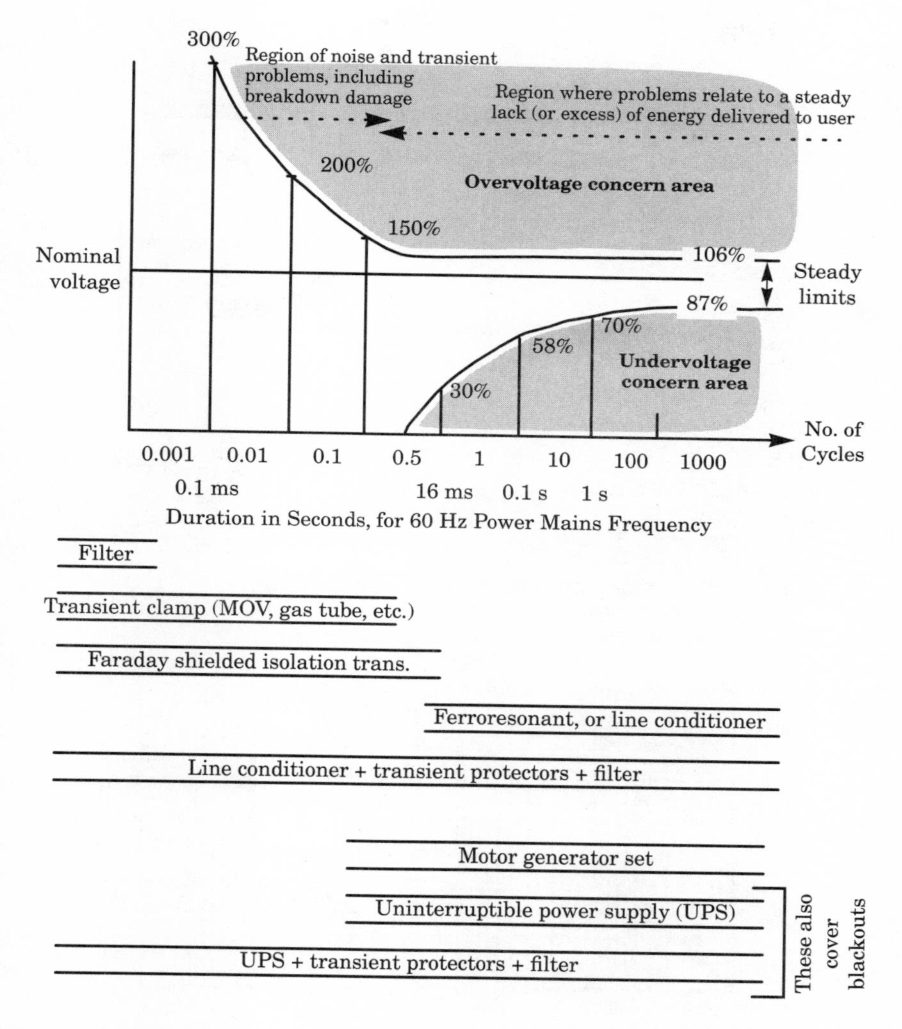

Figure 3.27 Template of line-voltage tolerances and solutions at installation level.

crease in the current probe's readout is a sure indication of an EMC improvement. In cases where conducted EMI exists also as DM currents, these too can be recognized by the probe, with the aid of a special wire arrangement (see Fig. 3.28).

To use the probe, you need to know the probe ratio in voltage/current (called *transfer impedance*), i.e., how many millivolts the probe output will deliver for 1 mA flowing through it. So, to find the unknown current, I, the transfer impedance Z_T of the probe is used as follows:

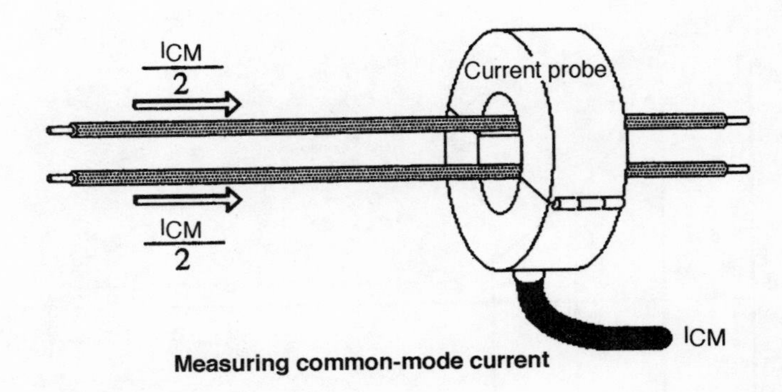

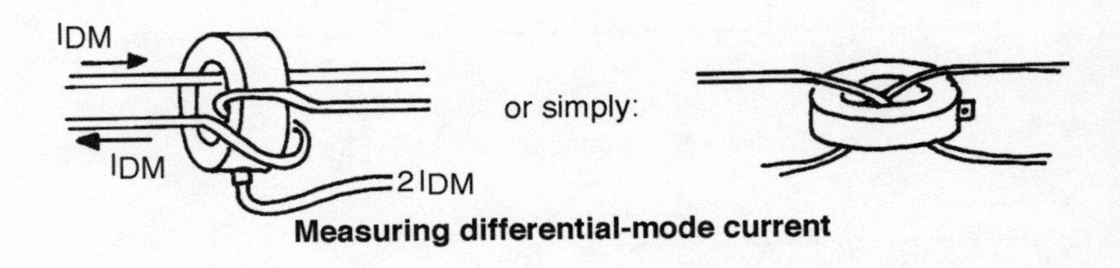

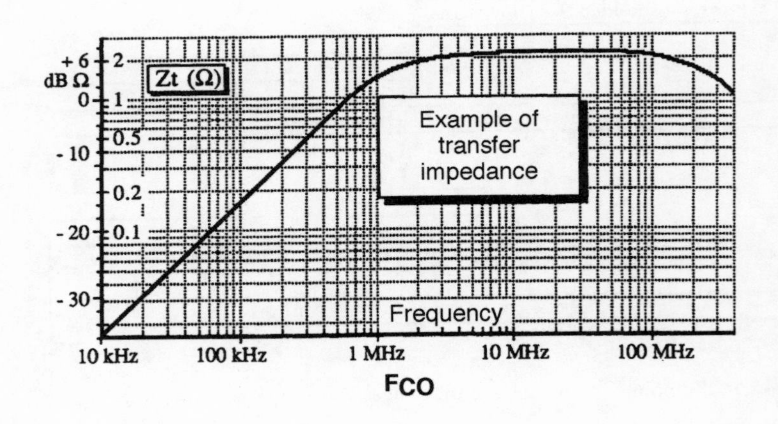

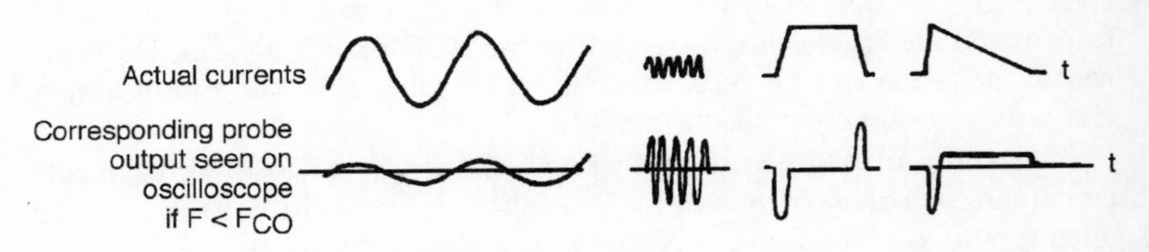

Figure 3.28 Current probe applications: example of distorted time-domain view delivered by an EMI current probe when used below frequency F_{CO}.

$$I\ \mathrm{dB}\mu\mathrm{A} = V_{(\mathrm{out})}\ \mathrm{dB}\mu\mathrm{V} - Z_T\ \mathrm{dB}\Omega$$

or

$$I\ \mathrm{dBA} = V_{(\mathrm{out})}\ \mathrm{dBV} - Z_T\ \mathrm{dB}\Omega$$

The Z_T parameter, of course, is not constant with frequency. The manufacturer provides a curve for Z_T versus frequency, which has the general shape shown in Fig. 3.28. Usually, below a few tens of kilohertz, the curve follows a 20 dB per decade slope. This means that if you use an oscilloscope, and the pulse width of the current corresponds to this low frequency range, beware that the scope picture is not a replica of the current waveform but its derivative, which will tend to amplify the highest frequency contents (see Fig. 3.28). This can be avoided, of course, by using a spectrum analyzer. Above its curve "knee," Z_T is usually flat such that the voltage output is a replica of the current. A homemade current probe is shown in Fig. 3.29.

3.5.2 Field Probes

While they belong to the general category of *antennas,* field probes are generally used to make proximity measurements near localized noise sources, leaky spots, seams, and so on. Although they can be calibrated

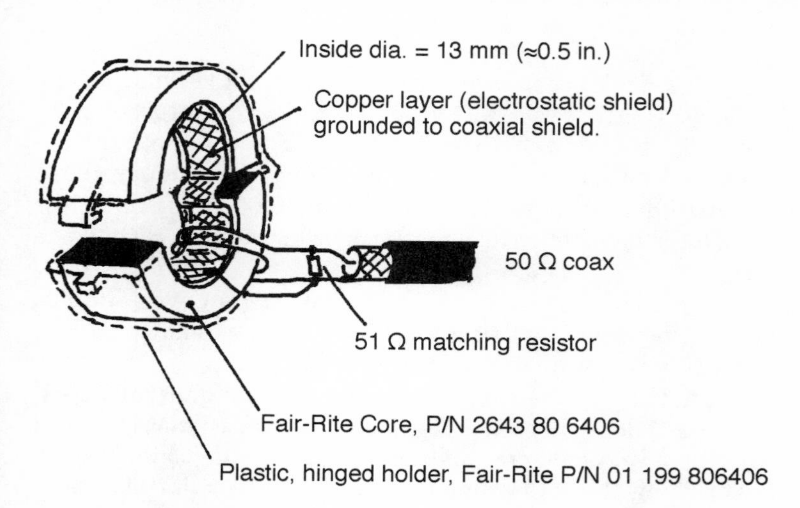

Useful range ≈ 0.1 MHz to 400 MHz
3 dB flat response = 3 MHz to 200 MHz, Z_t = 10 Ω (20 dBΩ)

Figure 3.29 Example of homemade current probe, for coarse current measurements (accuracy ±1.5 dB).

to a reference field, they are generally not used to make precise, absolute field strength measurements but instead to make quick assessments of the relative importance of one noise source versus another, or one seam leakage versus another.

The most broadly usable is the magnetic field sensor, made by winding a coil around an isolated mandrel (see Fig. 3.30). To keep it from acting as an electric antenna, a Faraday shield is built in the form of a nonclosed tube of copper or aluminum. One lead terminates on the center pin of a BNC plug, and the other lead is attached to the outer shell (ground) along with the probe shield.

The open voltage for any loop antenna at maximum field capture is

$$V_{volt} = 0.8 \times 10^{-9} F_{Hz} \times A \text{ cm}^2 \times N \times H \text{ A/m}$$

where A = loop area

N = number of turns

H = the incident, unknown, magnetic field

The 40-turn, 25 × 25 cm loop of Fig. 3.30, for the dimensions given, will deliver (if oriented for maximum pickup):

1. 80 mV per Gauss of field at 50 Hz (1 mV per A/m)
2. 1.6 V per Gauss of field at 1 kHz (20 mV per A/m)

Therefore, with a 1 mV scope sensitivity, it can detect H-fields as low as 1 A/m @ 50/60 Hz. With a 1 MΩ scope input, such a probe can be used up to at least 100 kHz. With a 50 Ω scope input, the output becomes constant (≈0.1 V per A/m) and independent of frequency above 5 kHz. This sensitivity is needed for detection of 50/60 Hz fields at low threshold (e.g., CRT monitors are sometimes vulnerable to H-fields as low as 1 A/m).

If the EMI frequencies are above a few megahertz, an even simpler H-field probe can be made by using a piece of coax, as shown in Fig. 3.30. The unterminated braid acts as the Faraday shield.

Since such H-field loops are very directional and are generally used in extremely near fields where the distance factor is $1/r^2$ or $1/r^3$, it is important to keep a very constant distance when making comparisons. An easy way of doing that is to tape a piece of plastic ruler along the probe diameter, cut to a length that will make it a spacer gauge. A typical distance range for close-in evaluation of box seam leakages or PC board hot spots is 2 to 10 cm from the rim of the loop.

One of the advantages of small "sniffer" probes is that they can be brought very close to the sources. This will

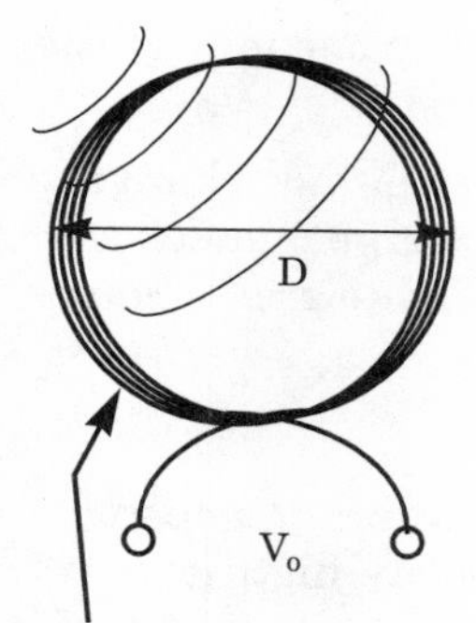

Loop open voltage:

$$V_{o(dB\mu V)} = -62 + 20\log(A_{cm^2} \times N) + 20\log F_{MHz} + 20\log H_{A/m}$$

When loaded by 50 Ω, the cutoff frequency is reached for $L\omega = 50\ \Omega$, i.e.,

$$F_{co(MHz)} \cong \frac{1300}{D_{cm}\left[\log_n\left(\frac{D}{d}\right) + 1.8\right]}$$

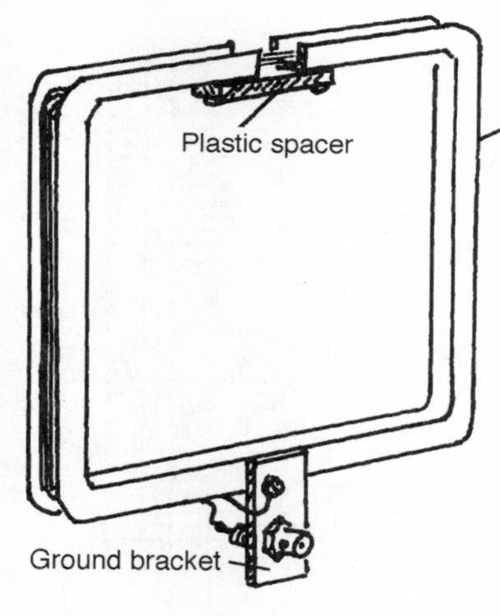

LF loop, for use at 50/60 Hz, up to ≈100 kHz
Example: 25 × 25 cm (10 × 10 in), 40 turns

$$H_{dB\mu A/m} = V_{dB\mu V} + 32 - 20\log F_{kHz}$$

up to $F_{CO} \approx 4$ kHz. Then,

$$H_{dB\mu A} = V_{dB\mu V(50\Omega)} + 20$$

HF loop, 1 turn
Example: at 1 MHz, 0.7 mV/cm^2 of open voltage, for 1 A/m of H-field

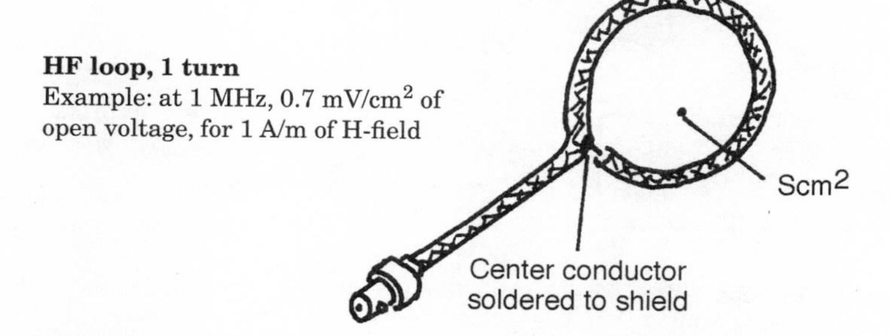

Figure 3.30 Examples of LF and HF H-field loops with electrostatic shields.

- prevent them from picking up radiations from other portions of the system, which could mislead your investigation, and
- enhance the reading from the localized source, while not changing the ambient reading. Your source emissions will emerge more clearly above the background clutter, CW stations, and other signals. Sometimes, you can even switch the spectrum analyzer to a lesser sensitivity, clearing out the ambient.

Although less useful than the H-field loop, an ordinary telescopic rod, fitted with a BNC connector, can serve to identify high levels of CW fields. An example is shown Fig. 3.31.

A 0.75 m (30 in) rod, loaded with a 50 Ω coaxial load, can at least give an idea of the nature of an ambient E-field, if it reaches typical

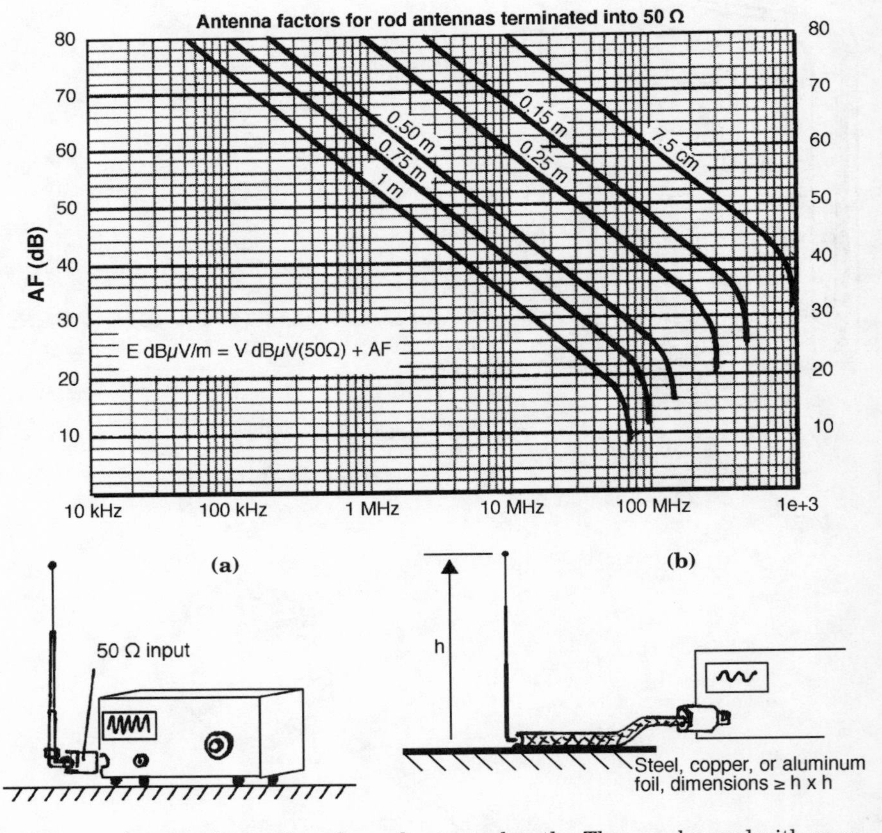

Figure 3.31 Antenna factors for a few rod antenna lengths. They can be used with an oscilloscope or spectrum analyzer. For better sensitivity and accuracy, a reference ground plane connected to a coaxial shield (b) is preferable to the vertical and undefined reference of (a).

values for EMI concerns. Such an antenna will produce approximately:

- 1 mV per V/m of field around 1 MHz
- 10 mV per V/m of field around 10 MHz
- 200 mV per V/m of field around 90 to 100 MHz

100 MHz is the upper frequency limit for using this 0.75 m rod, since it corresponds to its 1/4 wavelength resonance. It should not be used above that limit, or it should be shortened, with a corresponding change in characteristics.

For instance, in Fig. 3.31, using a 0.75 m rod, assume that a 10 mV p-to-p signal is measured at a frequency of 1.6 MHz. For 1.6 MHz, the rod delivers about 1.6 mV per V/m. Since the peak voltage is 0.5×10 mV = 5 mV, the corresponding ambient field is 5/1.6 = 3 V/m peak, or 2.1 V/m rms. This could indicate a powerful AM broadcast station, within a mile distance.

More commonly, the EMC community uses *antenna factor* to define its voltage-to-E-field conversion, when using a spectrum analyzer or EMI receiver, such that

$$E_{\mathrm{dB\mu V/m}} = V_{\mathrm{dB\mu V}} + \mathrm{AF}$$

3.5.3 More Hints for Troubleshooting Noise Problems

Noise measurements in the engineering lab or in the field do not always require specialized setups with EMI receivers, simulators, and the like. You sometimes have to deal with an elementary noise problem right on the corner of the workbench. Even for apparently trivial situations, few precautions are in order.

The Ground Plane. Any sound investigation of a noise problem (or simple measurement of a signal waveform, S/N ratio, etc.) should be made EMI-free. A reference plane (see Fig. 3.32) should always be in place. Do not rely on the earthing terminal of the ac outlet to quiet down HF noise since, to the contrary, several equipments powered from different outlets in the same room can give way to CM noise problems. A simple way is to use a steel or aluminum sheet metal or a simple aluminum foil. It will be the voltage reference for all your measurements. A paper overlay can be placed atop this plane if there is a risk of short circuits with your test jig. The plane is connected to the earth ground, for safety considerations.

The chassis of all your equipments, and the 0 V references of your circuit under test (unless it *has to* remain floating) must be connected

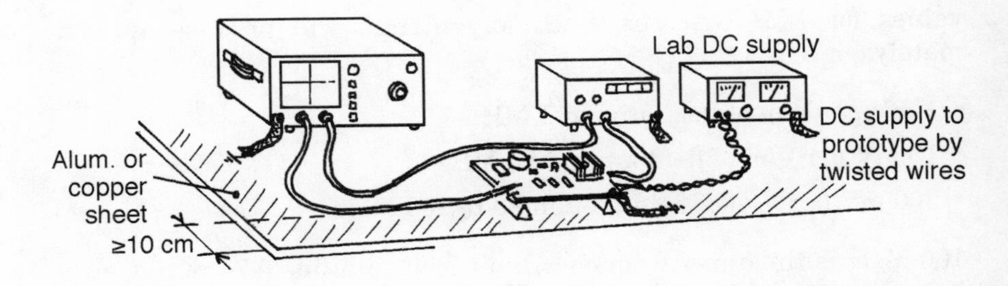

- All wiring and instruments are kept away from ground plane edges.
- Instruments and the prototype under test are grounded to reference plane using short straps.
- Hookup wiring and coaxial cables are run close to the reference plane (preferably attached to it by adhesive tape).
- The prototype 0 V reference is grounded to the reference plane near its I/O connector area.

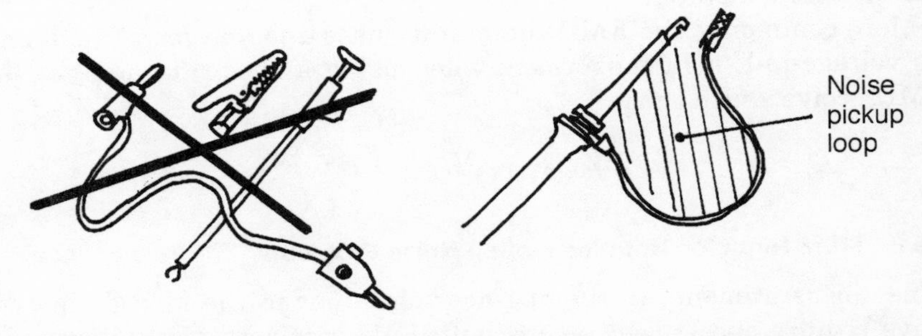

- A few items that should not have been used when troubleshooting EMI, or any measurements above a few megahertz.

Figure 3.32 Some practical guidelines for prototype measurements.

to the plane. The cables, loose wires, and coaxials must be laid on this plane at the lowest possible height. This is the simplest and best guarantee against induction, field pickup, and crosstalk.

The Instruments and Accessories. If a spectrum analyzer is not available, it is at least possible to make some broad EMI diagnostics using an oscilloscope, with the following precautions:

- The oscilloscope bandwidth must be sufficient to catch both the wanted signals you are seeking (which may be low-frequency analog) as well as the spurious noise. A 100 MHz BW is a minimum. (You still can use the BW limit switch when necessary.)

- It should have a selectable 50 Ω input, or you can use a separate 50 Ω coaxial load. Try to make most measurements via a 50 Ω input, if your circuit can tolerate this loading without affecting the wanted signals. This will stabilize the measurement impedances and make you less prone to capacitive couplings. For circuits that cannot tolerate a 50 Ω loading, use a highpass filter by inserting a capacitor, a few nanofarads in value, in series.
- If the signal (or noise) being measured is above a few megahertz, do not use loose wires, banana extensions, or alligator clips. Always prefer coaxial cables.
- With voltage probes, use the shortest probe cable that is compatible with your setup.
- Avoid lengthy ground clip leads (Fig. 3.32).
- FET amplified probes, or differential active probes, are very efficient and will enhance your ability to make low-level signals and noise measurements.
- Beware of line mismatches and VSWR when the cable lengths approach or exceed 1/4 wavelength. This would correspond in air, to length (m) = 75/F (MHz). But with the dielectric constant of your cables, the criteria is shifted lower, and it is prudent to check for $l \leq 20/F$ (MHz). If there are cable lengths that exceed this criterion, you must respect matching conditions (another reason for preferring 50 Ω inputs). In that respect, beware also of derivations with "T" junctions, which can add a mismatch by shunt loading if they too violate the criteria for length.

3.5.4 Recognizing Capacitive vs. Magnetic Crosstalk

Crosstalk can show up

- at development stage, between traces in PCB or backplanes, in connectors, in ribbon cables and flexprint, in multipair cables.
- in the field, between system cables and other customer's installation wiring (power or other)

Once you have reason to suspect possible crosstalk (see, for example, the flowchart in Fig. 3.24), several obvious solutions exist, such as:

- separating culprit from victim wirings
- twisting one of the two
- shielding at least one of the two
- reducing cable heights above ground

But they may not be practical at a certain stage. For instance, at a large customer site, the culprit and victim cablings are no longer accessible, or even visible, over 90 percent of their shared path. Here, for applying other solutions, more needs to be known about the capacitive or magnetic nature of the crosstalk.

All that is needed is that either the culprit circuit load or the victim circuit source be accessible and temporarily modifiable. Try one of the following checkups while keeping an eye on the victim's disturbed signal.

	V_{victim} decreases?	V_{victim} increases?
Short out the culprit's load, right across its terminals.	Crosstalk is capacitive.	Crosstalk is magnetic.
Open culprit circuit at load side.	Crosstalk is magnetic.	Crosstalk is capacitive.
Short out victim's line at the source side.	Crosstalk is capacitive.	Crosstalk is magnetic.

This identification will help select the most appropriate fix.

Against Capacitive Crosstalk

- Try reducing the HF contents of the voltage spectrum with decoupling capacitors across culprit (source-end side) or victim (load-end side).
- Use a shield or guard wire tight against the culprit wires (grounded *at least* at the source end).
- Use a shield or guard wire tight against the victim wires (grounded *at least* at the receiver end).

Against Magnetic Crosstalk

- Try reducing the HF current spectrum with ferrite beads in series in the culprit circuit, preferably at the source side.
- Use a shield or guard wire tight against the culprit wires (grounded *both* ends).

If the result is disappointing, check to see if you have neglected some contributor over the coupling length. Typical hard-to-catch contributors to crosstalk are the connectors (DIN, Sub-D, Mil-C, and so on) when the culprit and victim pins are close. Try to locate one or more unused pins between the two families and ground them on both sides of the connector.

Representative Vendors of Devices Mentioned in This Chapter

The U.S. address is always listed first, if applicable.

LISNs

Fischer Custom Communications (FCC)
2917 W. Lomita Blvd.
Torrance, CA 90505
Tel.: (310) 891-0635
Fax: (310) 891-0644
www.fischercc.com

Schaffner EMC, Inc.
9B Fadem Rd.
Springfield, NJ 07081
Tel.: (973) 379-7778
Fax: (973) 379-1151
www.schaffner.com

Schaffner-Chase EMC Ltd.
Broadwood Test Centre
Rusper Road, Nr Capel, Dorking
Surrey RH5 5HF
U.K.
Tel.: 44 (0) 1306 713333
Fax: 44 (0) 1306 713303
www.schaffner-chase.co.uk

Low-Noise Preamplifiers

Schaffner-Chase EMC Ltd.
(see above)

Current Probes

Schaffner-Chase EMC Ltd.
(see above)

Current Injection Clamps

AH Systems
9710 Cozycroft Ave.
Chatsworth, CA 91311
Tel.: (818) 998-0223
Fax: (818) 998-6892
ahsystems.com

Fischer Custom Communications (FCC)
(see above)

TEGAM, Inc.
10 Tegam Way
Geneva, OH 44041
Tel.: (440) 466-6100
Fax.: (440) 466-6110
www.tegam.com

Miniature Surface Current Probes (B. Danker)

Modelshop Philips/Novatronics
Building SFH-1
5800 JB Eindhoven
31 40 273 7050

CM/DM Splitters

Chauvin Arnoux, Inc.
d.b.a. AEMC Instruments
99 Chauncy St.
Boston, MA 02111
Tel.: (617) 451-0227
Fax: (617) 423-2952
www.aemc.com

French office:
AEMC
Immeuble St. Georges
Av de la Liberté
38180 Seyssins, France
Tel.: 33 (0) 4 76 21 2390

EFT Generators

Haefely Trench
1308 Devils Reach Rd.
Woodbridge, VA 22192
Tel.: (703) 494-1900
Fax: (703) 494-4597
www.trenchgroup.com

Keytek
One Lowell Research Center
Lowell MA 01852-4345
Tel.: (508) 275-0800
Fax: (508) 275-0850
www.keytek.com

Schaffner EMC, Inc.
9B Fadem Rd.
Springfield, NJ 07081
Tel.: (973) 379-7778
Fax: (973) 379-1151
www.schaffner.com

Main office:
Schaffner EMV AG
Nordstr. 11
4542 Luterbach
Switzerland
Tel.: (41) 32 6816 626
Fax: (41) 32 6816 641

ESD Simulators

Schaffner
(see above)

Keytek
(see above)

EMV-Schloder
Im Stockmade 7/1
76307 Karlsbad
Ittersbach, Germany
Tel.: (49) 72 48 6468
Fax: (49) 72 48 8560

Small Field Probes

EMC Test Systems, L.P. (ETS)
P.O. Box 80589
Austin, TX 78708
Tel.: (512) 835-4684
Fax: (512) 835-4729
www.emctest.com

Chapter

4

Conduction-Type Fixes

In a large majority of EMI situations, a conducted path is involved. This does not necessarily imply that the problem is a purely conducted one from one end to the other. It could just as well be a conducted emission that ultimately appears as a radiated interference or, conversely, an ambient radiation that ends up in a conducted mode on I/O wires.

4.1 Mode of Operation of Conduction Fixes

When the EMI manifests itself as a conductive path, regardless of whether we decide to fix it at the source output (emission problem), at the victim input (susceptibility problem), or somewhere along the coupling path (emission or susceptibility), the purpose is one of the following:

1. To reduce the EMI current by inserting a high impedance in series with the conductor
2. To divert the EMI current to ground, or to some return conductor, by inserting a low impedance in parallel
3. To block the EMI current by opening the path via some galvanic isolation device
4. To neutralize the EMI current by making it work against itself, for example, through mutual induction

The choice of one of the techniques listed above is, of course, based on the nature of the EMI problem, as indicated in Chap. 3, and on hardware constraints such as space availability, access, cost, and so on. But it also depends strongly on the impedance of the circuits involved. Trying to reduce the EMI current by inserting a high impedance—namely a choke or ferrite toroid—in a circuit that is already a high impedance will be for naught. Reciprocally, it is also useless to insert a low-impedance shunting device—a decoupling capacitor—for bypassing EMI current in a circuit that is already a low impedance. Figure 4.1 provides a broad overview of all the EMI reduction components acting at the conduction level.

Finally, another aspect to consider is the frequency domain in which the EMI reduction is needed. Some solutions, such as galvanic isolation, are excellent at low frequencies (say, below the megahertz region) but become progressively transparent as frequency increases. Figure 4.2 provides attenuation figures for all components and cabling techniques for ground-loop reduction. It shows, among other things, how floating the PCB (0 V) reference vs. chassis grounds becomes useless with a 10 m cable link above ≈7 MHz.

In addition, Fig. 4.3 shows an inventory of the principal common-mode fixes arranged according to two major families, namely

- Those based on galvanic isolation (opening of the ground loop, most applicable for low or medium frequency problems)
- Those based on HF filtering and decoupling (generally more appropriate above the megahertz region)

It is important to note that neither optical isolators nor fiber optics will be described in the subsequent fixes. This is not because we disregard their usage, but because they are typical engineering options that must have been incorporated in the original design, so they do not lend themselves easily to a "fix-it" situation.

4.2 Series Attenuation Devices vs. Shunt Attenuation

With filtering, the decision of using capacitors versus inductors, or a combination of both, is guided by the rule of maximum mismatch. Simply said, in any filtering circuit, capacitors should always "see" a high impedance on both sides, and inductors should see low impedances. Applying this maximum-mismatch rule, a truth table (Table 4.1) was established. For each condition, two possibilities are given, the "n" parameter being the number of L, C elements in the filter circuit.

When both source and load impedances are matched, or at least both are in the "high" or "low" category, the simplest filter is of one element. However, the n = 1 filter provides an attenuation slope of just 20 dB/dec of frequency (equivalent to 6 dB/octave) (see Fig. 4.4).

For instance, if the useful signal spectrum—that is, the frequency portion that the filter should not alter—is 8 MHz, the cutoff frequency of the filter (i.e., the 3 dB attenuation point) has to be selected a little

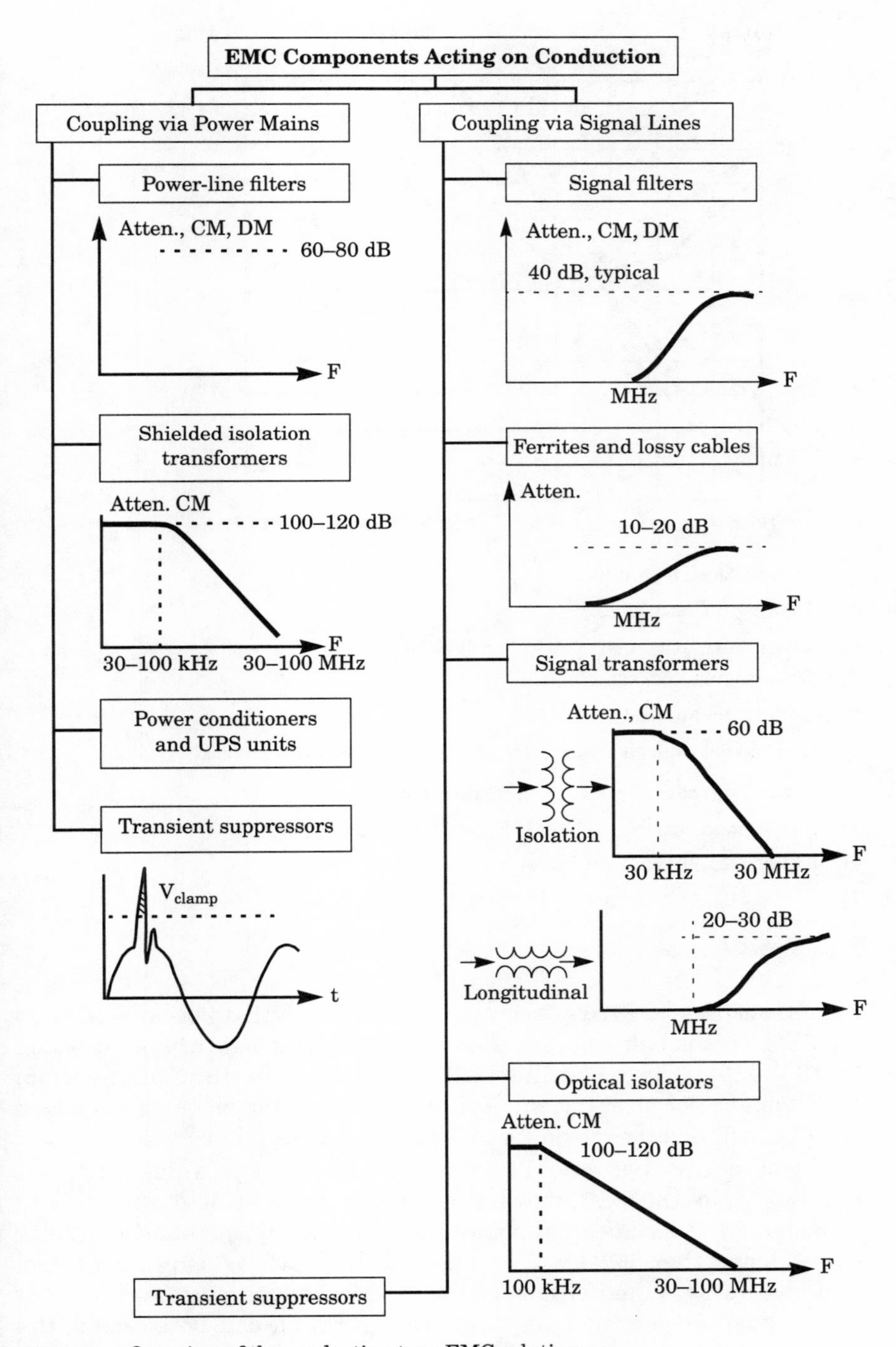

Figure 4.1 Overview of the conduction-type EMC solutions.

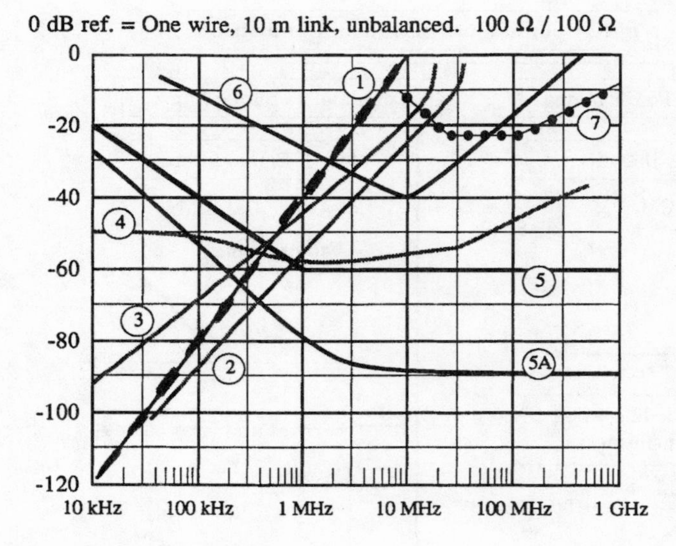

Loop isolation techniques

① Floating 0V (star ground)

② Differential signal transformer, low capacitance

③ Opto-isolator, low capacitance

Balancing technique

④ Balanced link with good symmetry (MIL 1553 Bus with Raychem Twinax)

Ground-loop cancellation & shielding techniques

⑤ Single-braid coaxial (5A) Double-braid coaxial

⑥ CM choke, 20 turns (≈ 0.5 mH)

⑦ Bifilar miniature Balun (D.I.L. package) for Hi-speed Bus

Figure 4.2 Ground-loop reduction techniques, 100 Ω/100Ω.

higher, such as 10 MHz. Assume now that the EMI to be filtered is at 30 MHz. This is half a decade above 10 MHz (on a logarithmic scale, log 30/10 ≈ 0.5), so an n = 1 filter will provide only 10 dB of attenuation, which might not be sufficient. A filter with a higher "n" term, such as a "pi" or "T" filter, corresponding to 60 dB/decade should be selected.

The borderline between the so-called "high" or "low" value for Z_{source} and Z_{load} is, of course, somewhat arbitrary. A good rule of thumb is to consider 50 Ω as an appropriate borderline. Any impedance significantly lower than 50 Ω will be regarded as "low" for using this table, and vice versa. When either one or both impedances are in the 50 Ω range (say ±50 percent), any L, C configuration can be selected, the choice being guided by volume or cost considerations.

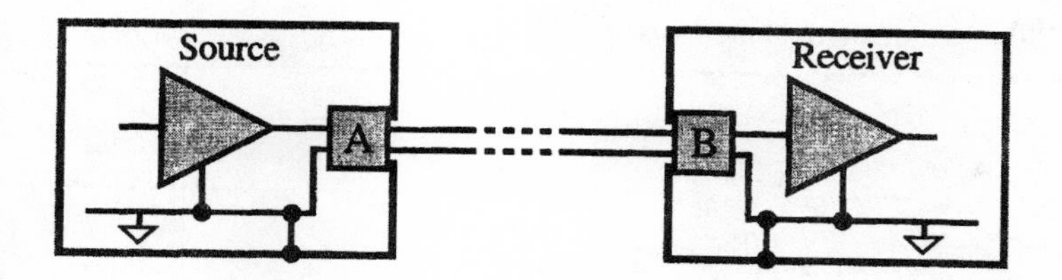

In LF : Opening the loop at source **OR** receiver side

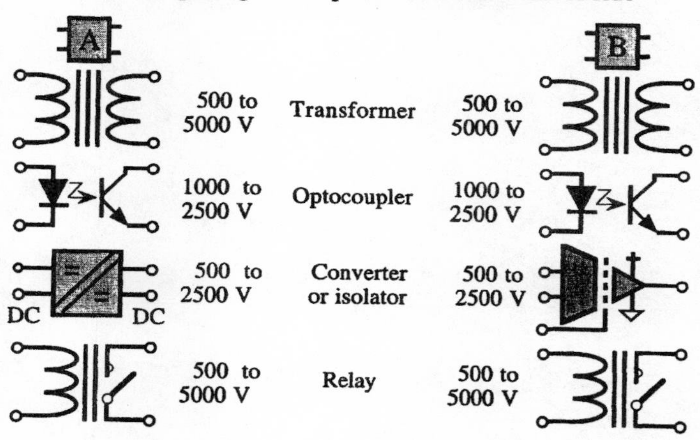

Decoupling **HF** CM noise at transmit **AND** receive side

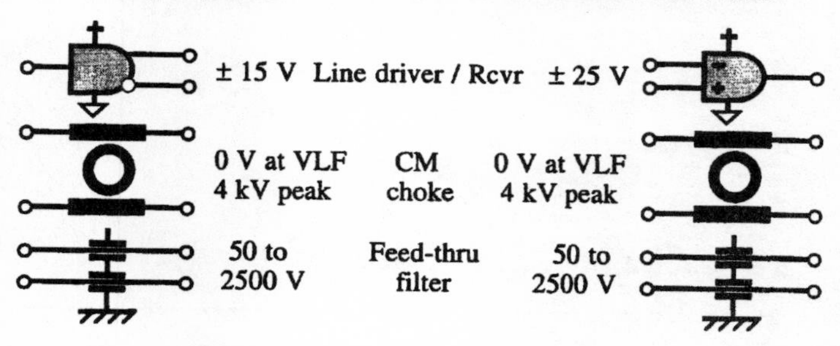

Figure 4.3 Common-mode reduction components, with their typical voltage-withstand range.

Table 4.1 Truth Table for Optimal Filter Arrangements

Z source	Filter	Z load
Low	n = 1 (20 dB/dec)  n = 3 (60 dB/dec) 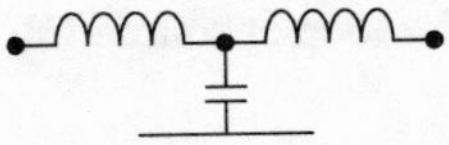	Low
Low	n = 2 (40 dB/dec) 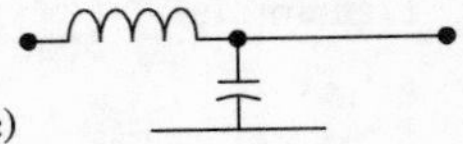 n = 4 (80 dB) 	High
High	n = 1 (20 dB/dec)  n = 3 (60 dB/dec) 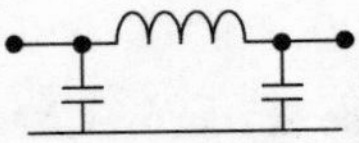	High
High	n = 2 (40 dB/dec) 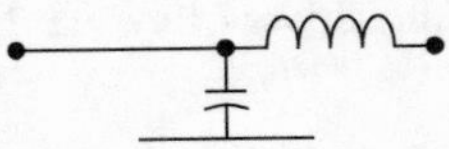 n = 4 (80 dB/dec) 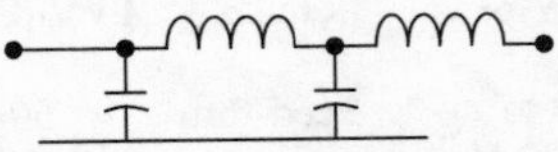	Low

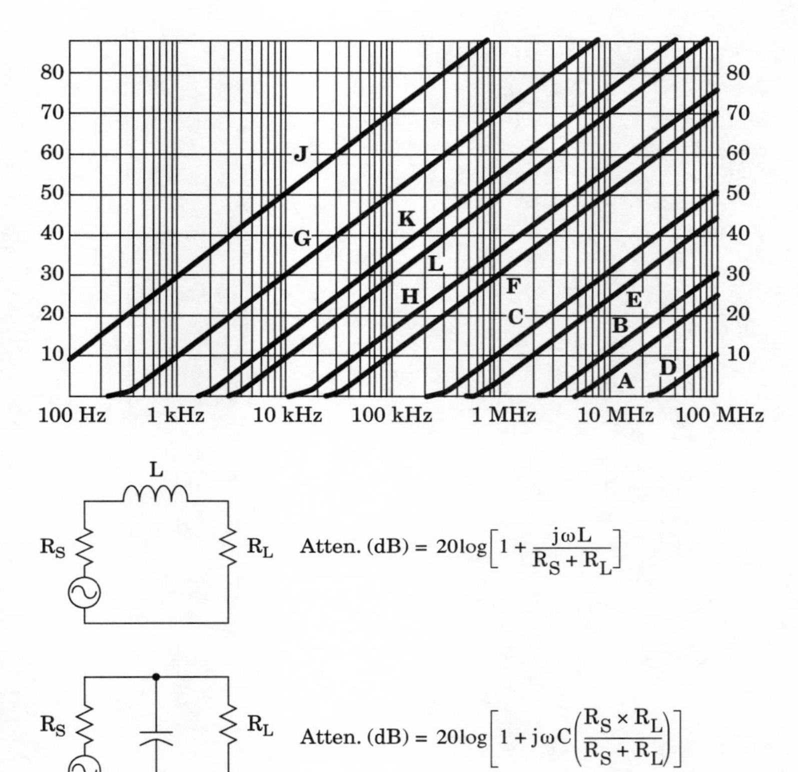

	L		C	
R_S, R_L	1 mH	10 mH	1 nF	10 nF
1 Ω, 1 Ω	G	J		D
50 Ω, 50 Ω	H	K	A	E
1 Ω, 50 Ω	H + 6 dB	K + 6 dB		D + 6 dB
100 Ω, 100 Ω	F	L	B	C
1 kΩ 1 kΩ	C	F	C	F
50 Ω, 1 kΩ	C + 6 dB	F + 6 dB	A + 6 dB	E + 6 dB

Figure 4.4 Ideal attenuation of a few simple L or C filters in typical impedance configurations.

Chapter

5

Capacitive Types of EMC Solutions

5.1 Theoretical Brief

A capacitor works by shunt-type attenuation (Fig. 5.1). That is, the capacitor's impedance

$$Z_x = \frac{1}{2\pi FC}$$

must be lower than the parallel combination of Z_s, Z_l. For instance, if one knows only that $Z_l = 1000\ \Omega$, a common mistake would be to assume that it is enough, if Z_x is smaller than 1000 Ω, to get some filter-

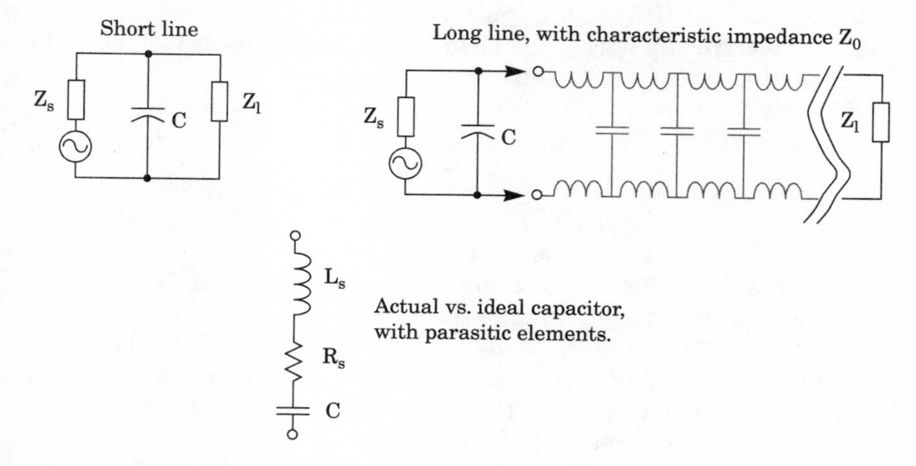

Figure 5.1 Filtering capacitor, actual configurations.

ing action. In fact, no definition of the filter can be attempted without knowing at least the modules of Z_s and Z_l, and preferably their complex value, if they are not pure resistances.

The cutoff frequency of such lowpass filter is given by

$$F_{co} = \frac{1}{2\pi R_p C} \tag{5.1}$$

where R_p = parallel combination of Z_s, Z_l.

This would be the exact 3 dB point on the attenuation curve if Z_s, Z_l were pure resistances, and an approximate value of F_{co} if Z_s, Z_l were a combination of resistances, inductances, and capacitances.

The attenuation at any frequency above F_{co} is given by

$$A_{dB} = 10\log\left[1 + \left(\frac{Z_s//Z_l}{Z_x}\right)^2\right]$$

$$= 10\log\left[1 + \left(2\pi FC \cdot \frac{Z_s Z_L}{Z_s + Z_L}\right)^2\right] \tag{5.2}$$

$$\approx 20\log 2\pi \cdot F \cdot C \times (Z_s//Z_L), \text{ for } F >> F_{co} \tag{5.3}$$

Using this formula, one must be careful to use the actual value of Z_s, Z_l at the calculation frequency, which may be different from their nominal value at low frequency, if Z_s or Z_l are not pure resistances.

A frequent situation occurs when one side of the filtering capacitors is not connected directly to a discrete impedance but sees a transmission line instead. This is the typical case of an I/O line whose length becomes electrically long, i.e., reaches or exceeds a 1/4 wavelength. In practice, this translates approximately into

$$l(m) \geq \frac{55}{F_{MHz}} \tag{5.4}$$

This formula takes into account an average propagation speed of 0.75 times the theoretical free space value.

In such a case (Fig. 5.1), the side of the capacitor that faces the far end does not "see" Z_l, but rather the transmission line characteristic impedance Z_0. In this case, the Z_0 value should replace Z_l in Eq. (5.3).

With capacitors, a very common limitation in the actual attenuation performance, compared to the theoretical one, is due to the parasitic R and L elements in series with the ideal capacitance C.

The value of R depends on the material and technology of the capacitor armatures (foils, discs, etc.). The value of L_s depends essentially on the lead length. For round, wire-type leads, a typical value for L_s is 10 nH per centimeter length.

5.2 Dielectric Materials and Tolerances

The majority of capacitors used for EMI filtering are the nonpolarized type. Various types of dielectrics and constructions exist, as described below.

1. *Paper foil.* Economical, ordinary quality. Limited use at HF because of parasitic inductance of the wound layer. Usable, in practice, up to a few tens of kilohertz.
2. *Metallized polyester or paper.* Economical. Lower inductance and smaller size than paper. Can survive short voltage breakdowns by self-healing. Efficient up to few megahertz. Typical application in class X* or Y† capacitors.
3. *Ceramic.* The typical EMI fix choice from megahertz to gigahertz, and even higher for the low-loss kind. Peerless for their wide temperature range and their high capacitance/volume ratio. Three grades of ceramic are available:
   - *NPO.* Best for stability, low loss, and precision tolerances (typically ≤10 percent from –55 to 125° C).
   - *X7R.* Average stability and tolerances.
   - *Z5U.* Mediocre dissipation factor and tolerances stability.

   Ceramics have wide use in low-voltage power and signal decoupling. For ac mains applications, due to voltage stress, values are limited to tens of nanofarads.

5.3 Capacitors for Differential-Mode (Line-to-Line) Filtering

Differential-mode (DM) here means that the capacitor is placed across the path of the intended current, i.e., line-to-line (two-wire configurations) or line-to-zero-volt reference (in most PCB configurations).

---

* Class X capacitors are qualified for use on ac mains where a failure of the component would not create a risk of electric shock—typically, line-to-line. They are tested for 100 hr @ 1.5 × rated rms voltage, plus 1 kV hi-pot testing.

† Class Y capacitors are mandatory for use where a failure could create a shock hazard—typically, line-to-chassis.They are tested at 1.7 × rated rms voltage, plus 2 kV hi-pot test, going to 4 kV if they are intended for class II products without an earth connection.

Capacitors placed across source output leads (emission problem) or victim input leads (susceptibility problem) will shunt the high-frequency contents of the EMI spectrum. Optimal filtering is that which leaves unaffected the frequency domain required for circuit performance while offering a low impedance for higher frequencies.

Indications

1. When a spectrum analysis on signal or power lines (using, e.g., a current probe) reveals that they carry unnecessary frequency components. This pollution, to some extent, can also be visible on an oscilloscope (Fig. 5.2).
2. When the victim's actual bandwidth is wider than what is strictly necessary for its purpose.
3. When a dc supply rail is coupling undesired frequencies into the circuit being fed.
4. When a victim is exposed to capacitive crosstalk.

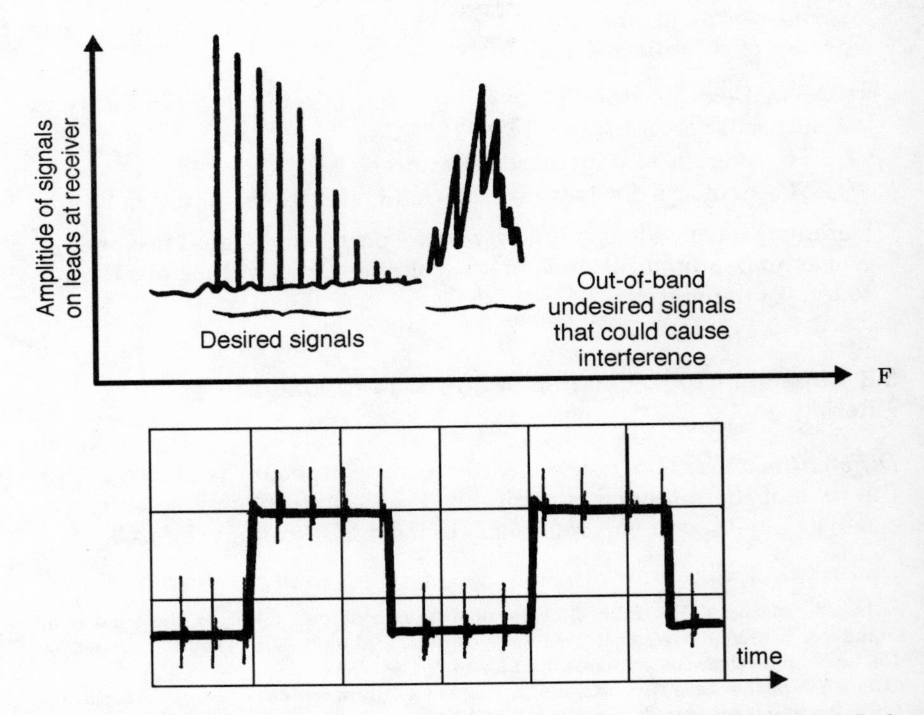

Figure 5.2 Example of unwanted high-frequency noise riding over a desired signal. Such noise can go unnoticed functionally but cause violations of radiated emission limits.

5. When the impedance of the circuit to be decoupled is greater than 50 to 100 Ω.

6. When the rise times of source components need to (and can) be slowed down.

7. On power entry, when the equipment offers a high power input impedance (for instance, power supply input with a front-end transformer).

8. On power entry, to improve the DM attenuation of an already installed line filter.

Prescriptions, Installation

1. The largest value of the filtering capacitor that is tolerable without significantly affecting signal integrity (the 3 dB point in the attenuation curve) is

$$C_{max} = \frac{1}{2\pi R_p F_{MHz}}, \text{ for C in microfarads}$$

where F = the highest useful frequency that should not be affected
R_p = parallel combination of R_{source} and R_{load}

If the rise time of the useful signal, t_r, instead of the bandwidth is known, then C_{max} is approximately equal to $(t_r/2R_p)$, with C and t_r in reciprocal units (i.e., microfarads and microseconds, nanofarads and nanoseconds, etc.).

2. Select a capacitor with good high-frequency behavior: use paper or polystyrene up to a few hundred kilohertz, and ceramic (ac) or solid aluminum electrolytic (dc) for higher frequencies.

3. Check the service voltage and surge withstand voltage against the application. For ac mains filtering, use class X capacitors (see Sec. 5.2).

4. For ac power-line filtering, beware that the capacitor should not draw too much 50/60 Hz regular current. This will degrade the power factor by wasting reactive energy and create inrush currents at turn-on. A good rule of thumb is that ac decoupling should not draw more than 1 to 3 percent of the 50/60 Hz nominal current. Translated into a convenient formula, this gives

$$C_{max} = 50\ I/V$$

for C in microfarads, with V and I being the rated rms current and voltage of the equipment at 50/60 Hz.

For example, we can calculate that, for a 120 V unit drawing 1.2 A

$$C_{max} = 50 \times \frac{1.2}{120} = 0.5\ \mu F$$

Larger values of rated current would lead to larger capacitors, but remember: this would no longer be cost-efficient, because it would correspond to system input impedances < 50 Ω.

5. Install the capacitor using the shortest leads possible, trimmed close to the capacitor can.
6. For improving the attenuation (DM) of an installed filter, install the capacitor across the filter terminals. Since attenuation of filters is seldom reciprocal, try to gauge which side (mains or load) is best for adding the capacitor. A good clue to this is to look at the filter schematic (from catalog data, or sometimes printed on the filter itself). The capacitor should always look at the side with the highest impedance.
7. Check that frequencies to be filtered are not significantly within the parasitic (inductive) region of the capacitor.

Limitations

As explained above, filtering capacitors:

1. Will not work in low impedance circuits (<50 Ω), which would require prohibitively large values of C,
2. Will exhibit parasitic resonance due to their lead inductance and will become progressively inefficient above this resonant frequency, and
3. Provide only a one-pole filtering (6 dB/octave, or 20 dB/decade). If the EMI is just slightly off the victim's 3 dB bandwidth, the attenuation will be meager. This can be improved by creating a steeper slope in combination with ferrites in an L or T arrangement.

Figure 5.3 is a universal reactance chart to help in the selection of bypass capacitors. To use it, locate the descending line corresponding to capacitor value, then locate the lead length on the right. The intersection of the capacitance line with the inductance line is the resonant frequency. For frequencies above resonance, the capacitor will behave as an inductance, and its impedance increases.

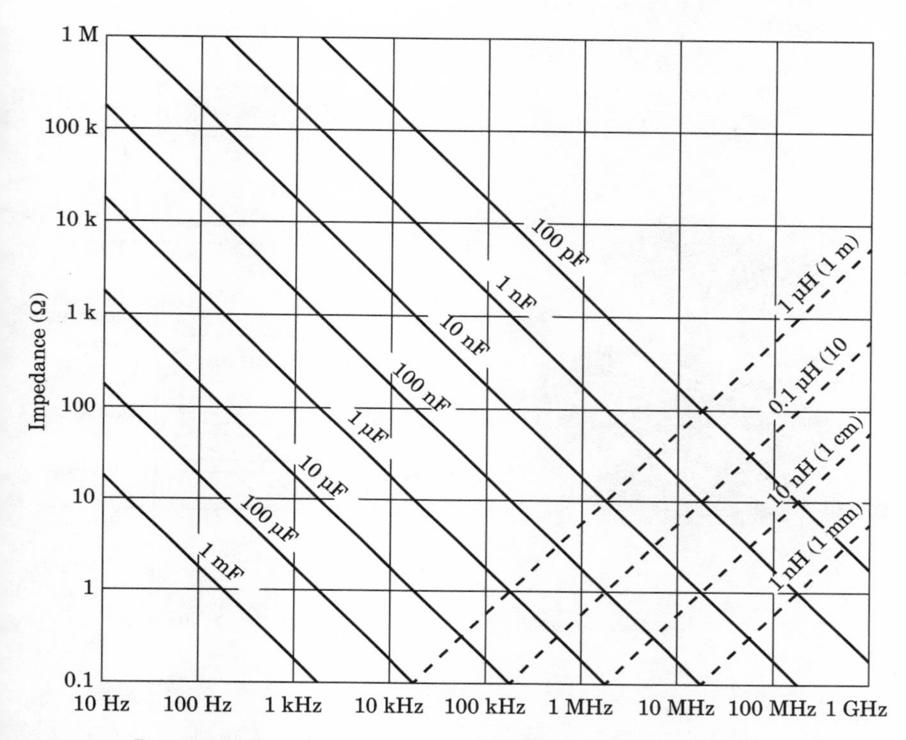

Figure 5.3 Reactance chart for determining impedance and parasitic resonance of discrete capacitors.

5.4 PC Board Capacitive Bus Bars and Flat Pack Capacitors

Flat bus bars and planar capacitors are noninductive power distribution components (Figs. 5.4 and 5.5). They consist of stacked copper foils with ceramic dielectric between. They can be used to distribute dc voltages across a whole card or locally to provide a capacitive decoupling, directly under a noisy IC module.

Indications

In general, it is used when there is high-speed logic (faster than ≈10 ns) and/or fast clock rates (>3 to 5 MHz), which results in (a) self-jamming and (b) out-of-spec emissions, typically in the 10–300 MHz range, in comparison with FCC or MIL-STD limits, and when one of the following constraints exists:

1. There is not enough room on a PC board to allow for large Vdc and 0 V copper lands, while a low-impedance distribution is needed.

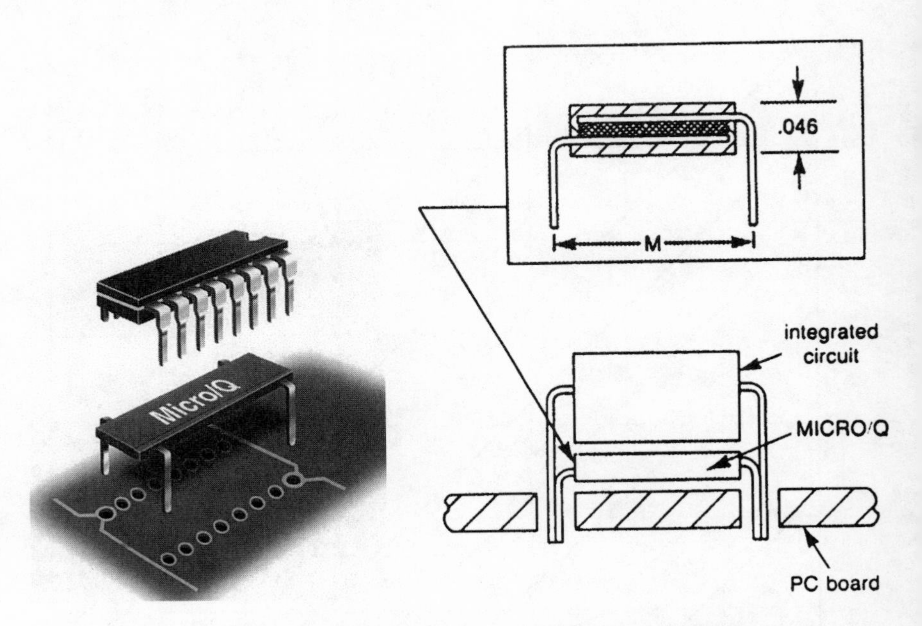

Micro/Q® decoupling caps are available in 14- to 18-pin ICs, rated 0.02 to 0.16 µF.

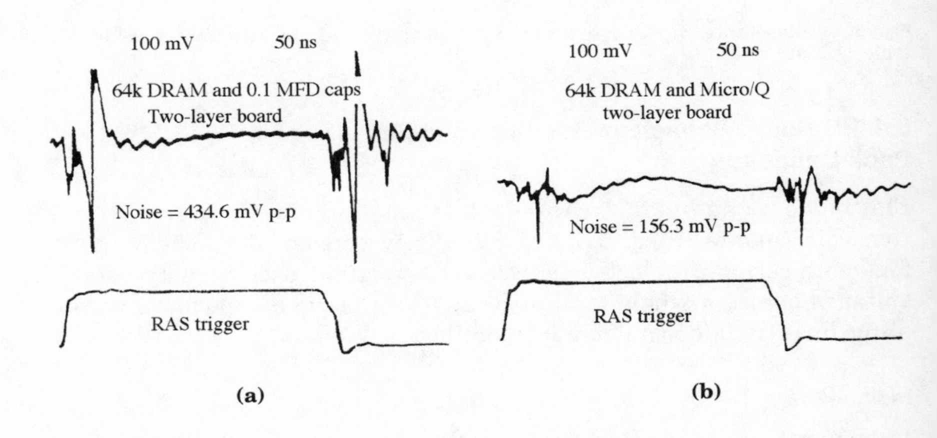

Figure 5.4 Examples of PC board V_{CC} decoupling capacitive bus bars, for DIP devices. They can be used as last-minute fixes to rescue a two-sided board when the multilayer option is not feasible. To demonstrate the effectiveness of various decoupling schemes in reducing IC noise, an array of 64 k DRAM chips was tested. Trace (a) shows a two-sided PC board with 0.1 µF MLC capacitors at every device. Peak-to-peak spike noise on V_{CC} was measured at 435 mV. Trace (b) shows the results of 0.03 µF Micro/Q 1000 decoupling capacitors on a two-sided board. The spike noise of 156 mVpp is the lowest due to much lower inductance. The sag level is measured at approximately 40 mV because of the lower capacitance value. *Courtesy of Micro/Q, C.C.I./Mektron.*

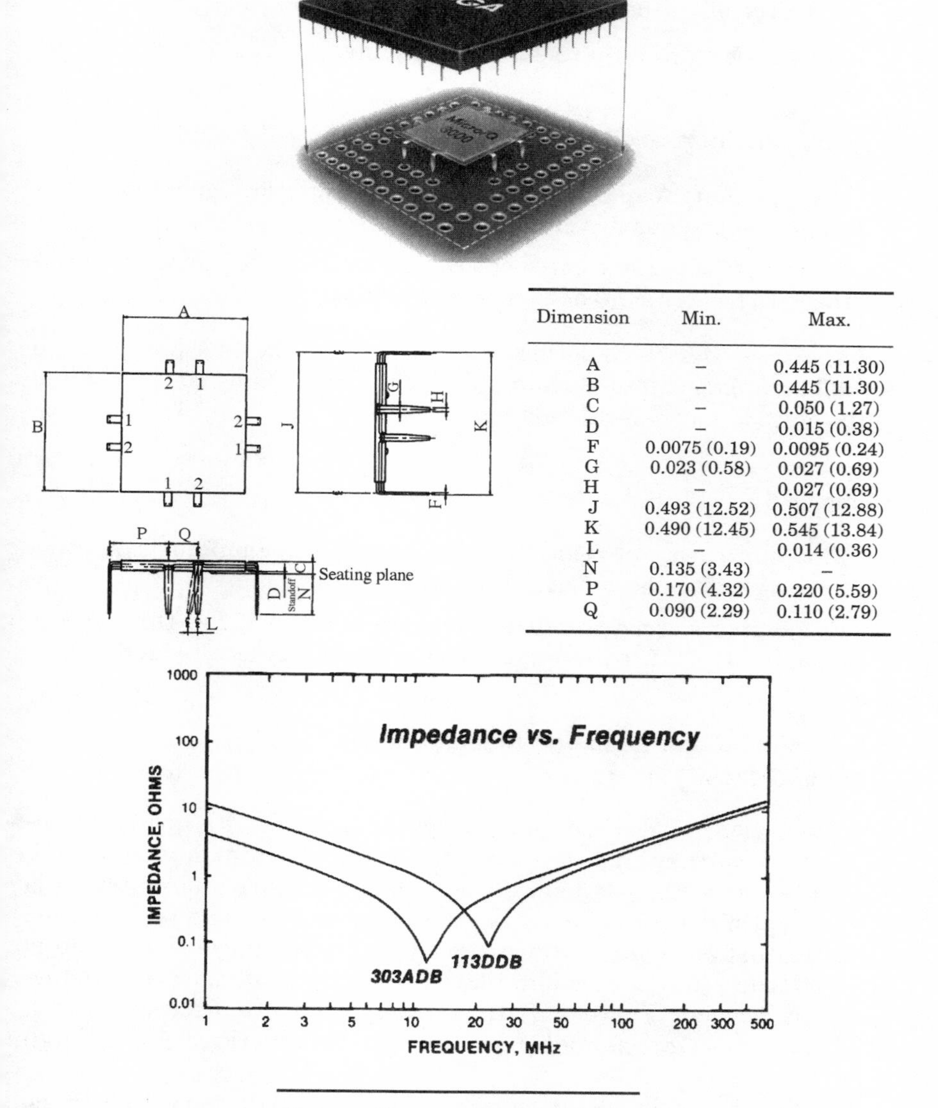

Dimension	Min.	Max.
A	–	0.445 (11.30)
B	–	0.445 (11.30)
C	–	0.050 (1.27)
D	–	0.015 (0.38)
F	0.0075 (0.19)	0.0095 (0.24)
G	0.023 (0.58)	0.027 (0.69)
H	–	0.027 (0.69)
J	0.493 (12.52)	0.507 (12.88)
K	0.490 (12.45)	0.545 (13.84)
L	–	0.014 (0.36)
N	0.135 (3.43)	–
P	0.170 (4.32)	0.220 (5.59)
Q	0.090 (2.29)	0.110 (2.79)

Part no.	Nom. capacitance	Dielectric
303ADB	0.030 µF	Z5V
113DDB	0.011 µF	X7R

Figure 5.5 Example of PC board V_{CC} decoupling capacitive pad, for PGA-style package. This, too, can be used as a last-minute fix, since it does not take board real estate and can be laid over existing traces. *Courtesy of Micro/Z, C.C.I.*

2. Multilayer boards cannot be used.
3. It is a retrofit situation, and the PCB cannot be redesigned.
4. Only a few chips require extra decoupling.

Prescriptions, Installation

1. Use one planar cap under each "hungry" chip (clock generators, microprocessors, bus drivers, etc.).
2. As a "dirty fix," the component can be mounted temporarily under the card (solder side) or even over the top of the IC.

Important: Before soldering, carefully check the bus bar pins. In some busses/capacitors, certain pins are not electrically connected and serve only for mechanical holding.

Limitations

1. This approach can reduce the Vdc-to-0 V noise only; it cannot be used against noise on signal traces.
2. By helping a stronger current sink from the 5 V rail through the chip, the H-field radiation of the chip itself may locally increase.

5.5 Capacitors for Common-Mode (Line-to-Ground/Chassis) Filtering

Common-mode (CM) decoupling is typically made with low-value capacitors (less than 100 or 10 nF) arranged to shunt unwanted HF currents to the chassis before they enter a sensitive circuitry (this for susceptibility) or before they get too far away from a noisy circuit (this against emissions). Since a good high-frequency attenuation is desired, the reduction or elimination of component parasitic inductance is crucial. Therefore, ultra-short leads are necessary, with leadless-type components being highly preferable (Figs. 5.6 through 5.8).

One favorite technology consists of feed-through capacitors (Figs. 5.9 and 5.10). These are more or less coaxial structures where the dielectric material is filled in between a center pin electrode and an outer tubular or peripheral electrode. This outer armature can be soldered, screwed, or pressed into the chassis or PC board ground plane. Because of this design, they exhibit no parasitic inductance; therefore, their frequency span is practically unlimited.

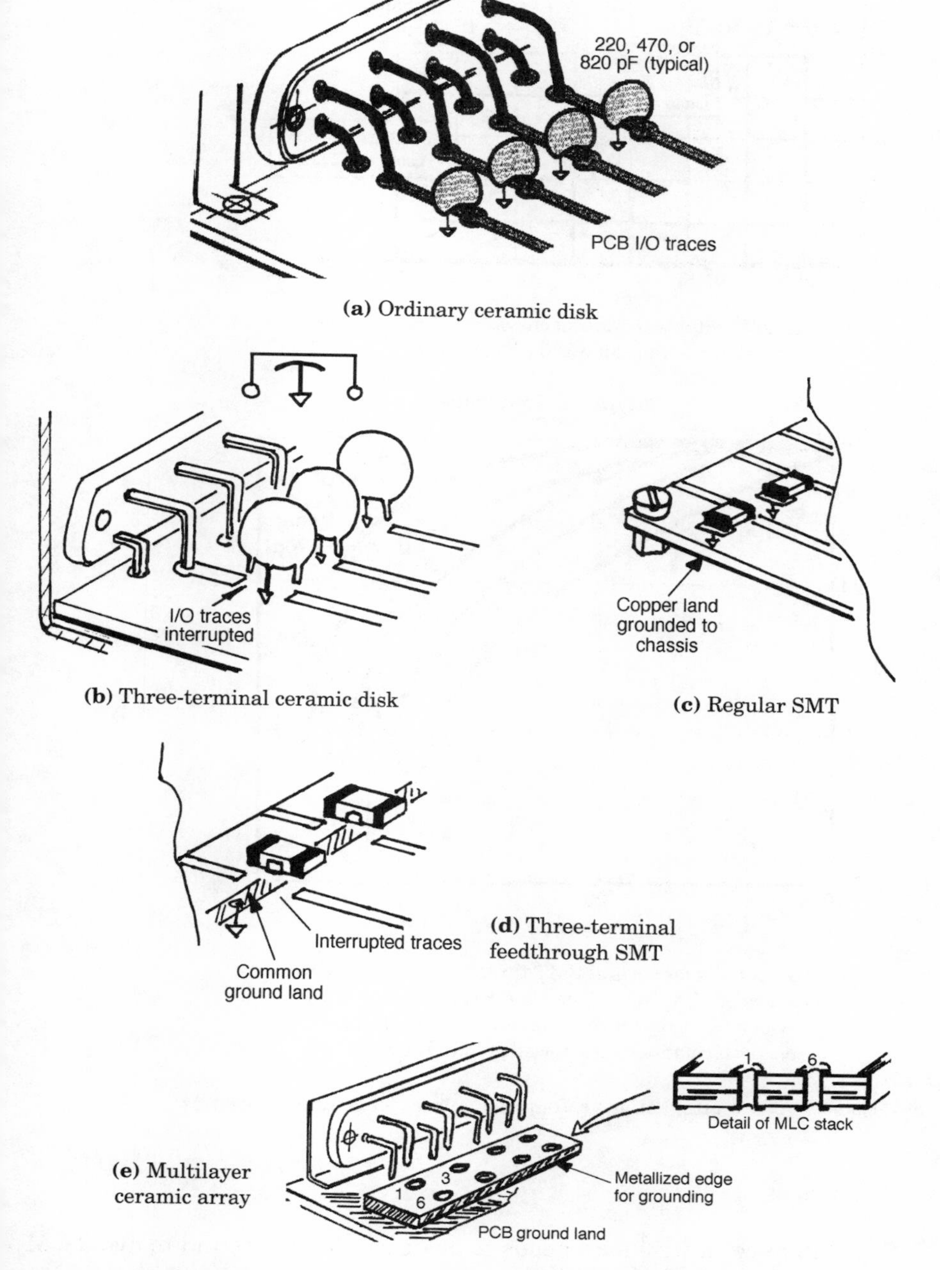

Figure 5.6 Practical mounting of I/O decoupling capacitors on PCBs. If there is some direct-contact ESD exposure for I/O pins, do not use X7R or Z5U dielectric; prefer NPO or voltage-limiting capacitors.

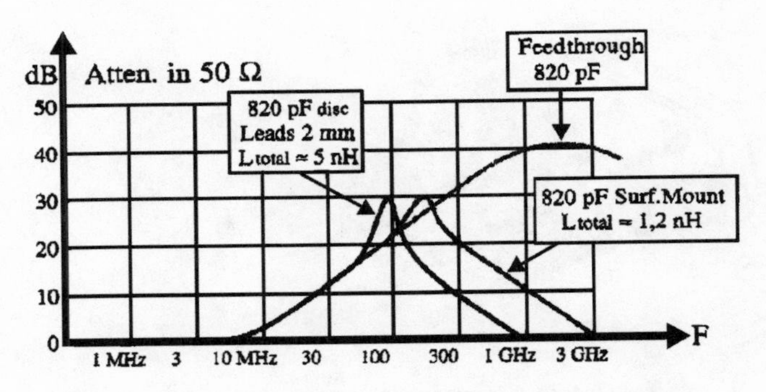

(a) Comparing filter attenuation curves for three packaging styles of the same 820 pF capacitor.

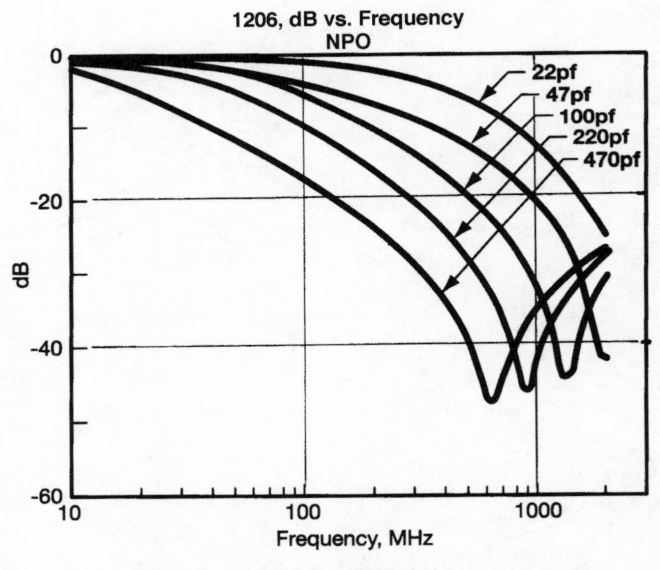

(b) Attenuations for several SMT three-terminal capacitors (courtesy of AVX Corp.). The 1206 SMT package, with a wider footprint for the grounding electrode, has better performance characteristics than the 805 size.

Figure 5.7 A few attenuation vs. frequency curves for ceramic capacitors.

Indications

1. Whenever a wire lead needs to be decoupled to ground to avoid CM emission or susceptibility.
2. For use with high-impedance (>50 Ω) circuits.

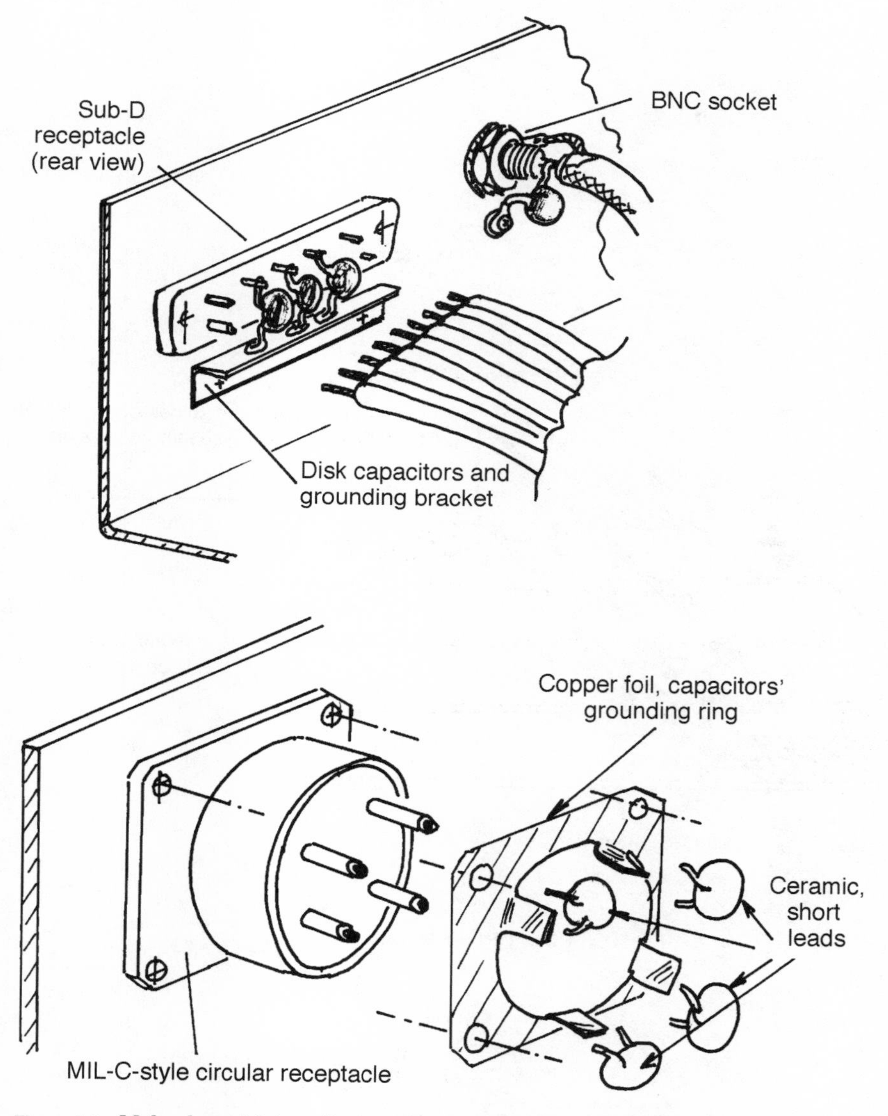

Figure 5.8 Make-do, quick installation of CM capacitors to filter connector areas.

3. To improve circuit symmetry to ground (i.e., CM immunity) at high frequencies.

4. Whenever wires cross a metal barrier that is relied upon for shielding effectiveness, feedthrough capacitors will prevent leakage at penetrations.

Press-fit or solderable capacitor

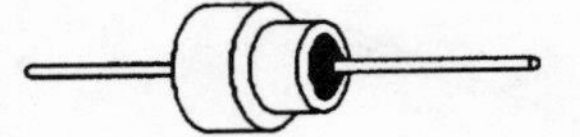

Threaded feed-through capacitor

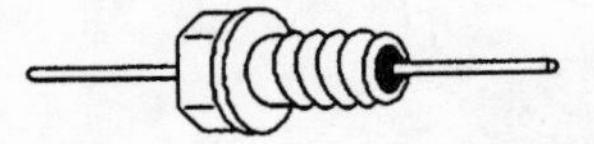

PCB, Horizontal Mount

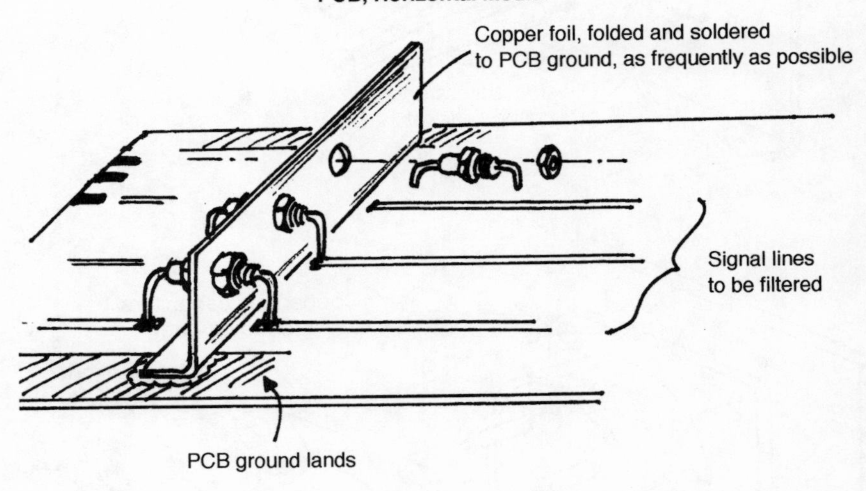

PCB, Vertical Mount

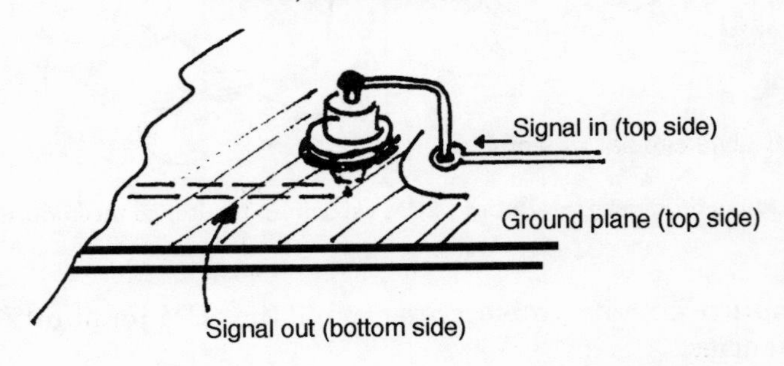

Figure 5.9 Although three-terminal SMT capacitors are a fair substitute, nothing can match the performance of real, wall-mounted feedthrough capacitors with 360° contact. Although their packaging is made for metal walls, it is possible, as a late fix, to mount them on a PCB. *(Continued on next page)*

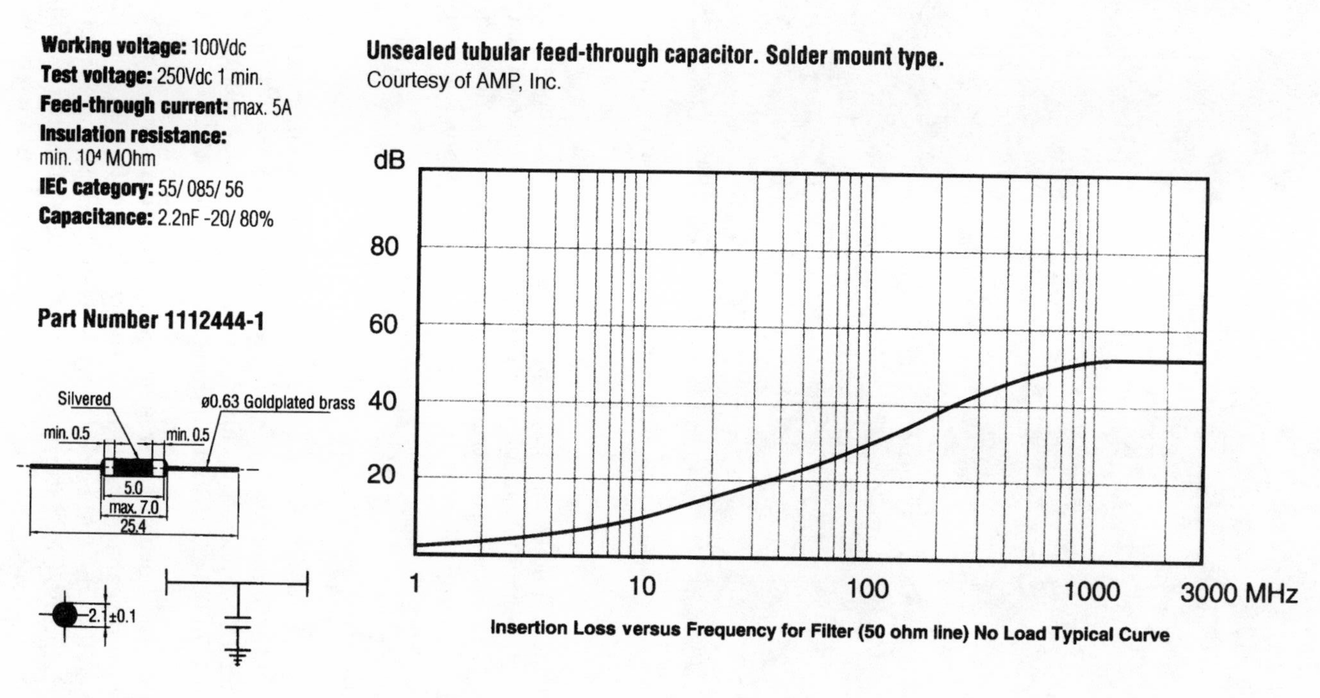
Working voltage: 100Vdc
Test voltage: 250Vdc 1 min.
Feed-through current: max. 5A
Insulation resistance:
min. 10⁴ MOhm
IEC category: 55/ 085/ 56
Capacitance: 2.2nF -20/ 80%
Part Number 1112444-1
Silvered
ø0.63 Goldplated brass
min. 0.5
min. 0.5
5.0
max. 7.0
25.4
2.1 ±0.1
Unsealed tubular feed-through capacitor. Solder mount type.
Courtesy of AMP, Inc.
dB
80
60
40
20
1
10
100
1000
3000 MHz
Insertion Loss versus Frequency for Filter (50 ohm line) No Load Typical Curve

Figure 5.9 *(Continued)*

Figure 5.10 Filtered terminal blocks with 2,500 to 5,000 pF feedthrough capacitors. *Courtesy of AMP, Inc.*

5. Every time the noise passing in or out has high-frequency contents (typically above a few megahertz), including ESD. In contrast, discrete capacitors or other traditional techniques (e.g., floating, single-point ground, RF ground chokes) could not work because of parasitic resonances.

5.6 Prescriptions, Installation

1. Install at the shield barrier. As a variation, the component can be installed through a PCB ground plane, preferably with this ground plane being connected within a short distance to a metal chassis (see, for example, "PCB Grounding Spacers," Sec. 11.2).

2. Equip all the incoming wires or leads with identical (same value, same brand) capacitors. Odd mixes could create line CM imbalance, spoiling the CM rejection by re-creating DM noise from what was originally CM interference. One single line left unfiltered (even if not noisy) can act as a capacitive detour that noise will take to bypass the adjacent filtered wires.

3. If each wire of a differential pair is decoupled to the ground by a capacitor of value C, the line sees C/2 across its conductors. This value C/2 needs to be compatible with the useful bandwidth (or rise time) that needs to be carried along this pair, as shown in Fig. 5.11. To avoid compromising the desired signal by feedthrough capacitors, refer to Table 5.1.

4. Carefully avoid recoupling of input wires with output wires: Install all the decoupling capacitors in the same zone (preferably at the I/O port).

5. The feedthrough capacitor can be efficiently combined with the ferrite fix in case of hybrid (high- and low-impedance) configurations. As shown in Fig. 5.7, the ferrite or inductor should look toward the low-impedance side of the line (generally the signal source).

Limitations

1. This approach is inefficient if a low-impedance, direct connection to CM ground (generally chassis) cannot be provided.

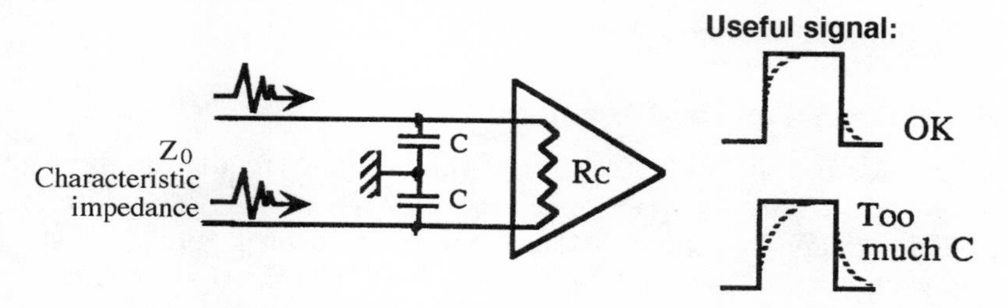

Condition, to avoid degrading a useful signal, with bandwidth F_{max}:

If X_C = impedance of one capacitor @ F_{max}
R_T = total impedance of victim line $R_{load}//R_{source}$ or $R_{load}//Z_0$ (long line)

We must keep $2 \times X_C > R_T$

$$\frac{1}{2\pi F_{max}\left(\frac{C}{2}\right)} \geq 3R_T \rightarrow C_{max} = \frac{100}{F_{max} \times R_T}, \text{ C in nF and } F_{max} \text{ in MHz}$$

or

$$C_{max} = \frac{0.3t_r}{R_T}, \; t_r = \text{rise time of useful signal in ns, for C in nF}$$

Figure 5.11 Choosing capacitor values for HF decoupling (CM).

Table 5.1 Maximum Permissible Values for CM Decoupling Capacitors

	Low-speed interface (typ.)	CMOS	TTL	HCMOS/AC
t_r	0.5–1 μs	100 ns	10 ns	3.5–1.5 ns
Bandwidth	300 kHz	3 MHz	30 MHz	100–230 MHz
Z*	100 Ω	500 Ω	100 Ω	50–30 Ω
C_{max} for good pulse integrity	2400 pF	100 pF	30 pF	20 pF
C_{max} for marginal pulse shape	7000 pF	300 pF	100 pF	60 pF

*Z = differential impedance (source and load in parallel)

2. Capacitors must be restricted to values small enough to avoid impairing circuit performance.
3. Capacitor tolerances must be as tight as possible. A balanced input decoupled to ground via capacitors having too wide a tolerance (e.g., –0/+50 percent) will become severely unbalanced.
4. Due to low capacitance (high cutoff frequency), this fix is not efficient against medium- to high-energy pulses, with pulse width longer than approximately 100 ns.

5.7 Filtered Connectors and Adaptors

Filtered connectors are an extension of the leadless, feedthrough capacitor concept, where a capacitor is incorporated with every connector pin (Fig. 5.12). In some filtered connectors, each pin is also equipped with a small ferrite tube such that an "L" type (C-L) or even a "pi" type (C-L-C) filter is created. Some components also use a multilayer ceramic array. Finally, some filtered connectors are offered as a male-to-female adaptor (5.11a) so that the effectiveness of the fix can be quickly evaluated.

An attractive option, which allows quick upgrading of an existing, nonfiltered receptacle, is the filter insert (see Fig. 5.13). This is a thin, flexible wafer with contact fingers on the rim, which can fit inside standard connector shells, provided they are *metallic* and *grounded*.

Indications

The indications are generally similar to those of ferrites and feedthrough capacitors. They are used

1. To reduce out-of-band emissions and out-of-band susceptibility from about 10 MHz to more than 1 GHz. This includes ESD.

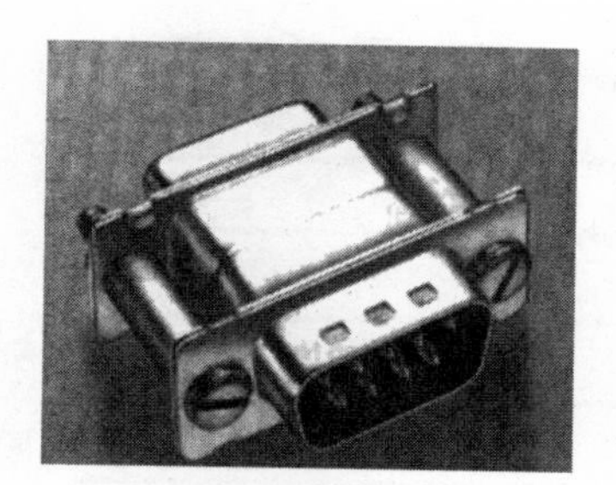

Filter Type	AMP Electrical Specification Number	Capacitance Range
CA	108-1139	4000 pf to 10,000 pf
CC	108-1132	1300 pf to 2500 pf
CD	108-1135	600 pf to 1000 pf
CE	108-1134	400 pf to 600 pf
CF	108-1133	240 pf to 360 pf

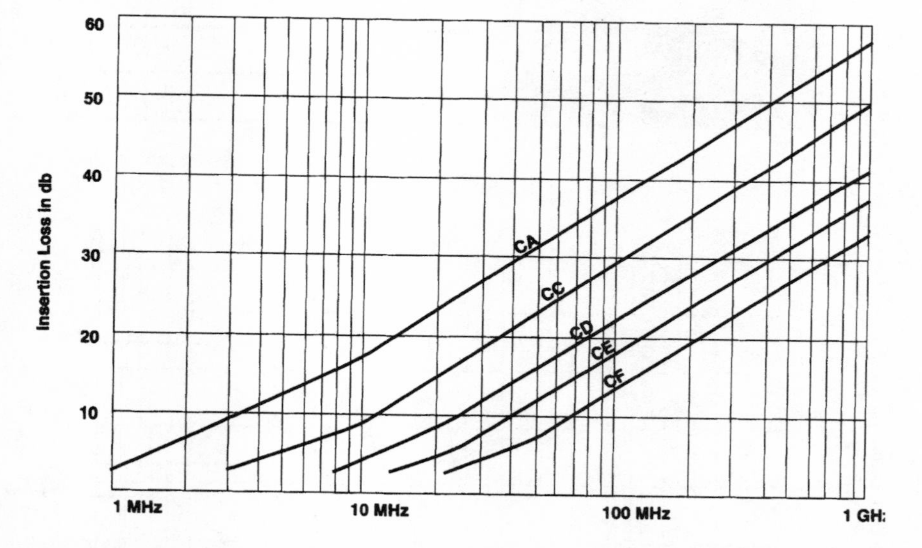

Figure 5.12a Examples of filtered connectors, AMPLIMITE series, less-expensive capacitive types. (Model shown is a male-female adaptor.) *Courtesy of AMP, Inc.*

2. To save space and weight. A single filtered connector will replace a set of one standard connector + discrete filter + filter mounting wall bracket + wiring.

3. When shielding effectiveness (SE) is critical, a filtered connector allows the realization of an ideal baffle mounting of the filter components.

Prescriptions, Installation

1. Select filter components, or attenuation curves, that are compatible with the useful bandwidth of the wanted signals (Fig. 5.13).

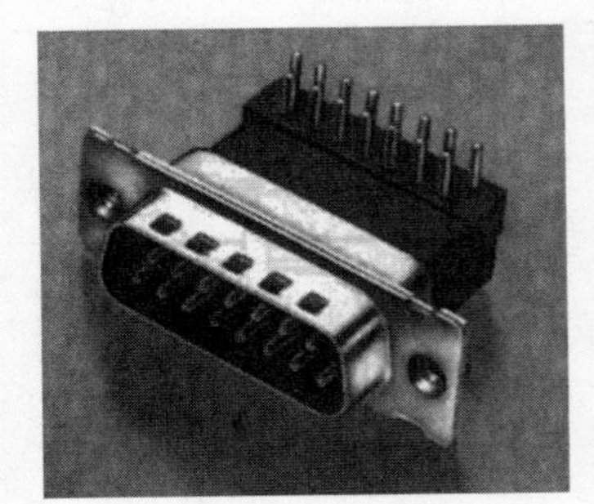

Filter Type	AMP Electrical Specification Number	Capacitance Range
DA	108-1116	3000 pf to 8000 pf
DB	108-1117	2000 pf to 5000 pf
DC	108-1118	1000 pf to 3000 pf
DD	108-1128	600 pf to 1000 pf
DE	108-1131	300 pf to 600 pf
DF	108-1124	130 pf to 400 pf
DG	108-1122	50 pf to 175 pf

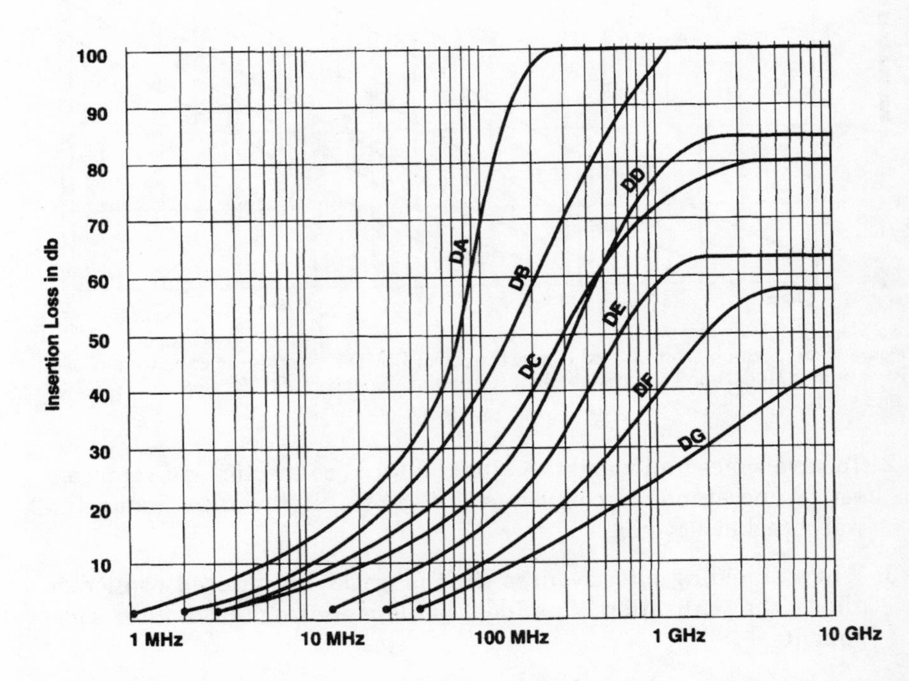

Figure 3.12b Examples of filtered connectors, AMPLIMITE series, distributed capacitor + ferrite types, which are more expensive but provide a steeper attenuation slope beyond cutoff. Both (a) and (b) types are available as a male/female adaptor, which perfectly suits troubleshooting and retrofit situations. The top performances rely heavily on a tight metal-to-metal contact of the connector housing to the equipment box. *Courtesy of AMP, Inc.*

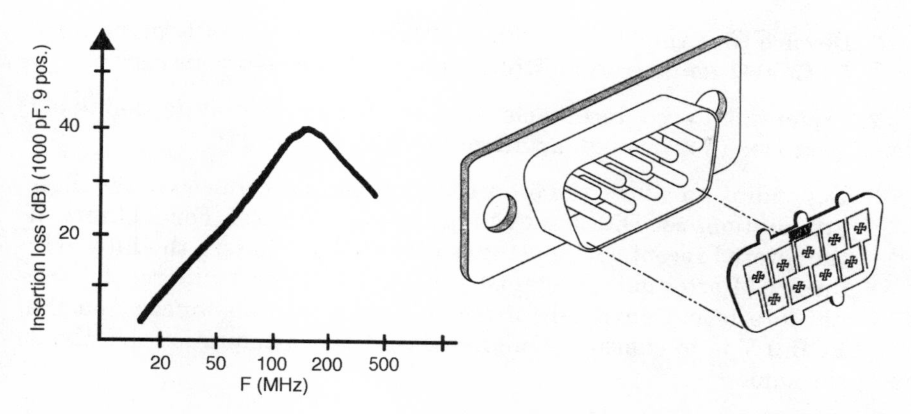

Filter Wafer Features

- Low-cost, high-performance alternative
- Available in all standard and high-density D-sub configurations
- Also available for MIL-C-38999 and MIL-C-26482 and other connectors
- Fits completely and unobtrusively inside mated connectors

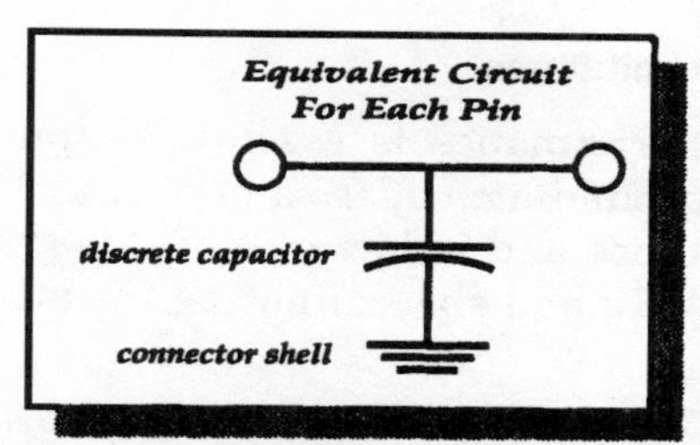

Standard Capacitances

Capacitance (pF) @ 1 kHz	Voltage rating (Vdc)
100	200
470	200
1000	200
1500	200
2000	100
4700	50
10000	50

Instructions for filter wafer installation

The filter wafer installs with the lettering facing out, over the connector pins.

- Set the filter wafer on the mouth of the connector with the pins aligned corresponding to the markings.
- Using your thumb or index finger, gently push the wafer on center until the head of the pins protrude through or are visible through the wafer.
- Using the face of the mating connector, seat the wafer to the back of the connector by mating the connector pair. The filter wafer should now be seated at the base of the connector pins. Ground tabs should rest firmly against the shell of the connector.

Figure 5.13 Capacitive filter wafer for quick upgrade of nonfiltered receptacles. *Courtesy of Microelectronic.*

Beware that vendor's attenuation curves are generally given for a 50 Ω/50 Ω configuration, which may not be true in your case.

2. Prefer a filtered receptacle version. Check that it is positively grounded (not through a wire or trace)

3. Depending on where the filter common plate is connected, the filter attenuations for DM and CM may not be identical. For instance, if the filtered receptacle housing is mounted on chassis, the filter will favor CM attenuation against a floated PCB, or a differential signal interface. Conversely, if the I/O lines are unbalanced and the PCB 0 V is to chassis ground, CM and DM attenuation are about the same.

Limitations

Performance is essentially limited by the quality of the receptacle grounding. Unless a perfect metal-to-metal bonding is achieved, resistance and inductance of this grounding can reach several 100 mΩ at VHF and above, limiting the attenuation to ≈40 dB (in 50 Ω).

Representative Vendors of Commercial Filter Capacitors

The U.S. address is listed first, when applicable.

Discrete

Murata Electronics North America, Inc.
2200 Lake Park Dr.
Smyrna, Georgia 30080
Tel.: (770) 436-1300
Fax: (770) 436-3030
www.murata.com
EMIFIL, NFM series

Philips Components
3200 N. First St.
San Jose, CA 95134
Tel.: (408) 570-5600
www.passives.comp.philips.com

P.O. Box 90050
5600 Eindhoven
Netherlands
Tel.: 31 40 2783 749
Fax: 31 40 2788 399

AVX Corp.
3900 Electronics Dr.
Raleigh, NC 27604
Tel.: (919) 878-6224
Fax: (919) 878-6218
www.avxcorp.com
Capacitor arrays W2F, W3F

Feedthrough

AMP Inc., U.S.A.
Box 3608
Harrisburg, PA 17105
Tel.: (717) 564-0100
Fax: (717) 592-3199
"QUIET LINE" series, filtered terminal blocks
www.amp.com

AMP, France
Tel.: 33 0 1 34 20 8888
Fax: 33 0 1 34 20 8600

Spectrum Control, Inc.
6000 W. Ridge Rd.
Erie, PA 16506
Tel.: (814) 835-4000
Fax: (814) 835-1600
www.spectrumcontrol.com

Filter Connectors

AMP
(see above)
"AMPLIMITE" series

Spectrum Control
(see above)

Tusonix, Inc.
7741 N. Business Park Dr.
Tucson, AZ 85743
Tel.: (520) 744-0400
Fax: (520) 744-6155
"Filter Wafer"

Micro-Electronics Mfg.
19191 Jasper Hill Rd.
Trabuco Canyon, CA 92679
Tel.: (949) 766-8596
Fax: (949) 766-8597
www.microem.com

PCB Flat Capacitors

Circuit Components Inc. (C.C.I.)
2400 S. Roosevelt St.
Tempe, AZ 85282
Tel.: (602) 967-0624
Fax: (602) 967-9385
Q/Pac® capacitive bus bars

Eurofarad
93 rue Oberkampf
F-75540 Paris Cedex 11
France
Tel.: 33 (0) 1 49 23 1000
Fax: 33 (0) 1 43 57 0533

Chapter

6

Inductive, Series-Loss EMC Solutions

6.1 Theoretical Brief

Inductors and ferrites are dual elements of capacitors. They work by increasing the loop impedance of the circuit (see Fig. 6.1), i.e., the device impedance, $Z_x = 2\pi F \times L$, must be larger than the series combination of $Z_s + Z_l$.

Therefore, just as for capacitive filtering (see Fig. 5.1), no value of L can be determined without knowing at least the modules of Z_s and Z_l, and preferably their complex values, if they are not pure resistances.

The cutoff frequency of such lowpass filter is given by

$$F_{co} = \frac{1}{2\pi R_s L} \tag{6.1}$$

where R_s = series combination of Z_s, Z_l.

The attenuation at any frequency above F_{co} is given by

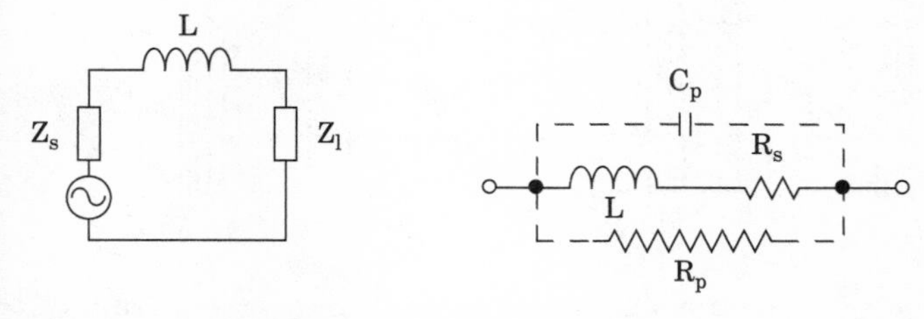

Figure 6.1 Inductive filtering and parasitic elements. R_s = coil wire resistance, and R_p = insulation resistance.

$$A_{dB} = 20\log\left(1 + \frac{Z_x}{Z_s + Z_l}\right) \approx 20\log\left(\frac{Z_x}{Z_x + Z_l}\right), \text{ if } Z_x >> (Z_s + Z_l) \qquad (6.2)$$

Again, as for capacitors, one must be careful to use the actual values of Z_s, Z_l at the calculation frequency, if these are not pure resistances.

Since inductors are dual elements of capacitors, they suffer a mirror kind of limitation: While with capacitors a resonance is created with the parasitic inductance of the terminal leads, an inductor will have a parasitic resonance produced by the interturn capacitance of its winding (Fig. 6.2).

The inductor being placed in series in a power or signal circuit, the device must be:

- Capable of carrying the functional current without overheating or unacceptable voltage drop.

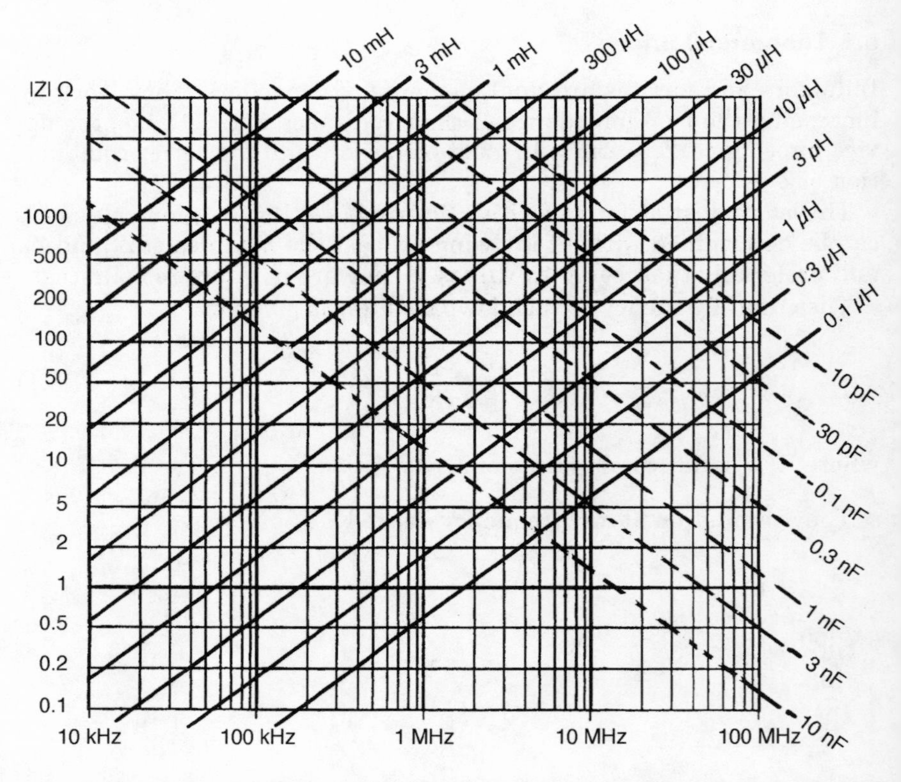

Figure 6.2 Reactance chart for determining impedance and parasitic resonance of discrete inductors. At the Lω = 1/c·ω resonance, the actual impedance is higher than the simple crossing of the two Lω, 1/c·ω lines. The peaking factor is $Q_o = R_p/(L\omega_o)$. Typical values of Q (for air cores) range from 10 to 60.

- Unaffected, i.e., not driven into magnetic core saturation, by the through current. This is especially true with high-permeability materials and, of course, irrelevant with air-core inductors.

6.2 Core Materials

EMI fixes that rely on series impedance, such as wound inductors and ferrites, are practically always applied above 30 to 100 kHz. At such frequencies, the laminated core materials used in ordinary 50/60 Hz up to kilohertz magnetic components are practically worthless.

A good relative permeability (μ_r) is needed up to the megahertz range and sometimes a great deal above. A high μ_r gives the benefit of a high value of inductance, L, per turn (termed the AL). Thus, fewer turns are required, which reduces the component volume and the parasitic turn-to-turn and turn-to-core capacitances.

Above tens of megahertz, where the value of μ_r unavoidably collapses, the material may become lossy, which is a blessing in EMI control (remember, turning noise into heat is one of best solutions), such that the material behaves as a frequency-dependent resistor and no longer as a true inductance.

The following core materials are typical of inductors and ferrites for EMI applications:

Iron Powder. With an iron-alloy powder core, the multitude of oxide-coated iron grains are isolated from each other, creating distributed gaps (instead of the localized slot in a classical air-gapped core). This prevents the core from going into saturation, at the expense of lower values for μ_r. Consequently, such cores can be used for single conductor, noncompensated chokes.

Soft Ferrites. Many alloys formulas are used, the most common being based on the following:

	μ_r	Resistivity (= lossy material)	Upper range of frequency
Mn-Zn	500–10,000	0.1–100 Ω•m	Megahertz
Ni-Zn	10–1,000	10^3–10^6 Ω•m	100 MHz

The Curie point is rather low, typically 150 to 200°C.

Nanocrystalline Materials. These are iron-based alloys, with particle size in the 10 nm range (≈1000 times smaller than standard magnetic materials). They have excellent magnetic properties, with high μ_r, typically ≥ 10,000, and Curie point ≥ 600° C. Core losses, for practical purposes, do not increase with temperature. They are usable up to 100 kHz, allowing large values of L for fewer winding turns. This, by decreasing the parasitic capacitance, results in a higher resonance frequency. Since it is a more expensive material, applications are typically for low-loss, high-inductance magnetic circuits. An example of such material data, the VITROPERM® from VAC Corp. of Germany, is shown on Fig. 6.3.

Basic properties			
saturation induction	Bs	[T]	1.2
saturation magnetostriction	λs	[ppm]	<0.5
resistivity		[Ω mm²/m]	1.15
curie temperature	T_c	[°C]	600
tape thickness		[μm]	20
Properties of tape-wound cores			
static coercivity	H_c	[A/cm]	0,005
hysteresis losses (100 kHz; 0,3 T)		[W/kg]	105
upper application temperature (continuous operation)		[°C]	120-150

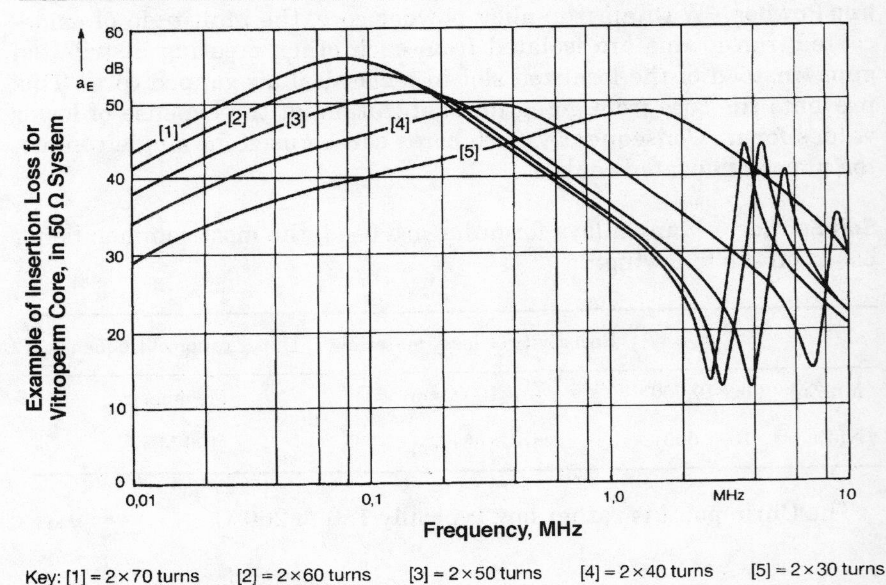

Figure 6.3 Example of material properties and EMI attenuation performance for nanocrystalline magnetic core (VITROPERM®). *Courtesy of VAC Corp.*

6.3 Ferrites and Ferrite-Loaded Cables

Ferrite toroids, beads, and sleeves (see Figs. 6.4 through 6.6) are often regarded as the miracle remedies that the EMC magician pulls out of his sleeve, resolving in seconds a problem that was lingering for months. Although the EMI/EMC storybook is replete with tales of this sort, ferrites deserve a better consideration than being the last-minute snake oil that could "save the day." Integrated in a PCB, they can be part of the circuit's EMC line of defense. And when it comes to troubleshooting and fixing, they ought to be used more intelligently than just as an empirical "cut-and-try."

For logic PC boards, seasoned EMC practitioners often use the "100 pF/100 Ω" rule of thumb whereby a 100 pF capacitor on the high-Z side (e.g., looking toward a gate input) teamed with a 100 Ω resistor

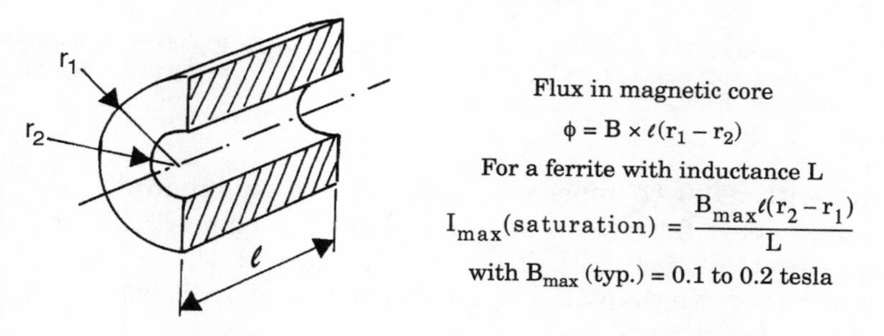

r_1
r_2
ℓ
Flux in magnetic core
$\phi = B \times \ell(r_1 - r_2)$
For a ferrite with inductance L
$I_{max}(\text{saturation}) = \frac{B_{max}\ell(r_2 - r_1)}{L}$
with B_{max} (typ.) = 0.1 to 0.2 tesla

Figure 6.4 Basic parameters of a ferrite toroid.

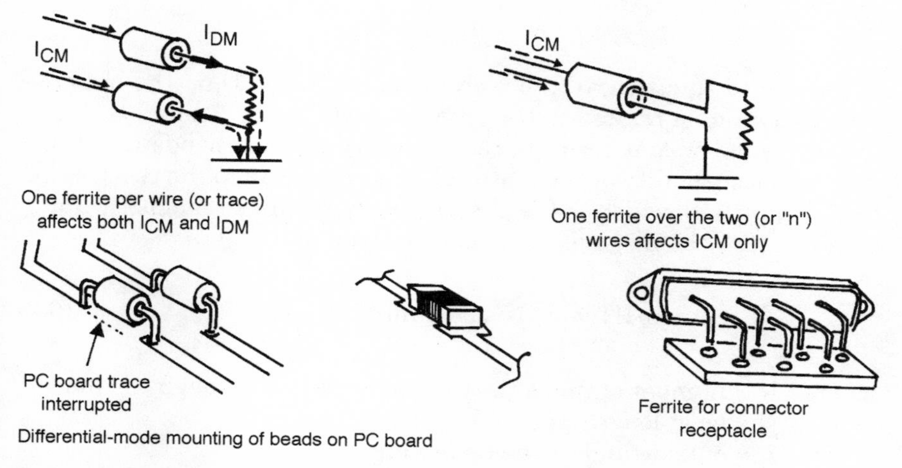

Figure 6.5 Typical applications of ferrite beads.

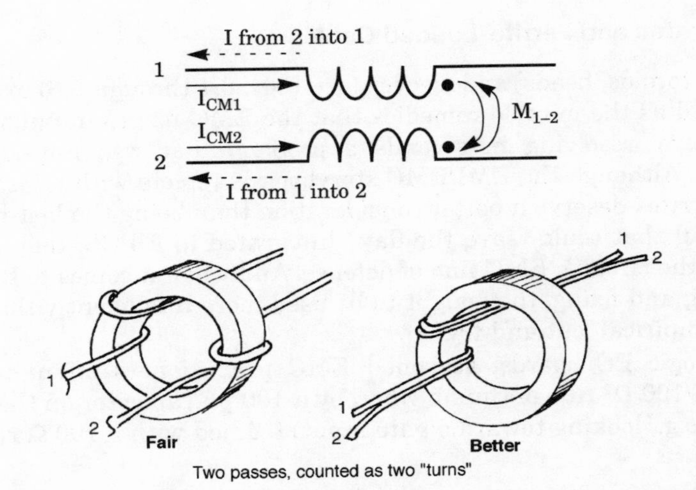

Figure 6.6 CM mounting of a ferrite toroid over two (or more) wires. Keeping the wires very close improves the symmetry and provides tighter mutual inductance, which is critical at high frequencies.

or a ferrite with ≥100 Ω impedance, will solve most immunity and emission cases above 50 MHz. A simple resistor can be an economical choice, e.g., with CMOS circuits drawing only 1 mA. However, when a PCB trace carries more than 10 mA, the ferrite is preferred, because it has a near-zero dc voltage drop, so no dc power is wasted.

The most interesting characteristic of an EMC ferrite bead or toroid is its impedance vs. frequency curve (Fig. 6.7). The impedance of the ferrite bead is a complex one, which can be expressed as

$$Z_b = R + j2\pi FL$$

In the above equation, R, the resistive term, depends on the material resistivity and is related to the eddy currents losses. Therefore, it is a frequency-dependent term. L, the inductive term, depends on μ_r, the relative permeability of the material. It corresponds to the reactive behavior (Fig. 6.8), which is the one primarily sought in magnetic applications below the megahertz level. It is given by

$$L(\text{nH}) = 0.2N^2\mu_r \, l(\text{mm}) \, \text{Ln}\left(\frac{D_1}{D_2}\right) \tag{6.3}$$

where N = number of times the conductor passes through the hole
l = bead length
D_1, D_2 = outside and inside diameters
Ln = natural logarithm

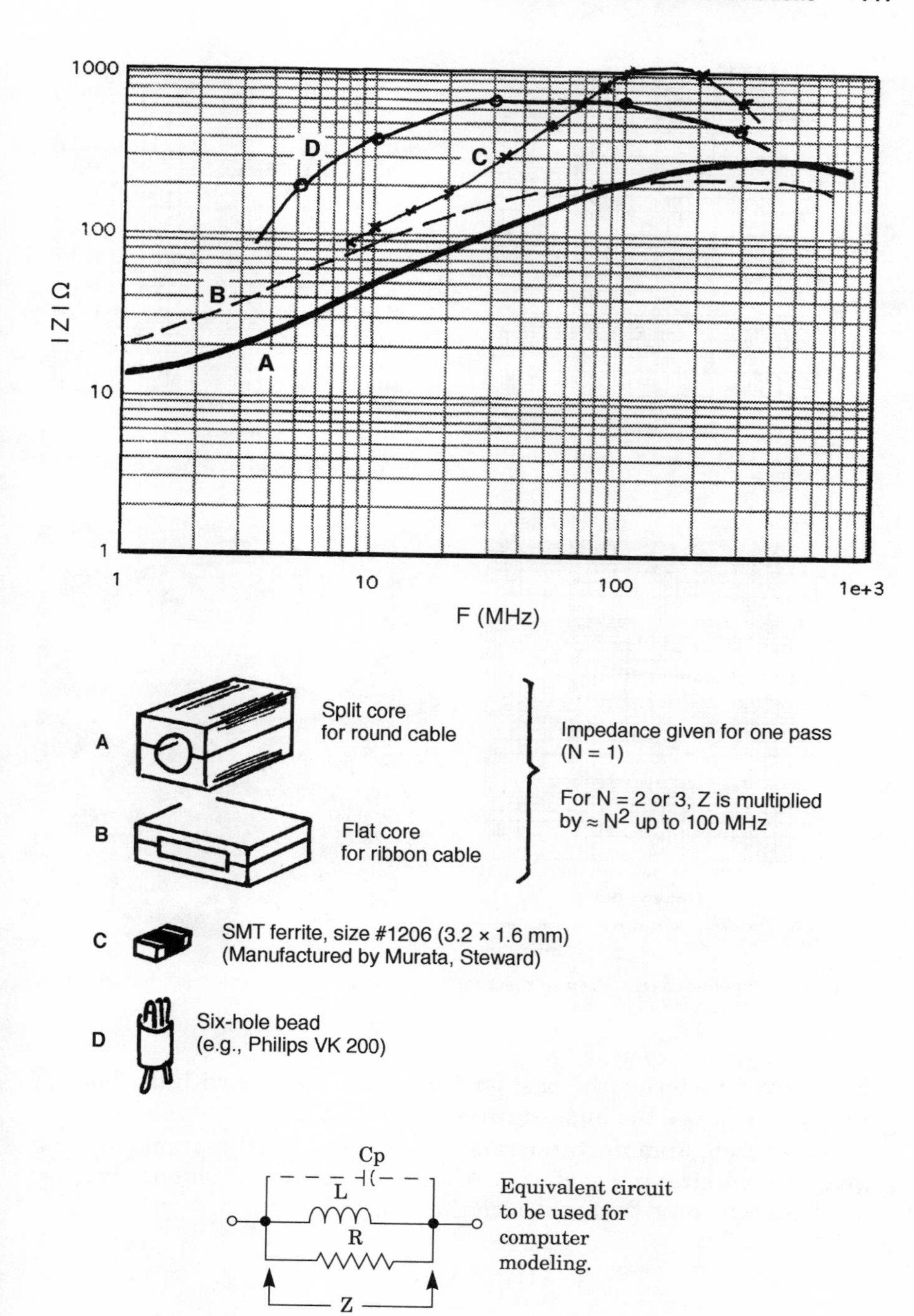

Figure 6.7 Performance of typical ferrites.

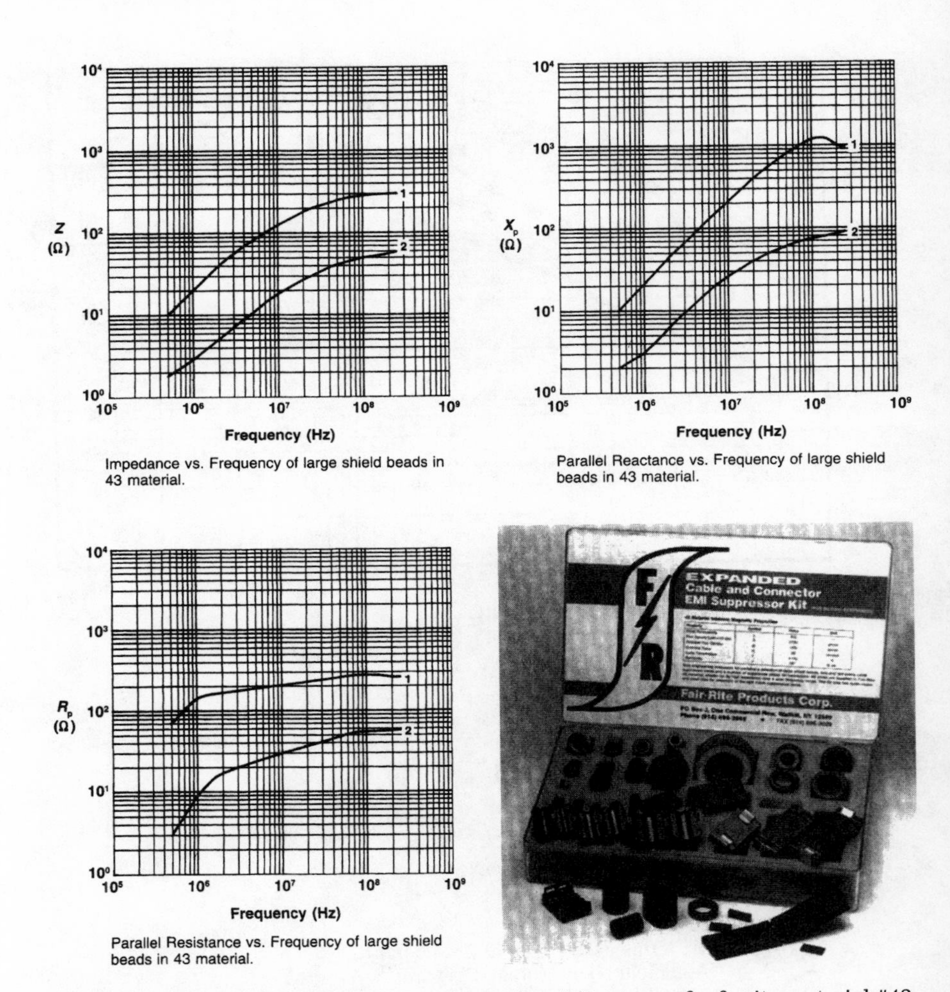

Figure 6.8 Impedance (R + jX) and reactive and resistive terms for ferrite material #43. *Courtesy of Fair-Rite Corp.*

For a given material, the best ferrite bead is the one with the highest l × S product, i.e., the highest ratio for l × (OD/ID).

Quite often, Manufacturer catalogs provide the AL parameter for a given toroid size and material. AL is a measure, in nanohenries, for one turn (one pass through) such that, for N turns,

$$L(\text{nH}) = \text{AL} \times N^2 \tag{6.4}$$

This N^2 dependency is a theoretical ideal that generally is not met except at near-zero current. More realistically, practitioners found that

$$L \approx AL \times N^{1.5 \text{ to } 1.8}$$

Combining Eq. (6.3) and (6.4):

$$AL(\text{nH/turn}^2) = 0.2\mu_r \, l(\text{mm}) \, \text{Ln}\left(\frac{D_1}{D_2}\right) \qquad (6.5)$$

The value of μ_r, in general, is more modest than for a purely magnetic component such as a transformer or a choke. Typical values of μ_r for EMC ferrites are in the 50 to 1000 range. But, in contrast with magnetic materials used at 50/60 Hz and up to few tens of kilohertz, the μ_r for ferrites keeps a stable value across a very wide frequency range, e.g., 0.1 to 10 MHz, and even beyond 100 MHz for some materials. The upper limit of Z_b is reached by either core saturation (too many ampere-turns) or too high an EMI frequency where parasitic capacitance starts to bypass Z_b.

As the current (and therefore the magnetic field) increases, L tends to decrease, but so does the "R" term. Typically, for a small or medium size bead or ring, this decrease starts for I in the amp range, with a 0.5 times decrease in Z_b for a few amps of dc bias.

It is interesting to note that the upper region of the Z_b vs. F curve is dominated by the resistive term R; i.e., the upper-frequency portion of the EMI spectrum is dissipated into heat.

6.3.1 Estimating the Attenuation of a Ferrite Bead

Slipped over a conductor, the ferrite performs as a lossy inductor, whose insertion loss (Fig. 6.9) is approximately

$$A(\text{dB}) \approx 20\log_{10}\left(1 + \frac{Z_b}{Z_s + Z_L + Z_w}\right) \qquad (6.6)$$

where Z_s, Z_L = source and load impedances of the circuit

Z_w = wire impedance if there is some length of conductor between actual source and load

Note: If the distance between the source and the far-end load becomes electrically long [i.e., typically, length (in meters) > 55/F (in MHz) (see Sec. 5.1)], the combination of $Z_L + Z_w$ should be replaced by Z_0, the characteristic impedance of the line (assuming that Z_L is the far-end side).

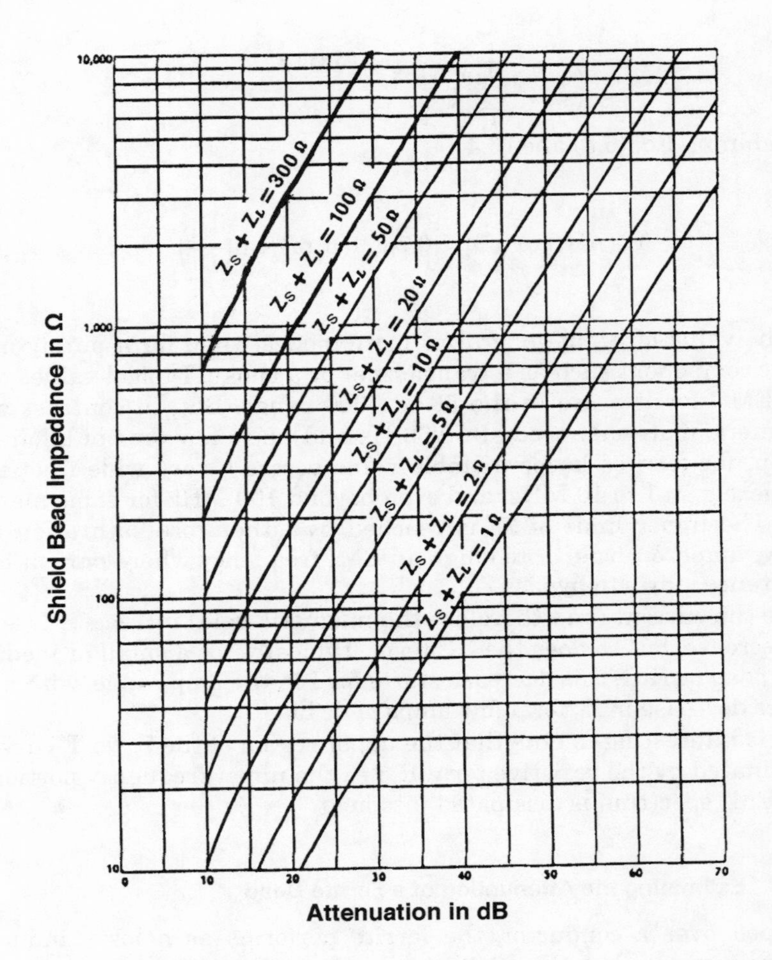

For utilization in a circuit with source and load resistances other than 50 Ω, insertion loss has to be recalculated by

$$IL_{dB} = 20\log\left(\frac{Z_b}{Z_s + Z_L} + 1\right)$$

Z_b Z_s Z_L

Figure 6.9 Estimation of ferrite bead attenuation vs. circuit impedances.

Conversely, if source and load are close to each other—say, less than few tens of centimeters, Eq. (6.6) becomes simply

$$A(dB) \approx 20\log\left(1 + \frac{Z_b}{Z_s + Z_L}\right) \tag{6.7}$$

These equations reveal several things that should be considered when choosing the ferrite as a possible solution:

1. *Ferrites have little effect in high-impedance circuits.* The best ferrite toroids or beads available in the mid 1990s are reaching Z_b values of 300 to 600 Ω in the VHF band (30 to 300 MHz). In a circuit whose configuration averages Z_s/Z_L, such as 150/150 Ω, the best possible attenuation, with a 300 Ω bead, will be

$$A(\mathrm{dB}) = 20\log\left(1 + \frac{300}{150 + 150}\right) = 6\ \mathrm{dB}$$

   That is, a 2× decrease of EMI current results.

2. *Increasing N,* the number of turns, ideally will multiply Z_b by N^2. In the above example, passing the conductor twice through the toroid will produce the following attenuation:

$$A(\mathrm{dB}) = 20\log\left(1 + \frac{300 \times 2^2}{(150 + 150)}\right) = 14\ \mathrm{dB}$$

   This is a 5× decrease in EMI current.

3. *If the wire impedance* connecting Z_s to Z_L is already large (very resistive conductor, high self-inductance), the attenuation may be disappointing because of the Z_w term.

Figures 6.10 through 6.12 illustrate attenuation measurement techniques and results for ferrites.

6.3.2 Ferrite-Loaded Cables

An interesting offspring of the EMI-suppressing ferrite technology is the ferrite-loaded cable. Instead of inserting discrete ferrites over its conductors, the whole cable length is turned into a lossy line. The basis is a fine ferrite powder, mixed at a high dosage (90 percent) with a polymer binder that can be extruded around the wires.

The attenuation is proportional to the coated-cable length, so it is expressed in decibels per meter length. By avoiding the abrupt discontinuities that a cascade of beads would create, there is a more even attenuation above hundreds of megahertz. The impedance and insertion loss of such lossy cable is shown in Fig. 6.13. Such cable can be used on a segment of given length to get the desired attenuation. Just like ferrite beads, they can be made in DM (each wire individually coated) or CM fashion (one coating around the whole bundle). This latter causes very little disruption of the useful DM signals.

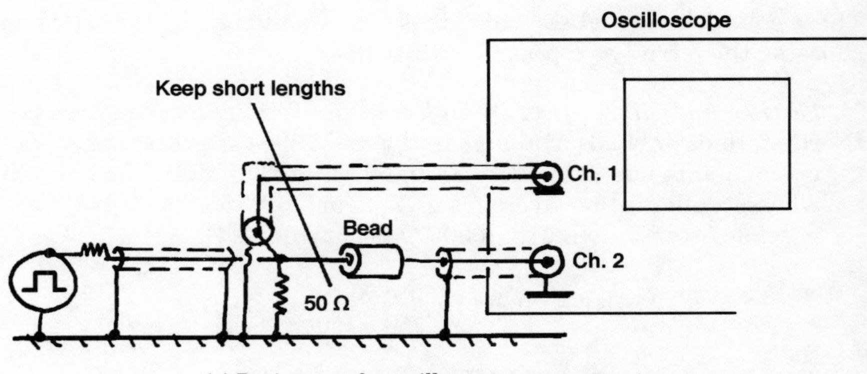

(a) Basic setup for oscilloscope measurements.

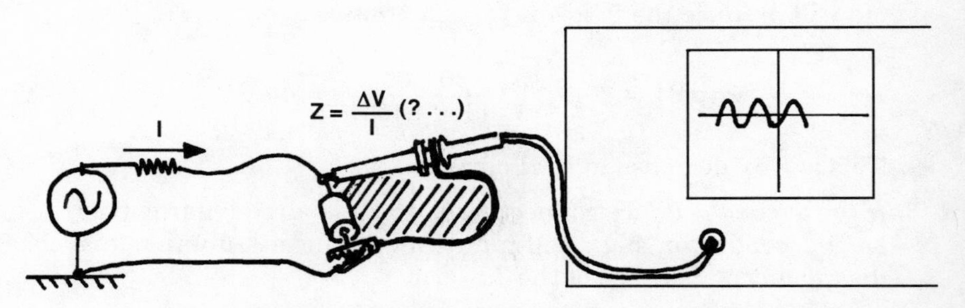

(b) A common mistake: People measure the voltage drop, ΔV, to estimate the ferrite impedance. Misleading results are caused by the pickup loop with the ground strap.

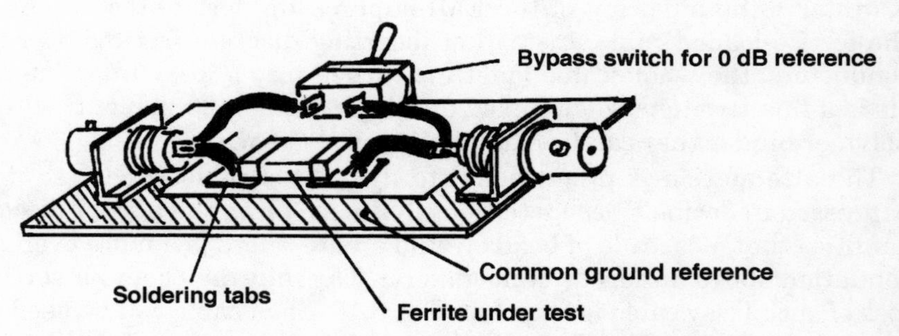

(c) Simple test jig for measuring SMT components. It is easily realized on a 7.5 × 5 cm (3 × 5 in) piece of PCB.

Figure 6.10 Measuring the attenuation of ferrite beads.

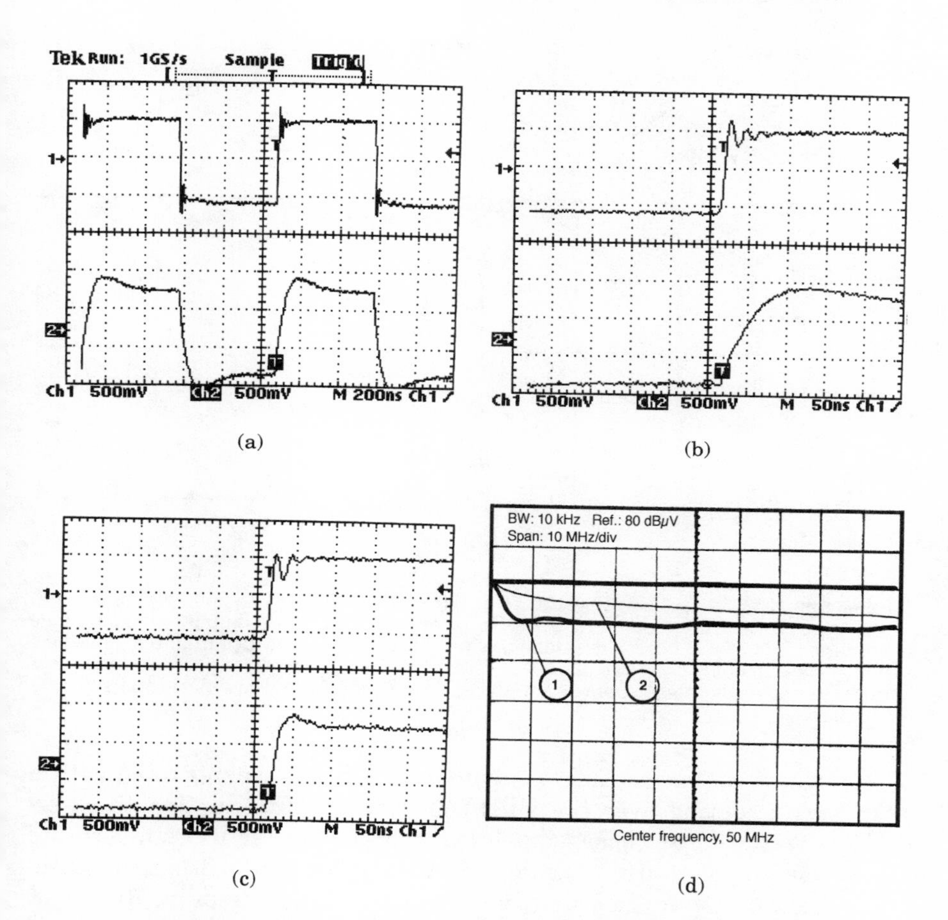

Figure 6.11 Measured attenuations of ferrite beads (setup per Fig. 6.10). (a, b) With ferrite bead type 1 (Fair-Rite material #43), the rise times of this 1 MHz square pulse train are significantly softened, changing from 10 ns (top) to 50 ns (bottom). The 50 MHz oscillations caused by scope input mismatch also disappear. (c) With ferrite type 2, the series impedance is lower than with type 1, and rise time is barely affected. Still, 50 MHz ringing is suppressed. (d) Frequency domain attenuation with the same ferrites, using a spectrum analyzer and tracking generator, the lowpass/HF-stop behavior of ferrites is clearly visible. Ferrite 1 provides 10 dB of insertion loss above ≈ 5 MHz, indicating a series impedance of 200 Ω (from the graph of Fig. 6.9) in a 50 Ω/50 Ω configuration. Ferrite 2 lags behind.

Among their many applications, there are two remarkable ones that can provide an instant relief in emergency situations:

1. *Lowpass filter power cord.* This is a ferrite-loaded power cable (Fig. 6.14) with standard VDE, UL, and CSA approvals. It incorporates an aluminum foil plus a small inductance and a 0.1 μF capacitor inside the ac plug.

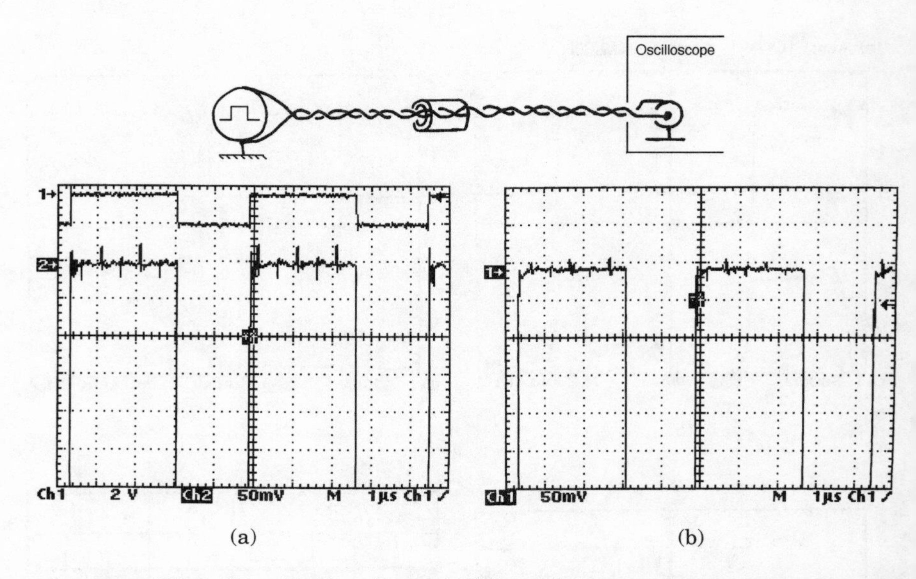

Figure 6.12 CM ferrite, selective effect. (a) Without the ferrite, the 200 kHz pulse train (top trace) is carrying residues of a 1 MHz repetitive signal, picked up by common-mode coupling. This 25 mV peak noise (magnified trace, bottom) can cause radiated emissions from this unshielded twisted pair. (b) With a ferrite toroid (three-pass, material #28 from Steward), the noise spikes are reduced to ≈10 mV, while the differential signal is unaffected.

2. *Coaxial cables for EMC measurements.* RG58 coaxial cable with a ferrite coating over the shield provides an attenuation to externally picked up noise—much better than the regular single braid cable. The CM attenuation in 50 Ω is at least 20 dB higher, somewhere between the ordinary RG58 and the double-braid RG214 or RG55. When a double-shielded coaxial is not available or practical (because of stiffness, for instance), this solution will improve the dynamic range and accuracy of emissions measurements and reduce the errors and uncertainty caused by the cable picking up some of the EMI ambient. And, since the performance is in part due to absorption loss, the quality of the coaxial connectors is not so crucial as for double braid.

Indications for Ferrites

1. As a high-μ device to create inductive drop in the 0.1 to 10 MHz range.

2. As a lossy device, when the EMI frequencies are in the VHF–UHF (30 MHz to a few gigahertz) range, which is where most out-of-spec emission arise.

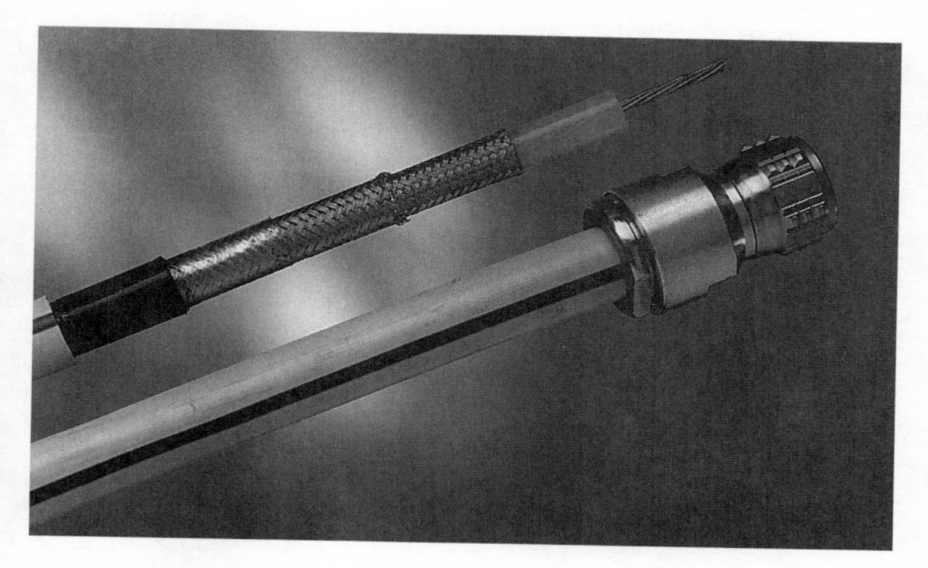

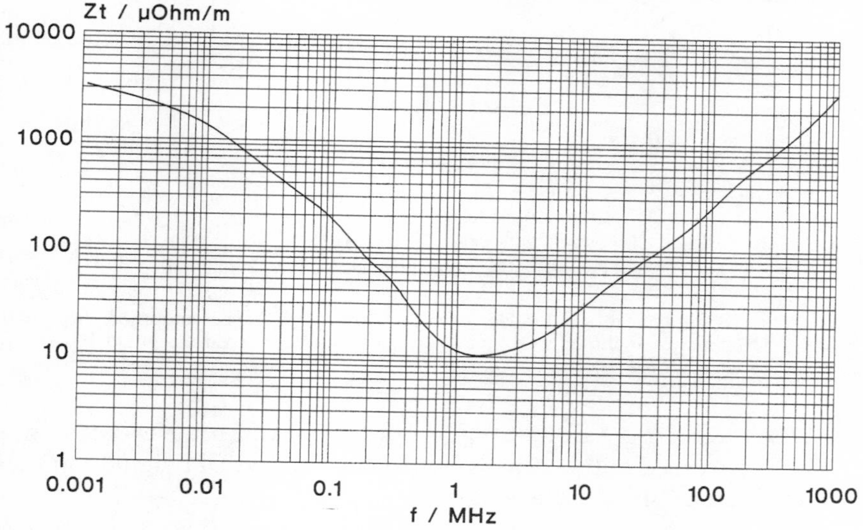

Figure 6.13 Coaxial cable type RG58 (50 Ω), superscreened with ferrite loading. *Courtesy of EUPEN KabelWerke, Belgium.*

3. When a current probe or other technique reveals that EMI currents are of the common-mode type (see Current Probes, Chap. 3).

4. When the traditional "floating" (or single-point ground) approach is inefficient for opening a ground loop, because of the high-frequency nature (typically > 1 MHz) of the problem.

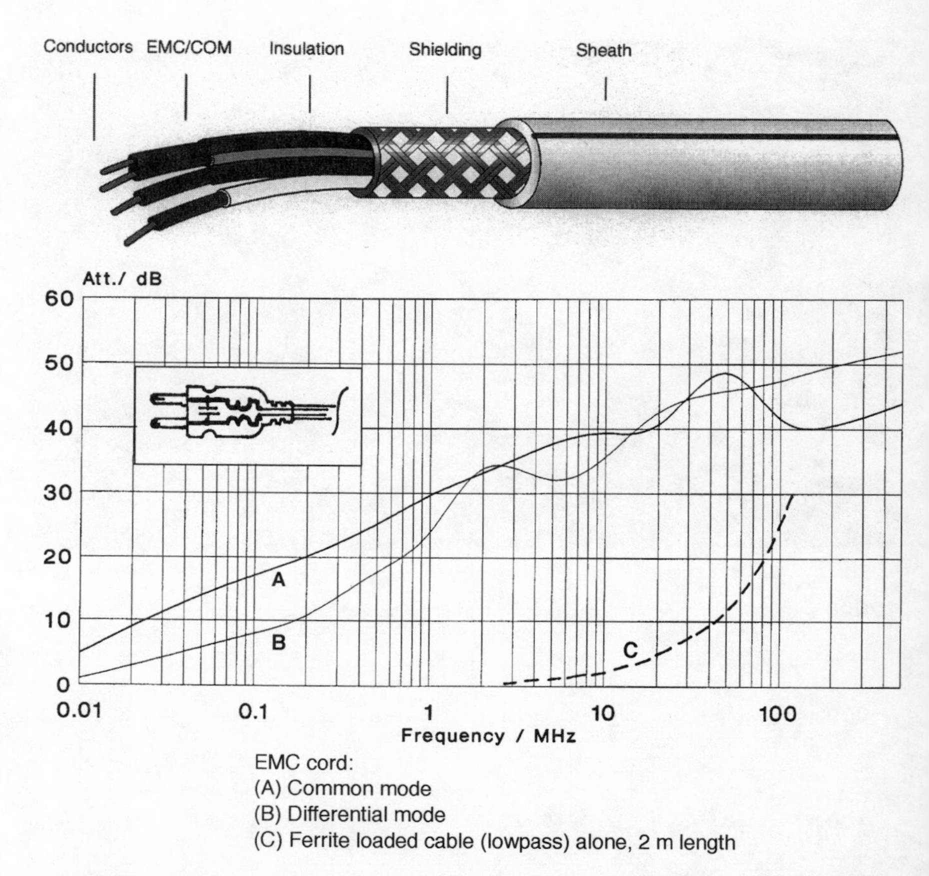

Figure 6.14 Example of ferrite-loaded "lowpass" cable (top). When combined with small L and C components, included in the ac plug, it becomes a filtered power cord (bottom). *Courtesy of EUPEN KabelWerke, Belgium.*

5. When, within the same frequency range, a CM interference has to be suppressed but, at the same time, a desired DM signal must go unaffected through the cable.
6. When cable shield currents have to be reduced without interrupting the shield.
7. When a single-braid coax needs to be improved by 6 to 12 dB without incurring the cost and complexity of a double-braid coax.
8. When the EMI source spectrum (DM or CM), or the victim rejection band, has to be cleaned up above a few megahertz.
9. More generally, with many emissions or susceptibility problems above a few megahertz, including ESD, when there is a conducted

process in the EMI phenomenon before (or after) it becomes a radiation coupling; ferrites reduce the antenna effect of I/O cables.

Prescriptions, Installation

1. For DM attenuation, install ferrites on each input lead (for victims) or output lead (for sources) of the component to be filtered. Select the ferrite with the smallest possible bore for the wire or lead size. "Long, fat beads are better." (There is less air and more magnetic material around the wire.)
2. For CM attenuation, install a ferrite around all the wires, including the 0 V or returns. The ferrite must be solidly fixed on the PC board, the cable, or the chassis using a spring clamp, nylon tie, or a spot of hard glue. To combat emission problems, install the ferrite at the source side of the cable. To combat susceptibility problems, install it at the victim side of the cable.
3. Avoid recoupling of outgoing with incoming leads. Several turns through the ferrite may help, since the ferrite equivalent reactance increases as N^2. However, more than three or four turns may tend to saturate. Many turns also aggravate the problem of parasitic capacitance.
4. Beware of mechanical abuse: Ferrites are fragile.
5. When in operation, check for possible heat evacuated by the ferrite.
6. With split-type ferrites, beware of any possible air gap. Mating faces must be pressed together perfectly.

Limitations

1. A small bead with only one turn generally is inefficient below 10 to 30 MHz.
2. A ferrite bead's series attenuation is predicated from the ratio

$$\frac{Z_{bead} + Z_{circuit}}{Z_{circuit}}$$

   Therefore, the bead typically cannot provide more than 10 to 15 dB attenuation in a 50 to 100 Ω circuit.
3. It is inefficient with high-impedance (>300 Ω) circuits.
4. It can be saturated by the intentional current if mounted in DM mode.

5. Performance drops significantly when core temperature increases toward the Curie point (typically 150 to 200° C).

6.4 Inductors, DM and CM

Inductors used for EMC purposes are low-loss, core-wound components, for use typically up to a few megahertz, beyond which self-resonance makes them progressively less efficient. Their most typical application is for filtering the ac input of switch-mode power circuits, the output of switch-mode speed drives and light dimmers, or as ripple filters for the dc outputs of switch-mode power supplies (SMPSs). Beside their use as stand-alone devices, they are also commonly incorporated in most EMI filters. Just like their dual element, the capacitor, they can be used against DM or CM interference (Fig. 6.15).

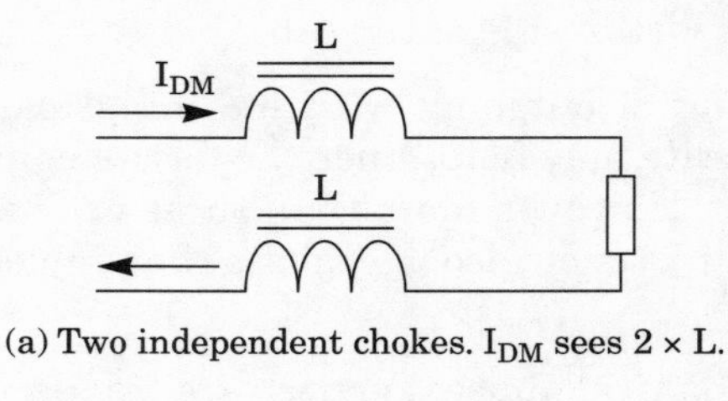

(a) Two independent chokes. I_{DM} sees 2 × L.

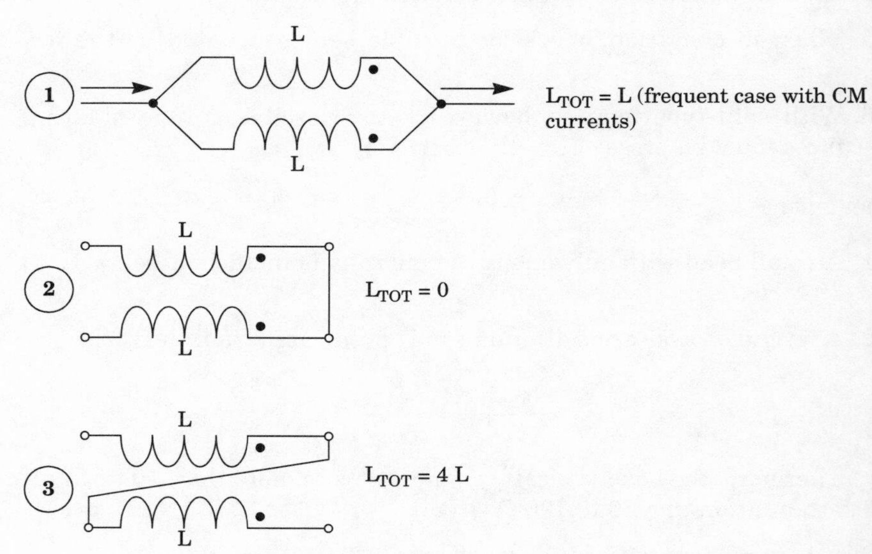

(b) Two coupled chokes (two windings, one core)

Figure 6.15 Specific arrangements with wound inductors.

1. *For DM suppression,* there must be one inductor per wire, therefore the winding carries all of the dc or low-frequency components and the undesired higher frequencies. As a result, the magnetic core can easily saturate and, for staying within a reasonable volume and cost, the value of L is rather limited, as soon as the normal DM current exceeds a few amperes.

   Since the stored energy in an inductor of given material and volume is 1/2 LI^2, a convenient rule of thumb with typical DM inductors of the present technology gives an approximate value for LI^2, related to the core volume and A_2

$$LI^2 \approx \frac{100\ \mu H \cdot A^2}{cm^3} \quad \left(\approx \frac{1600\ \mu H \cdot A^2}{in^3}\right)$$

   For instance, for a magnetic core with a bare volume (mean length × cross section) equal to 3 cm^3 (0.2 in^3), typical values obtainable with commercial chokes will be

   for I = 1A L = 300 μH
   for I = 4A L = 20 μH

   Open-core style DM chokes are less expensive and do not saturate easily, but this is at the expense of a high-leakage field.

2. *For CM suppression,* the inductor has a double winding. The four terminals are to be connected in such a way that the DM current produces opposing magnetic fluxes in the core, hence the other name for these chokes: *flux-compensated.*

   To the contrary, as the CM currents flow in the same direction, their fluxes also have the same direction, and the core behaves as a single inductance with "two-wires-in-hand" (another nickname for CM chokes). Since the normal DM current does not saturate the core, this allows more filtering capability (more L) in a given core size. Values of LI^2 in the range of 3,000 $\mu H \cdot A^2/cm^3$ (50,000/in^3) are easily obtained.

3. *Combined DM + CM chokes.* By letting a certain leakage flux take place in the device, a CM inductance with a certain DM value is created (see Figs. 6.16 and 6.17). Typically, with such chokes, $L_{DM} \approx$ 1 to 5 percent L_{CM}.

Indications

1. For conducted EMI suppression (DM and CM) in all sorts of switch-mode power circuits (SMPSs, variable speed drives, dc/ac inverters, etc.) on mains input side

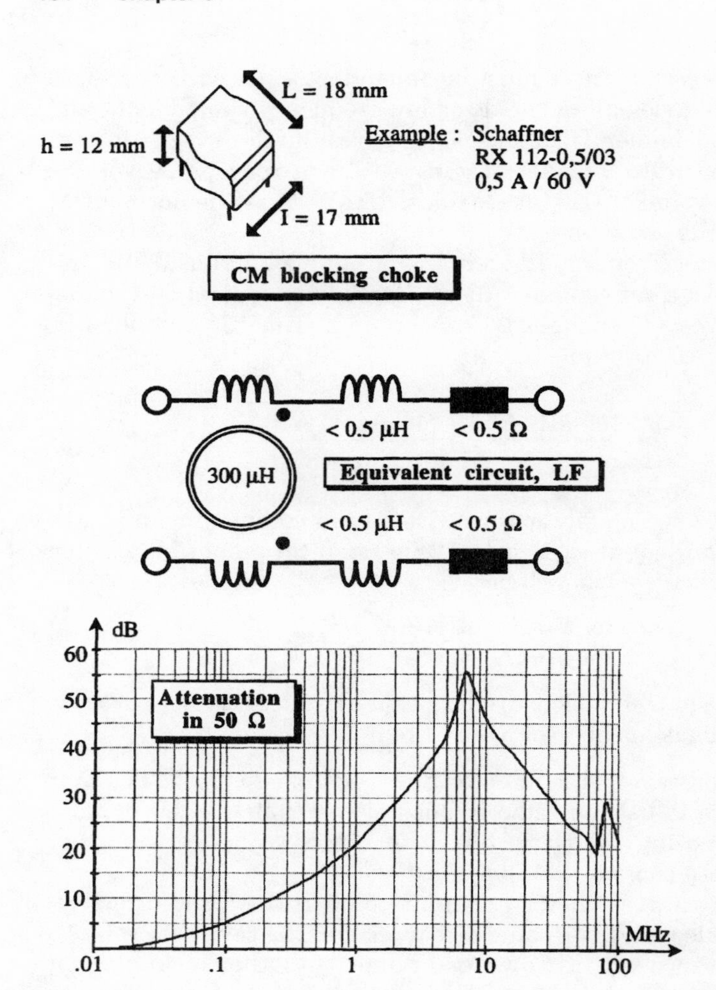

Figure 6.16 Common-mode choke with associated leakage inductances acting as DM chokes. *Courtesy of AEMC.*

2. For DM chokes, when the EMI source is a low-Z, high-current (typically above 0.5 A) circuit

3. For ripple and EMI suppression on the output side of switch-mode power circuits

Prescriptions, Installation

1. Select an inductor with a generously-rated current handling (dc bias or ac). Selection based on rms current can be misleading when the peak current has a high crest factor.

Part no.	CM L, typ. (mH)	DM L, typ. (μH)	DCR typ. (Ω)	Isolation (Vrms)	Current rating (A)	Self-res., MHz
P3717-A	25.0	1000	0.25	1500	3	0.1
Q4007-A	4.5	100	0.05	1500	5	0.6
Q4018-A	1.5	35	0.02	1500	10	1.0

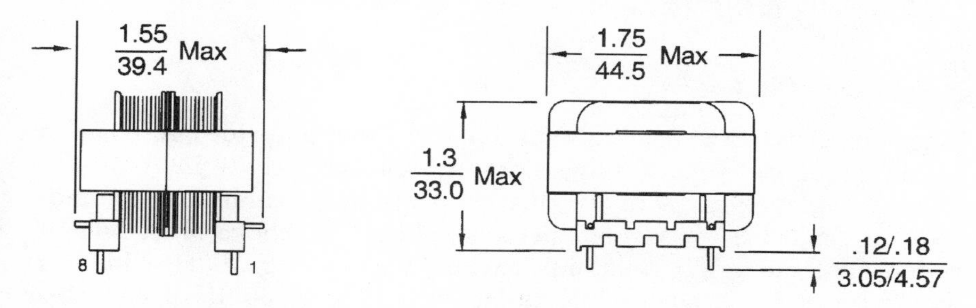

Figure 6.17 Example of combined CM + DM filtering choke. *Courtesy of Coilcraft.*

2. Install as close as possible to the switching source.
3. Check that the parasitic resonance frequency, F_{res}, is sufficiently above the DM or CM noise frequencies and definitely well above the SMPS switching frequency, F_s; typically $F_{res} > 5\ F_s$.
4. Prefer the packaging style with terminal lugs on opposite sides rather than side by side. This will shift the parasitic resonance to a slightly higher frequency.

5. Do not install near susceptible circuits or components (e.g., sensors, A/D converters, feedback loops, op amps, and analog instrumentation amplifiers).

6. Beware of the inductive kick at power-off. It may require a voltage transient protection device if there are fragile components on the same line.

7. Check that the resistive voltage drop (for *CM and DM* chokes) caused by the normal dc or 60 Hz ac current stays within acceptable limits. In a same way, *for a DM choke* only, check the inductive voltage drop ($\omega L \times I$). A good rule of thumb is that the combined IR and ωLI drop for the two-way trip does not exceed 0.5 to 3.0 percent of the nominal mains voltage. For a 50/60 Hz supply, this translates into an approximated rule, for the total DM inductance ($L_1 + L_2$)

$$L(\mathrm{mH}) \leq \frac{0.05\,U(\mathrm{volts})}{I(\mathrm{amps})}$$

Limitations

1. As in the case of any inductive, series component, it will not work in high-impedance circuits

2. Typically, it will not work above a few megahertz.

6.5 Ground Chokes

A ground choke (Fig. 6.18) is a special, low-value inductor put in series in the green (or green/yellow) safety wire to artificially increase the ground-loop impedance at HF, thus reducing the circulation of ground-loop currents via the interconnecting cables. The choke must have a value small enough to remain practically a short at 50/60 Hz (this is a safety concern). Typical values range from 20 to 1,000 µH. These special chokes are generally wound on a lossy ferrite core to decrease their "Q" at resonance.

Indications

1. Use against EMI in the range of a few kilohertz to a few megahertz when either of the following is the case:
   - The current probe shows similar CM currents in interconnecting cables and green safety wire or ground straps.
   - Interconnected equipments are grounded to safety line at different points that may not be equipotential in terms of HF.

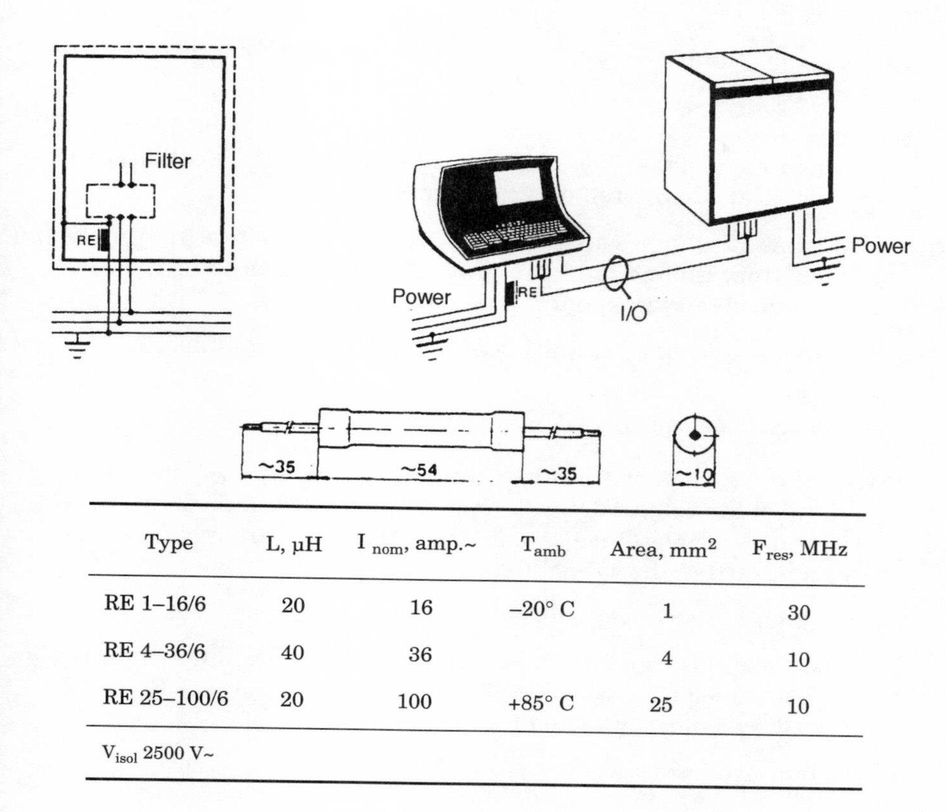

Type	L, µH	I_{nom}, amp.~	T_{amb}	Area, mm^2	F_{res}, MHz
RE 1–16/6	20	16	–20° C	1	30
RE 4–36/6	40	36		4	10
RE 25–100/6	20	100	+85° C	25	10
V_{isol} 2500 V~					

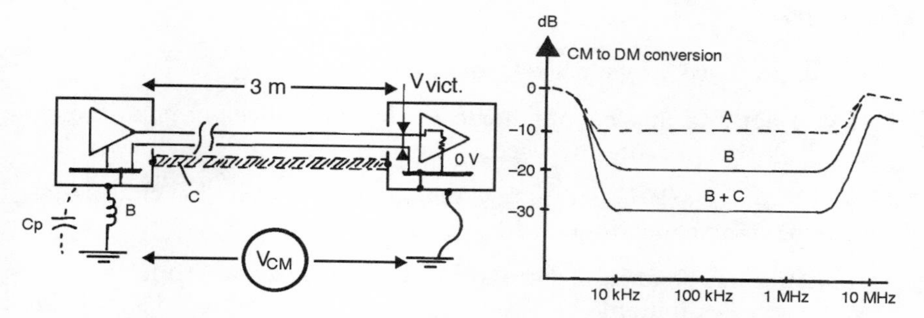

A. CM to DM conversion ratio. Unbalanced link, 3 m long, 100 Ω terminated at each end.
B. One grounding choke (40 µH) added, with stray capacitance box-to-earth reference approx. 100 pF.
C. One ground choke plus large braid between the two boxes, running close to the wire pair.

Figure 6.18 Application example of a ground choke.

2. Use when the building or system safety ground bus is very polluted.
3. Use to decrease ground currents injected by the first harmonics of a switch-mode power supply. This can help to meet FCC/VDE/MIL conducted specs and relieve filter requirements.
4. Use when it is impossible to float either the PCB zero volt or the chassis from ground, and the ground-loop problem is in the kilohertz to megahertz region.
5. Use as a temporary fix, waiting for a better, safer solution.

Prescriptions, Installation

1. Install only when the interconnected equipments (i.e., the "from" and "to" of the cables) are well under control, within the same zone. Do not use if cables have an outdoor path (an RF ground choke creates a hazard in the case of lightning surge).
2. Install within or as close as possible to the equipment housing.
3. The fix works better if the two equipment cases are interconnected by a low-impedance, wide braid or similar connector (see Sec. 11.3, on raceways and companion braids).
4. The improvement will range from 0 to ≈ 30 dB.

Limitations

1. It will not work at very low frequencies.
2. It will not add any benefit if the PC board is already floated or if optoisolators are already used.
3. It will not work above a few megahertz, where the chassis-to-ground stray capacitance will bypass the choke.
4. It should not be used if there are CM surge arrestors (line-to-chassis) in the equipment—or the surge arrestors' grounding should be rearranged such that they bypass the choke during a lightning-induced surge.

6.6 Common-Mode Bifilar Chokes (Longitudinal Transformers)

Section 6.3 described how two (or more) conductors passing in the same ferrite toroid were selectively attenuating the CM currents. This principle is extended to a multiple-winding choke where 2, 3, and typically up to 8 conductors can be filtered against CM by a single component.

Figure 6.19 shows the typical attenuation performance of an eight-winding (for four balanced pairs) CM transformer. More than 10 dB reduction of CM current is obtained from 3 to 500 MHz, mostly by the mutual cancellation effect between windings. By contrast with the isolation transformer (a dc-blocking device—see Chap. 9), the bifilar CM choke (also named *longitudinal transformer*) provides no isolation at dc or low frequencies.

This type of component is especially useful with high-speed parallel digital busses, where CM reduction is needed, with no prejudice to the DM useful signals. It can be regarded as artificially bringing a good symmetry into an unbalanced or poorly balanced link, hence the name *balun* (for balanced/unbalanced) sometimes used for such a device.

6.7 Combined L,C Elements

By combining small inductors with capacitors in two elements (L + C) or three elements (pi or tee), one can obtain (see Chap. 4) more efficient filtering with steeper attenuation slopes. Also, the performance limitations caused by the inherent weaknesses of each component (e.g., parasitic shunt capacitance for inductors and parasitic series inductance for capacitors) are shifted to higher practical values. HF insertion losses greater than 40 dB, which are very difficult to get by a single C or L filter, are obtainable by a C-L-C or L-C-L arrangement. DIP or SMT components are commonly available wherein the manufacturer has optimized the packaging for minimum parasitic effects (Fig. 6.20).

The pi (C-L-C) arrangement is preferred when both source and load sides are high impedances (see Sec. 6.4). However the pi scheme carries one typical vulnerability: It is very sensitive to a nonideal mounting of the ground terminal. Figure 6.20 shows that, if some parasitic impedance exists between the C common and the actual chassis ground, it immediately spoils the attenuation of the filter circuit. If this parasitic element is inductive (and it usually is, e.g., a PCB trace, a piece of wire, a metal stud, or thin bracket), the impedance Z_p increases with frequency, F, while, simultaneously, the impedance of the capacitors, C, is decreasing. A high-frequency corner is reached where the CM current, instead of being drained to ground at point G, is using the two capacitors to bypass the inductor, making the filter more and more useless as F increases. For this reason, the pi scheme must be used only when direct, leadless connection of capacitors to ground can be achieved (see Fig. 6.21).

A quite common arrangement for tee type (L-C-L) is shown in Fig. 6.22. The center conductor (2) is the capacitor ground lead, whereas

Specifications

Part Number*	Lines	Max. Current (mA)	L/winding (μH)	DCR Max. (mOhms)	Isolation (Vrms)
DLF8000	8	100	28	100	300
DLF8500	8	500	25	45	300
DLF4000	4	100	28	100	300
DLF4500	4	500	24	45	300
DLF3000	3	100	28	100	300
DLF3500	3	500	24	45	300
DLF2000	2	100	28	100	300
DLF2500	2	500	24	45	300

* For optional cover add "C" to part number. Not available on 4-line parts.

16 15 14 13 12 11 10 9 / 1 2 3 4 5 6 7 8

8 7 6 5 / 1 2 3 4

6 5 4 / 1 2 3

6 4 / 1 3

Figure 6.19 Example of longitudinal (CM) multifilar filters. (Continued next page). *Courtesy of Coilcraft.*

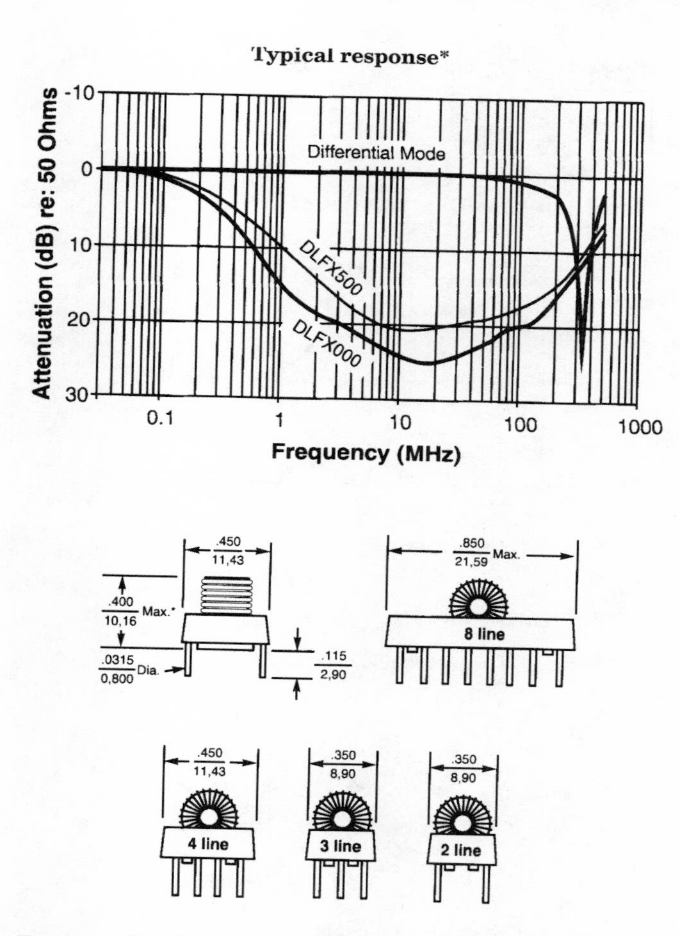

Figure 6.19 Example of longitudinal (CM) multifilar filters. (Continued)

terminals (1) and (3) are the inductive input and output, with ferrite beads inductance typically in the range of hundreds of nanohenries.

Although it is normally best suited for low values of Z_{source} and Z_{load}, the tee network is sometimes preferred with medium values of Z (e.g., 100 to 300 Ω) because it does not suffer from the vulnerability described for the pi scheme, that is, if connection 2 is not ideally grounded, the two inductances will still perform correctly against HF currents.

Indications

Indications are the same as for simple L or C filters except that they are used when the attenuation slope has to be steeper, such as 40 or 60

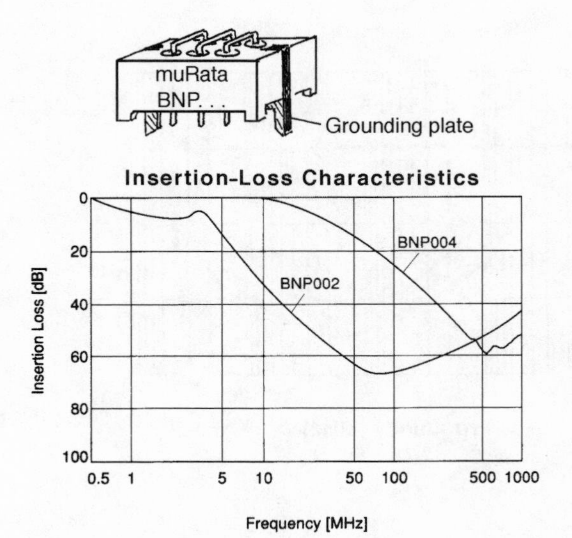

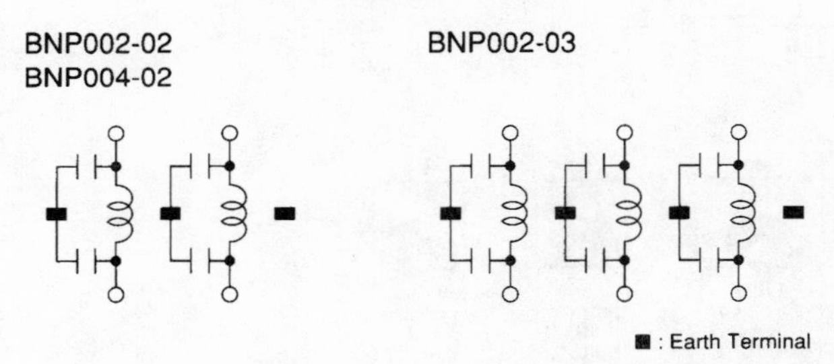

(a) Example of pi filter for signal applications (Murata, EMI-FIL series).

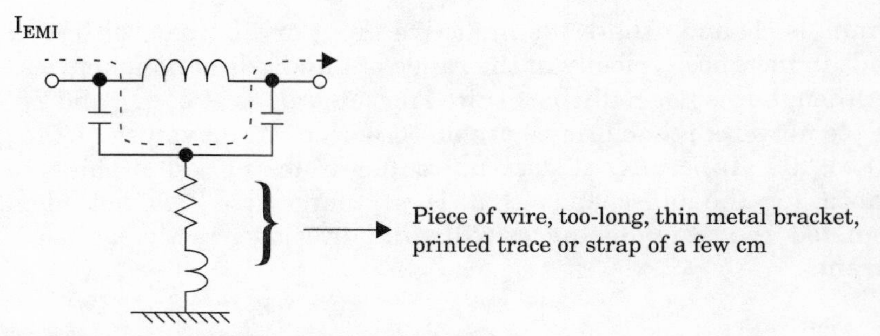

(b) Spoiling the efficiency of a pi filter by parasitic impedance in the grounding connection.

Figure 6.20 Small LC filters for signal applications.

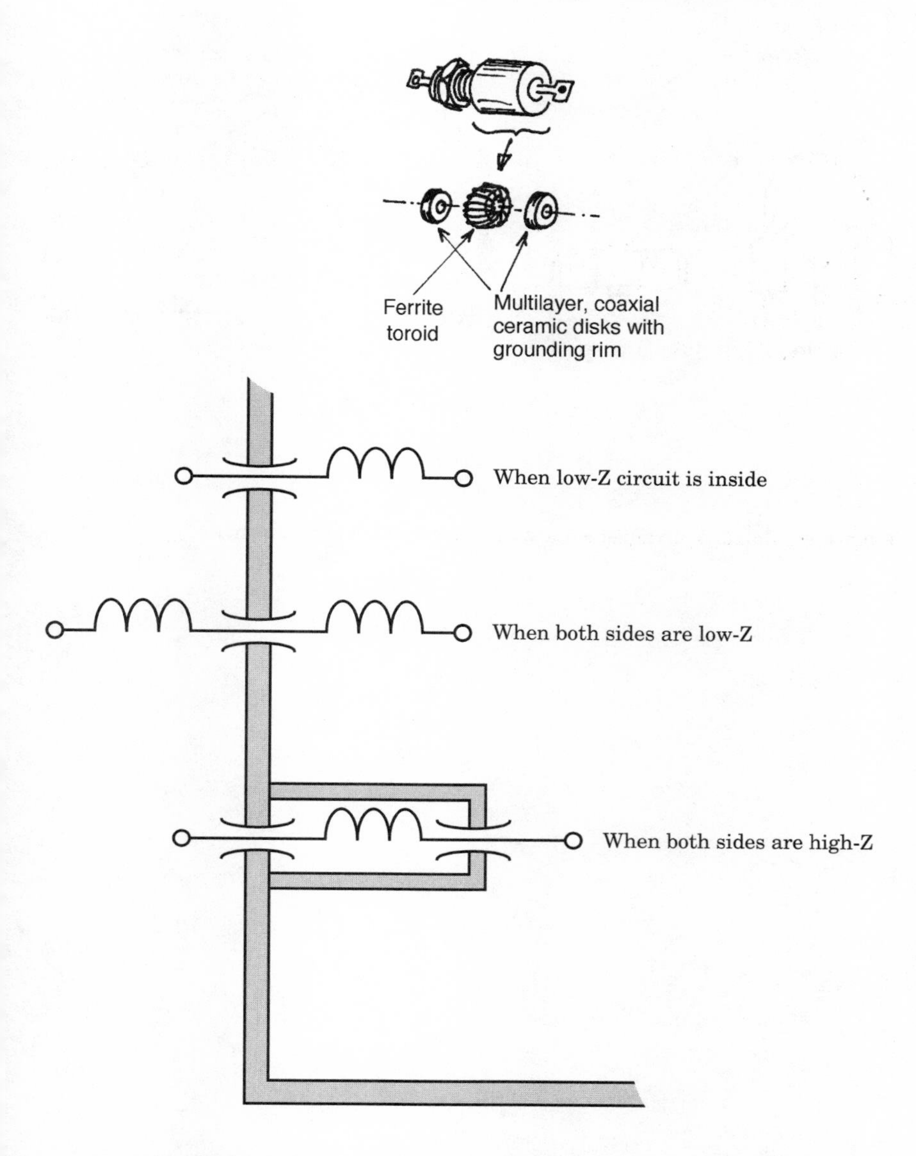

Figure 6.21 Feedthrough filter arrangements for L-C, L-C-L (tee), or C-L-C (pi) styles.

dB/decade, than a first order; for instance, if one needs 30 dB of filtering at 50 MHz while keeping intact a useful bandwidth of 5 MHz.

Prescriptions, Installation, Limitations

These are the same as for L and C filters (see Secs. 5.3, 5.5, 6.3).

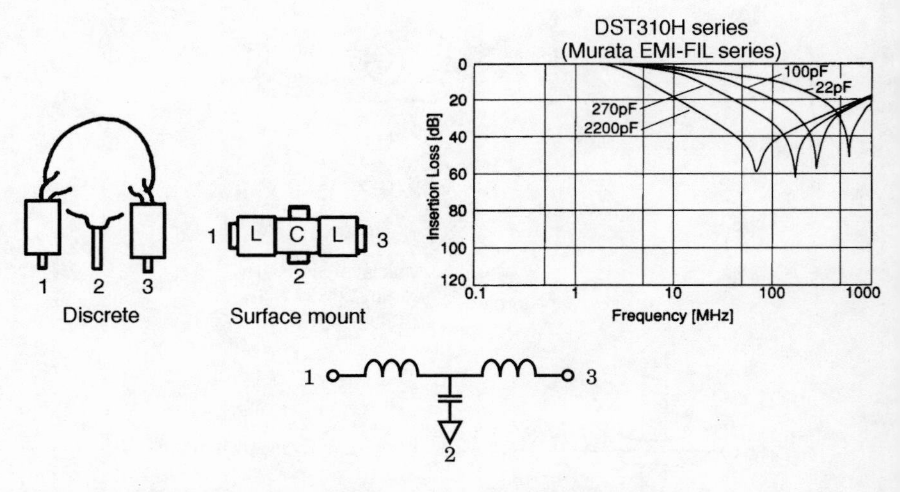

Figure 6.22 Common arrangements of tee filter for PCB applications.

Representative Vendors of Commercial Inductors and Ferrites

INDUCTORS

Coilcraft, Inc.
1102 Silver lake Rd.
Cary, IL 60013
Tel.: (847) 639-6400
Fax: (847) 639-1469
www.coilcraft.com

Schaffner EMC, Inc.
9B Fadem Rd.
Springfield, NJ 07081
Tel.: (973) 379-7778
Fax: (973) 379-1151
www.schaffner.com

Main office:
Schaffner EMV AG
Nordstr. 11
4542 Luterbach
Switzerland
Tel.: (41) 32 6816 626
Fax: (41) 32 6816 641

Vacuumschmelze GmbH (VAC)
Gruner Weg 37
D-63450 Hanau
Germany
Tel.: 49 6181 38-0
Fax: 49 6181 38-2645
www.vacuumschmelze.de

FERRITES

Fair-Rite Products Corp.
P.O. Box J
1 Commercial Row
Wallkill, NY 12589
Tel.: (914) 895-2055
Fax: (914) 895-2629
www.fair-rite.com

FerriShield, Inc.
Empire State Bldg.
350 Fifth Ave.
New York, NY 10118
Tel.: (212) 268-4020
Fax: (212) 268-4023
www.ferrishield.com

Murata Electronics North America, Inc.
2200 Lake Park Dr.
Smyrna, Georgia 30080
Tel.: (770) 436-1300
Fax: (770) 436-3030
www.murata.com
EMIFIL, NFM series

Steward
P.O. Box 510
1200 E. 36th St.
Chattanooga, TN 37401
Tel.: (423) 867-4100
Fax: (423) 867-4102
www.steward.com

Würth Elektronik GmbH
Rieden Strasse 16
D-74635 Kupferzell
Germany
Tel.: 49 (0) 7944 9193-50
Fax: 49 (0) 7944 9193-51
www.wuerth-elektronik.de

FERRITE-LOADED CABLES

Cablerie Eupen
Malmedyer Str. 9
B-4700 Eupen
Belgium
Tel.: 32 87/597000
Fax: 32 87/597100
www.piap.ch/cable/ceupen.html

Chapter

7

Power-Line Filters

Modern power-line filters contain an L, pi, or T arrangement of line-to-line (DM) and line-to-chassis (CM) high-frequency capacitors, combined with CM and/or DM chokes. The individual properties of these discrete components have been described previously in Chaps. 5 and 6, and the information will not be repeated here. The optimal choice among L, C, Pi, or T styles is based on the decision factors explained in Chap. 4 (see Table 4.1).

DM capacitors in power-line filters are generally in the microfarad range, while CM capacitors, because of safety aspects (leakage to earth), are generally limited to a few nanofarads. Some filters incorporate an earth choke of few hundred microhenries to block HF ground current circulation into the system (see Sec. 6.5, "Ground Chokes").

Indications

1. A power-line filter (Fig. 7.1) is practically a must in all modern electronic equipment for conducted emissions or susceptibility aspects, or both. To some extent, a good filter may also help to reduce radiated EMI coming in or out via the power cord.
2. A power-line filter is a must every time a switching power supply is used (unless the regulator is self-filtered). (See Fig. 7.2.)

Prescriptions, Installation

1. Select a line filter with a generous rating for the current consumption of the product. Beware of the rms value, which can be an underestimation in the case of pulsed currents with low duty cycle (high peak value, small average value).

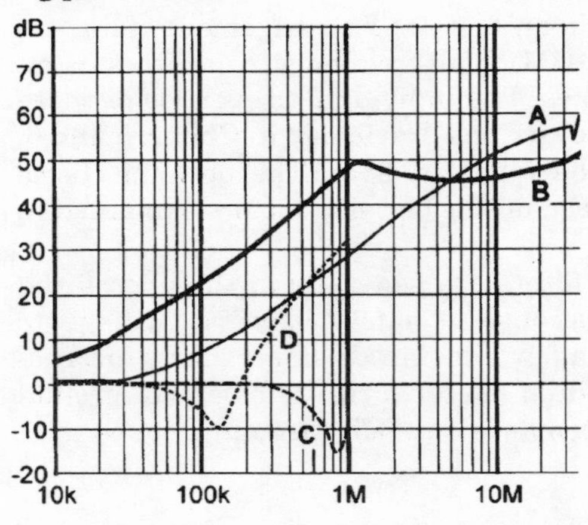

Figure 7.1 Example of a compact filter, with a good balance among size, cost, and performance. In addition to the usual attenuation curves for DM (curve A) and CM (curve B) in a 50 Ω/50 Ω configuration, the manufacturer provides the performance data for odd impedance arrangements, per CISPR 17: Z_{mains} = 0.1 Ω, Z_{load} = 100 Ω (curve D) and Z_{mains} = 100 Ω, Z_{load} = 0.1 Ω (curve C). *Courtesy of Schaffner Corp.*

2. Select a filter that has a combination of CM and DM filtering components.

3. Check that the line-to-ground capacitors can withstand abnormal voltage during temporary surges or line-to-ground faults. (Capacitors should be 250 V rated for single-phase power in the U.S.A. and for 115 V mains in general, and rated 400 V for single-phase power in Europe and for 230 V mains in general.)

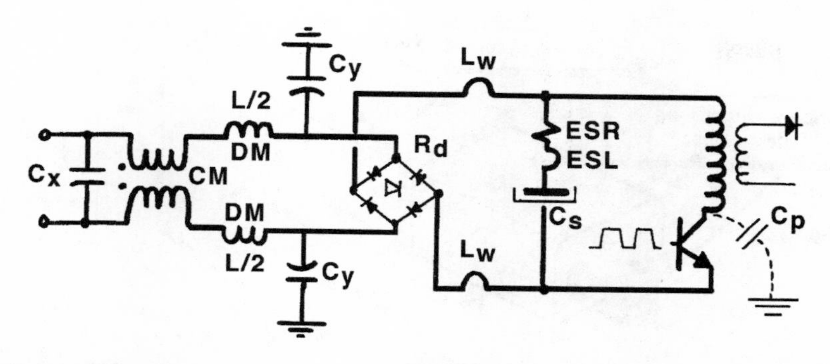

Designation		Typical values
ESR, ESL	Equivalent series resistance and inductance (parasitics)	100 mΩ, 15 nH
C_p	Parasitic capacitance of switching transistor hot electrode (collector or drain) to chassis ground	10 to 200 pF
L_w	Wiring inductance	10 to 50 nH (×2)
R_d	Dynamic resistance of rectifiers in "on" state	50 mΩ (×2)
L/2	DM inductance of EMI filter	10 to 300 µH (×2)
L_{CM}	CM, bifilar inductance	0.3 to 30 mH
C_y	CM filtering capacitors, class Y	2.2 to 47 nF

The DM inductance sees the right-hand side of the low impedance (typically < 1 Ω) of the tank capacitor C_s with ESR, ESL. On the left-hand side, the DM inductances see the low impedance of DM capacitor C_x.

The CM choke sees, on its right, the low impedance of CM capacitors C_y, which form a divider bridge with C_p. On its left, the CM choke sees the low/medium impedance of the power mains source (or the LISN, during a test).

Figure 7.2 Proper mounting of CM/DM chokes and capacitors for a switch-mode power supply filter.

4. Check that CM capacitor values are low enough to avoid creating safety issues and triggering the installation ground-fault detectors.

5. Install the filter in a feedthrough style or using a "dog house" right at the point where the power cord penetrates the housing (Figs. 7.3 and 7.4). If this is impossible, at least keep the portion of the power cord that is between the filter and the box wall as short as possible, and preferably shielded.

6. The "input" and "output" sides of the filter, as labeled by the manufacturer, correspond to the best attenuation for most typical situa-

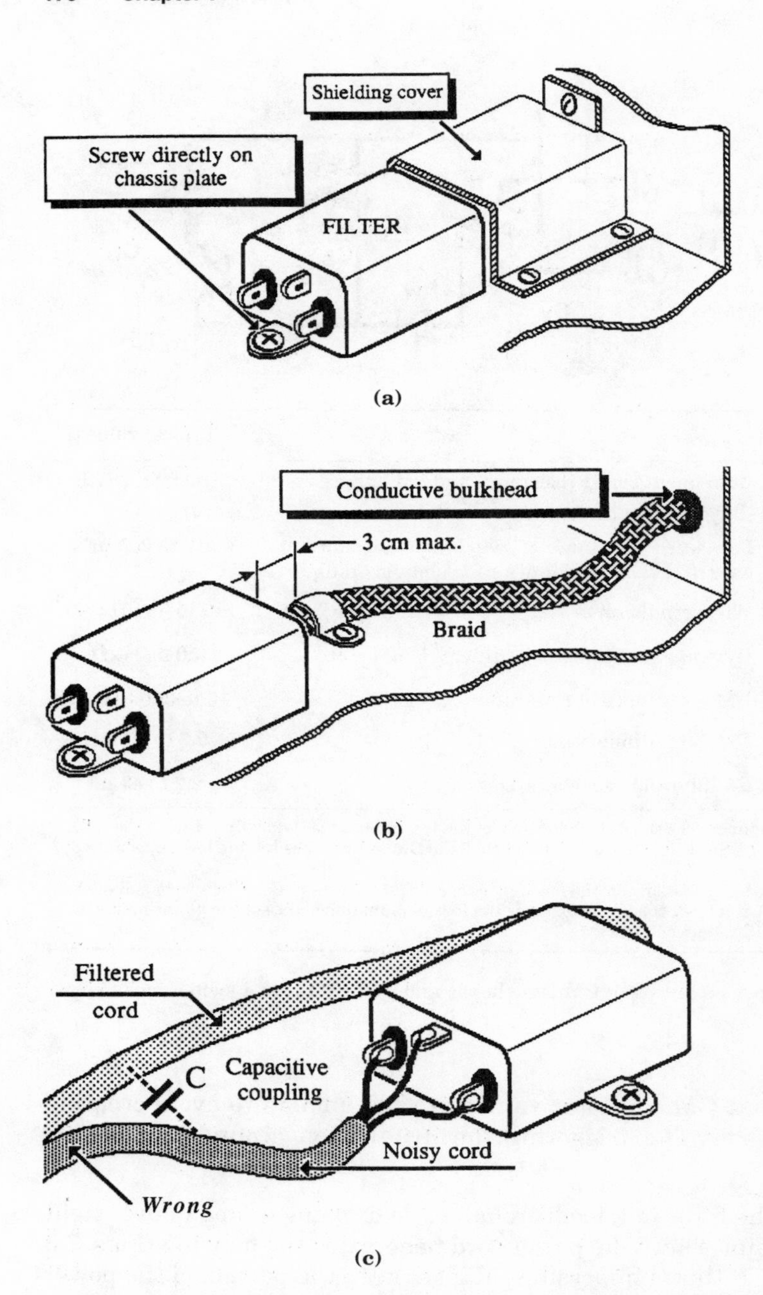

Figure 7.3 Mounting precautions with power-line filters: (a) filter inside the equipment, with "dog house" shield for input lead area, (b) filter inside the equipment, with a shield for the inner portion of the power cable coming in, and (c) parasitic recoupling between input and output wiring. *Courtesy of AEMC.*

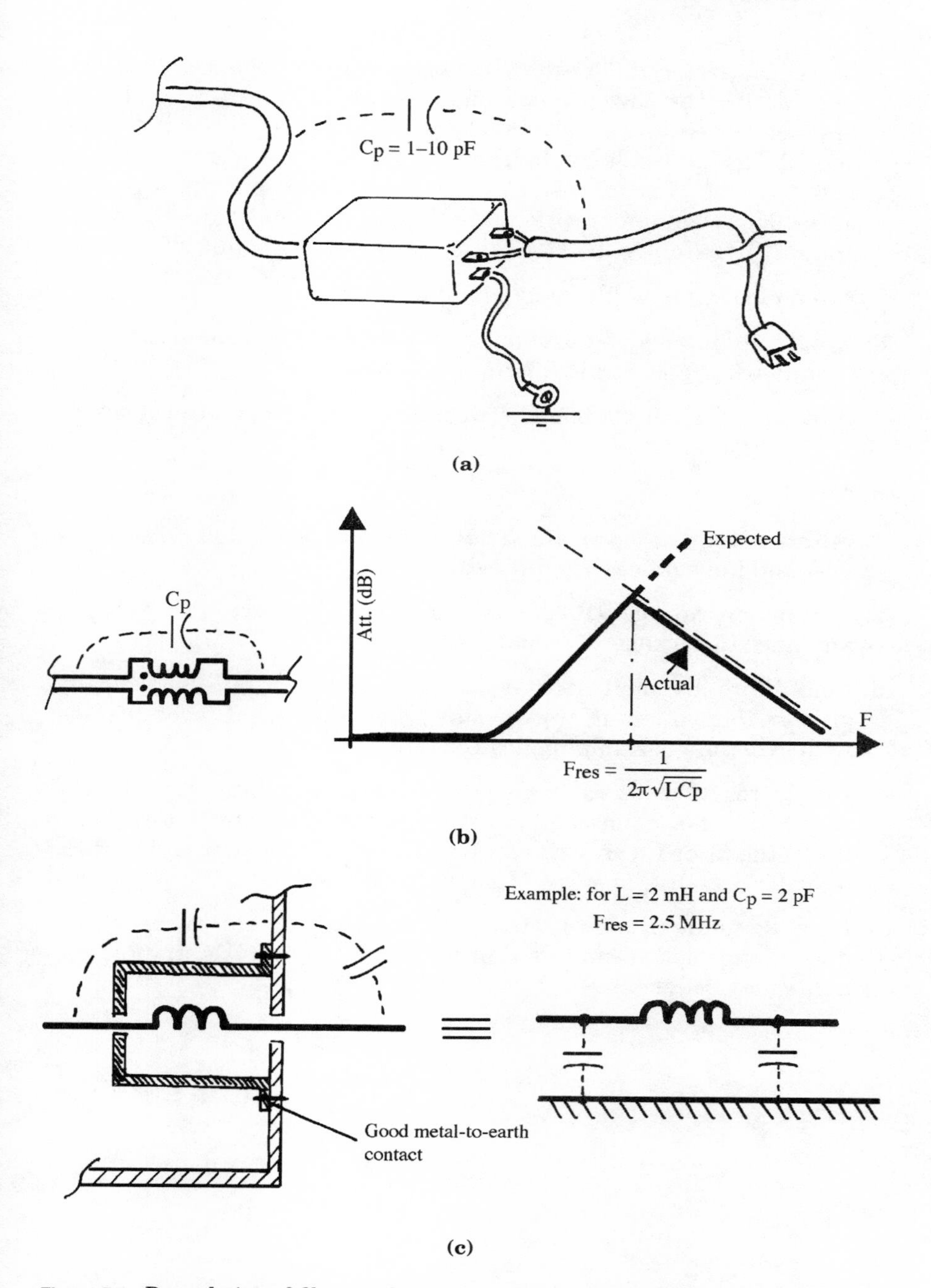

Figure 7.4 Degradation of filter performance by incorrect mounting. Even if incoming and outgoing wires are not tangled, the "open space" mounting (a) leaves a stray capacitance C_p, which is shunting the filter inductance, typically above a few megahertz (b). The through-wall mounting (c) makes this shunting impossible.

tions. However, compare the filter schematic with your application. Considering the power mains impedances (CM and DM) and your product power input impedances (CM and DM), filter inductors should face the side with lower impedance. As a general rule, power source impedances are lower than equipment input impedances. However, apparatus that use a power regulator with front-end rectifiers/tank capacitor exhibit very low DM input impedance.

7. Avoid recoupling of filtered and nonfiltered leads.
8. Avoid coupling inside the equipment between wires going to the filter and wires going to I/O ports.
9. Provide a good, direct bond between the filter can and the chassis.

Limitations

1. Filters are qualified in 50 Ω/50Ω configurations. Real-life source and load impedances are different.
2. For current ratings >10A, power-line filters are often heavy, bulky and hard to package as "come-late" fixes.
3. Line filters are inefficient against long-duration pulses (typically greater than a few microseconds) where they need to be teamed with overvoltage clamping devices (Fig. 7.5).
4. Purely reactive filters (high Q) can be driven into oscillation if power mains or equipment generate strong harmonics corresponding to the filter L-C resonance. Because of this, some filters contain lossy ferrites or Q spoiling (damping) resistors.
5. Because manufacturers often optimize their filters to meet conducted emission standards that stop at 30 to 50 MHz, the components may be less effective above those frequencies.

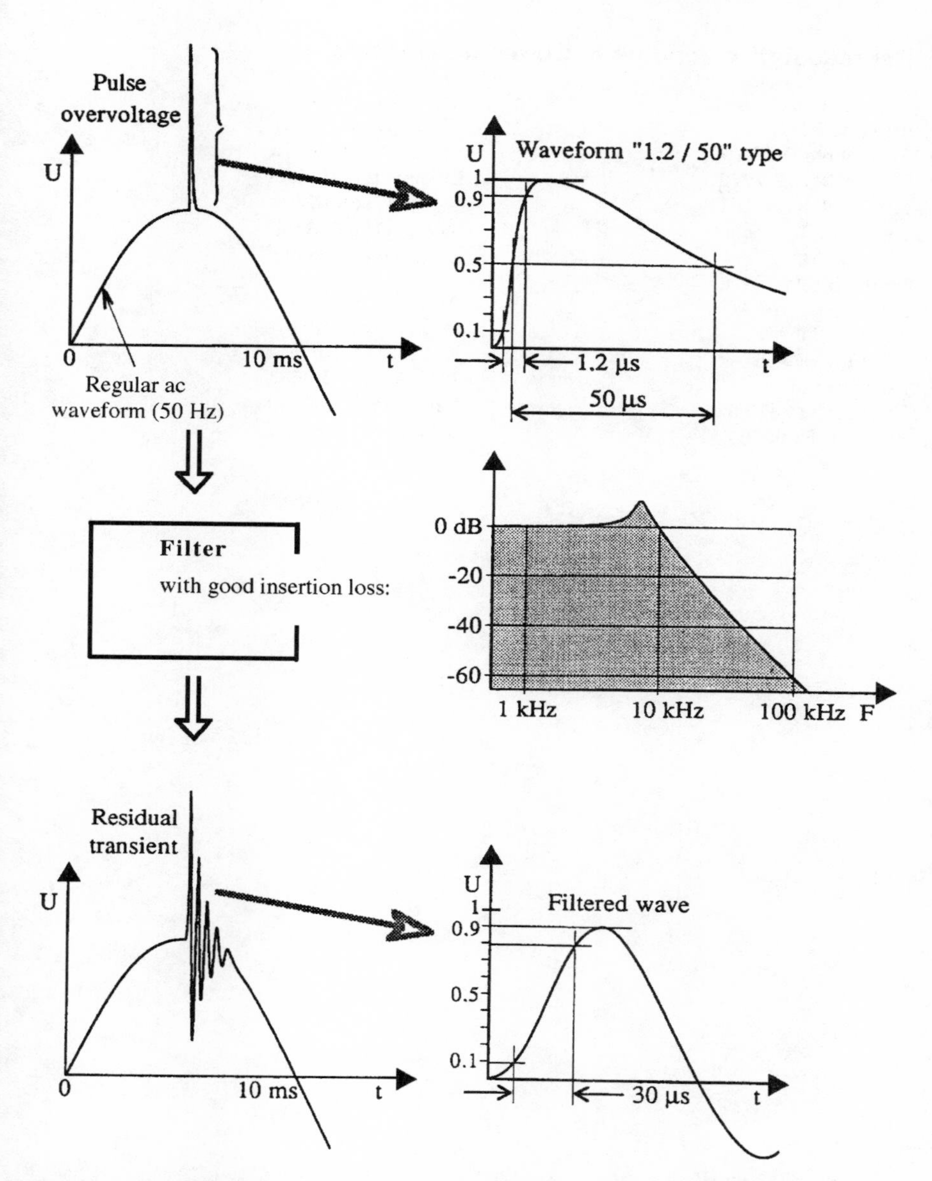

Figure 7.5 An RFI filter is not a good solution against long-duration pulses. For instance, the typical 50 μs surge, resulting from lightning strike coupling, is barely reduced when processed by a filter with a 10 kHz cutoff frequency. The pulse is simply "stretched" with a longer rise time due to the high-frequency attenuation of the filter. *Courtesy of AEMC.*

Representative Vendors of Commercial Filters

Schaffner EMC, Inc.
9B Fadem Rd.
Springfield, NJ 07081
Tel.: (973) 379-7778
Fax: (973) 379-1151
www.schaffner.com

Main office:
Schaffner EMV AG
Nordstr. 11
4542 Luterbach
Switzerland
Tel: (41) 32 6816 626
Fax: (41) 32 6816 641

Corcom Division
CII Technologies
844 E. Rockland Rd.
Libertyville, IL 60048
Tel.: (847) 680-7400
Fax: (847) 680-8169
www.corcom.com

Aerovox Inc.
9C Founders Blvd.
El Paso, TX 79906
Tel.: (915) 772-0555
Fax: (915) 772-0030
www.aerovox.com

Chapter

8

Power-Line Isolation Transformers, Power Conditioners, and Uninterruptible Power Supplies

A transformer with distinct primary and secondary windings, termed an *isolation transformer* (in contrast to the *autotransformer*) creates a discontinuity in the CM ground loop of a power or signal cable. The primary-to-secondary magnetic coupling permits the normal transfer of the DM voltages, while the galvanic isolation prevents the CM voltages from passing across the transformer.

Further improvements to the transformer have also resulted in more complete power-line isolation devices such as power conditioners and uninterruptible power supplies (UPSs).

8.1 Power-Line Isolation Transformers

An ordinary isolation transformer is an adequate device for interrupting low-frequency (50/60 Hz up to few kilohertz) ground loops occurring via the power mains. When frequency increases, the high value of the galvanic isolation (typically $> 10^6\ \Omega$) is soon bridged by the primary-to-secondary parasitic capacitance, C_{1-2} (Fig. 8.1).

For an ordinary power transformer with an "E" type core, C_{1-2} can vary from 50 to 100 pF for small transformers (100 VA or less) up to several nanofarads for transformers in the kilovolt-amp realm. Therefore, with such a transformer, the isolation barrier against CM loop currents is efficient up to a few kilohertz, dropping to nearly zero

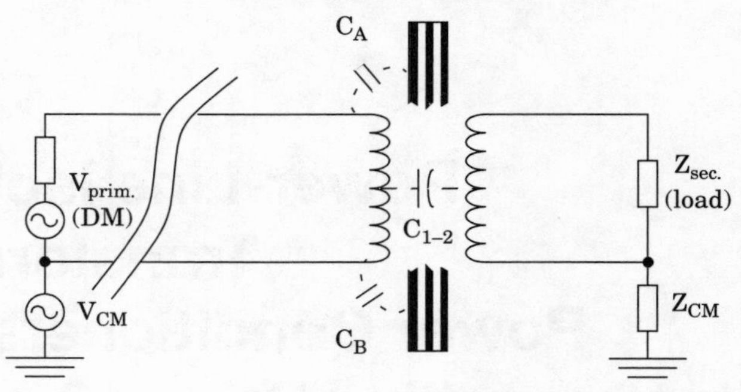

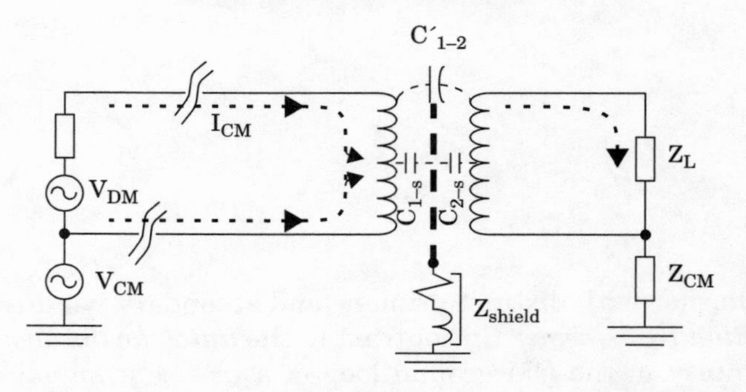

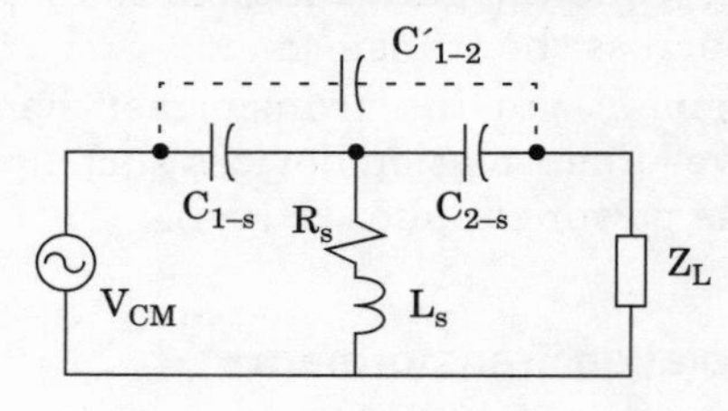

For CM attenuation, assuming $C_{1\text{-}s} = C_{2\text{-}s}$, $\dfrac{V_{sec}}{V_{CM}} \approx C^2_{1\text{-}s} \cdot \omega^2 \cdot (R_s + jL\omega)Z_L$

The rejection deteriorates as F^2 or F^3.

Example:
With typical values, $C_{1\text{-}s} = C_{2\text{-}s} = 300$ pF, $L_s = 10$ nH, $C'_{1\text{-}2} = 0.1$ pF,

For $Z_L = 50\ \Omega$, $V_{sec}/V_{CM} =$ –116 dB @ 150 kHz (due to $C'_{1\text{-}2}$)
–86 dB @ 1.5 MHz
–26 dB @ 15 MHz

Figure 8.1 Primary-to-secondary EMI coupling in transformers.

(typically 6 dB) around a few megahertz. Although the CM attenuation is the most prominent effect, a certain degree of DM coupling also exists at high frequencies through C_{1-2}, whereas the magnetic coupling no longer occurs at such frequencies.

To reduce the effects of C_{1-2}, special winding sequences and separate (instead of concentric) stacking of primary and secondary windings are used, which can divide the parasitic capacitance by a factor of 3 to 10. To get better isolation, another technique is used, as described next.

8.2 Faraday-Shielded Transformers

In a Faraday-shield transformer, an isolated aluminum or copper foil is wrapped between the primary and secondary coils, with the foil edges not touching, to avoid forming a shorted loop. This Faraday, or electrostatic, shield is connected to ground, thus replacing C_{1-s} by two new capacitances: C_{1-s} and C_{2-s}. A CM voltage appearing on the primary side will inject a current in the grounded shield via C_{1-s}. Provided this shield is grounded, no shield voltage will appear; therefore, no voltage is injected into the secondary via C_{2-s}.

In reality, since the shield resistance and inductance are not null, the flow of CM current in this shield impedance causes a small shield voltage, which in turn injects a noise current into the secondary, through C_{2-s}.

The equivalent circuit in Fig. 8.1 shows that, because of shield impedance Z_s, the attenuation rolls off quite rapidly above few hundred kilohertz with a slope of $1/F^2$ or $1/F^3$, depending on the mostly resistive or inductive nature of Z_s (Fig. 8.2).

Indications

- In power-entry rooms or power distribution frames as a simple isolation transformer with a 1:1 ratio, for breaking 50/60Hz ground loops.
- In the same application to locally re-create a grounded-neutral ac distribution, isolated from the general site ac distribution (which can be isolated-neutral, for instance).
- To prevent a too-frequent tripping of ground fault detectors (GFDs) in systems that have a significant earth leakage current. Typical examples are shielded rooms with large input filtering capacitors, and line impedance stabilization networks (LISNs).
- To be combined with a power-line filter whose attenuation starts at only above a few tens or hundreds of kilohertz.

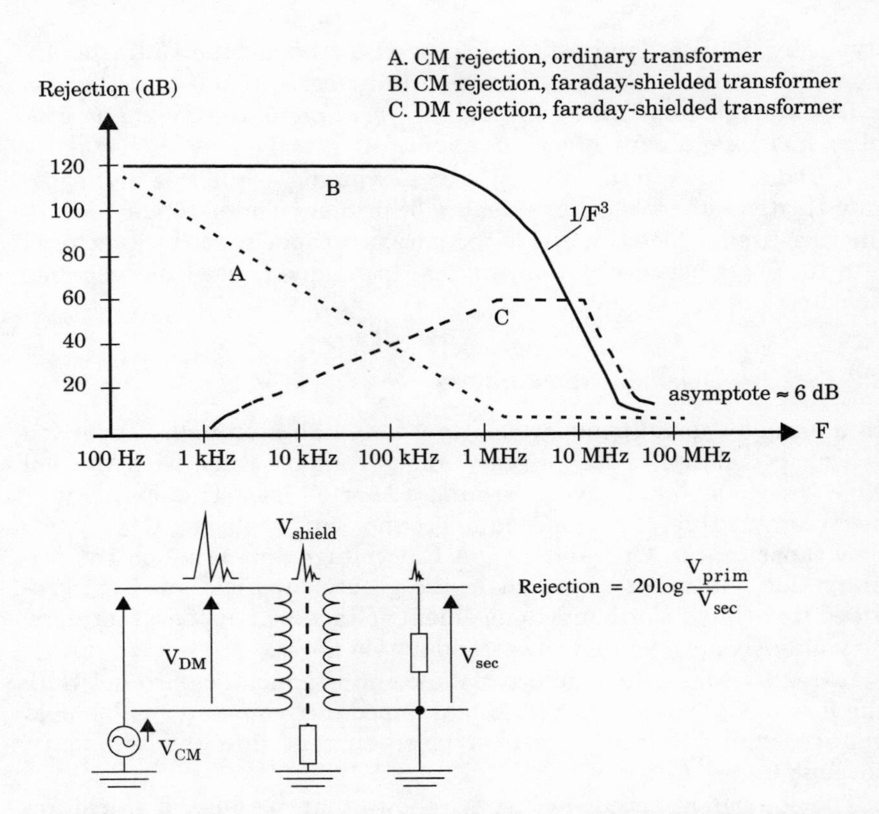

Figure 8.2 Typical values of noise rejection for unshielded and shielded transformers. The $1/F^3$ slope is due to the small parasitic inductance of the shield and its grounding connection.

Prescriptions, Installations

- Install in such a way that input cabling is kept separated from output cabling. Do not run in the same conduit. The best approach is to use shielded cables, distinct metallic raceways, or flexible metallic hoses.
- Mount the transformer on a ground reference plane (GRP): cabinet wall, shielded room collector-plate, transient plate (see Sec. 11.5), or on the local HF ground reference grid.
- If it is a shielded-type transformer, connect the shield terminal to the above GRP with very short, wide strap or braid—preferably to a jumper wire.
- For the secondary-side ac branch, make sure the earthing conductor (green or green/yellow) is connected to the transformer GRP. This GRP, in turn, is tied to the general site earthing network.

- The best solution, but not easily applicable as a late fix, is to install the transformer in a through-wall mounting, the transformer laminated core flanges being bolted through the cabinet sheet metal.

Limitations

- An isolation transformer, shielded or not, is not a regulator. DM voltage fluctuations at 50/60Hz, or 400Hz, or low-order harmonics up to a few kilohertz, will not be attenuated.
- CM interference above a few hundred kilohertz (for the ordinary transformer) or a few tens of megahertz (for the shielded type) will not be attenuated.
- It is absolutely worthless if the transformer structure and shield are not bonded to a GRP. (The CM noise is carried through via the green or green/yellow wire.)
- As any wound core, a transformer radiates a strong magnetic field locally. Do not install near H-field-sensitive components (magnetic read/write devices, CRTs, and so on).

8.3 Power Conditioners and UPSs

When the EMI problems are caused by voltage fluctuations, ripple, or even outages of the regular ac mains, a common solution is to locally re-create a better regulated ac delivery using line conditioners or to provide interrupt-free delivery via an *uninterruptible power system* (a.k.a. *uninterruptible power supply,* or *UPS*), based on battery backup.

At one time, these devices were regarded as a panacea for any power mains-related problem. However, because of their cost and size, they should be used specifically to compensate or substitute for an unreliable power distribution system. Against high-frequency EMI, many other solutions, previously discussed, are more appropriate in terms of volume and economy.

8.3.1 Line Conditioners

A line conditioner is a special transformer with a regulation circuit on its secondary, essentially needed to correct slow variations (i.e., undervoltages or overvoltages) on the 50/60Hz mains. For such irregularities, having durations ranging from one period to quasipermanent, a line conditioner typically will hold an output voltage around the nominal 120 or 230 V value within a ±5 percent margin, against ±20 percent fluctuations of the input voltage, and this for a wide range of secondary loads. Of course, because the line conditioner has no energy storage (with the exception of the ferroresonant type, which has a short, one-cycle, autonomy), it cannot compensate for a 50 percent drop or a full outage.

The transformer secondary voltage can be regulated in three ways:

1. It can employ a set of secondary taps, staggered in 10-volt steps, with an automatic tap changer operated by a stepper motor or, in more recent devices, by solid state switches. This type can handle large overloads during start-up sequences (3 to 15 times the nominal current).
2. It can use an auxiliary winding that operates in the ferroresonance regime. This is a robust, inexpensive technique, but it cannot handle more than a 50 percent overload.
3. It can incorporate an auxiliary saturable winding with an adjustable dc current.

Some conditioners that use an isolation transformer also incorporate a Faraday shield, thus combining the advantages of the two devices.

Indications

- When there is evidence that the ac mains voltage is very irregular, with excursions outside the normal tolerances of the user's equipment.
- When these ac voltage fluctuations do not exceed ±20 percent, and there are no outages.
- When the installation contains electrical or electronic equipment of ancient generations, which cannot tolerate more than a few percentage points of ac voltage shift.
- For conditioners with a shielded isolation transformer, the indications combine with the CM and DM attenuation properties of these devices.

Prescriptions, Installation

Installation is simple, according to manufacturers' instructions.

Limitations

1. A line conditioner does not absorb harmonic distortions, which will be carried over the secondary.
2. There is significant H-field radiation with the ferroresonant type.

3. It cannot make up for a temporary dropout.
4. If the transformer is the autotransformer type, there is no CM attenuation.

8.3.2 Uninterruptible Power Supplies

In contrast with the line conditioner, a UPS is an ac/dc/ac converter coupled with a battery bank. Thus, it can provide a well-regulated ac output, even during a momentary interruption of the normal ac mains. Two basic philosophies are used: *offline* (or *standby*) and *online*.

Offline. The loads are normally fed by the ordinary ac mains, while the inverter is off or in standby mode, charging its own backup batteries (Fig. 8.3). In case of ac mains failure, the privileged ac branch is automatically switched on the inverter. The main advantages of this scheme are its simplicity and the good long-term efficiency, since the inverter losses are only incurred when needed. The drawbacks are as follows:

- In the "normal" ac mains mode (i.e., most of the time), all the power mains disturbances (HF noise, transients, and so on) are seen by the loads.
- The switchover from normal to emergency is not smooth, with generally a short voltage interruption (typically 5 to 10 ms).

This scheme is typically suitable for small installations such as private offices, shops, and so forth where the equipments such as PCs are

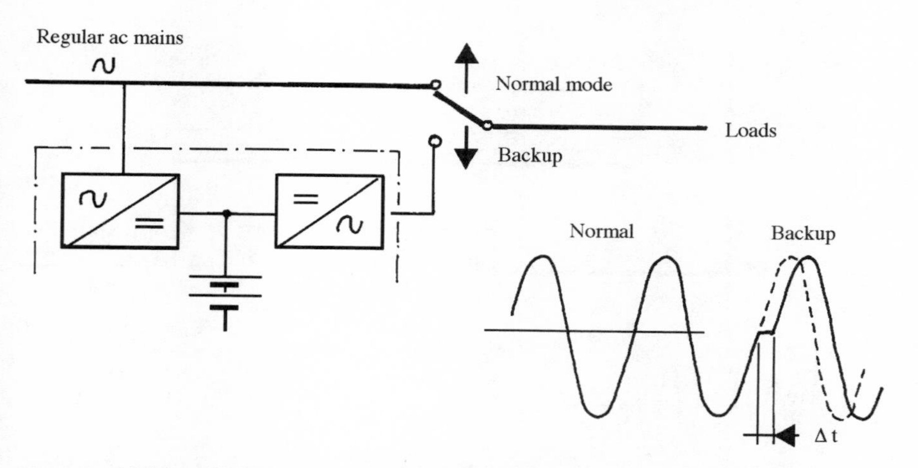

Figure 8.3 Offline UPS arrangement for low-power loads. The normal-to-backup transfer switch causes a small phase shift, Δt.

normally immune to power mains EMI noise and can tolerate a half-cycle voltage drop.

Online. The loads are always supplied from the inverter, and hence decoupled from the regular ac mains, which serve only to charge the battery bank through the ac-dc charger (Fig. 8.4). In case of inverter failure or overload, a static switch transfers the load to the ac mains. This static switch is a frequent cause of EMI trouble since, even in the *off* state, it provides only a mediocre isolation barrier against high-frequency coupling. The main advantage of the o-line UPS is the regularity of the ac supply. If the UPS is correctly designed with *input* and *output* EMI filters, the protection extends down from voltage fluctuations and frequency jitter up to CM and DM EMI noise and transients.

The drawbacks, besides the permanent waste involved in inverter and charger losses, are as follows:

- EMI noise can be generated by the switch-mode inverter itself (which can be solved by efficient filtering).
- It offers mediocre isolation, in terms of HF, of the static switch when in *off* mode (which is solved by some manufacturers who add an electromechanical switch in series).
- The neutral conductor generally is not interrupted; therefore, the CM noise carried on the neutral (neutral vs. earth noise) can reach the protected side if there is no efficient filtering.
- The internal source impedance is significantly higher than the regular ac source impedance. Therefore, some loads with high inrush

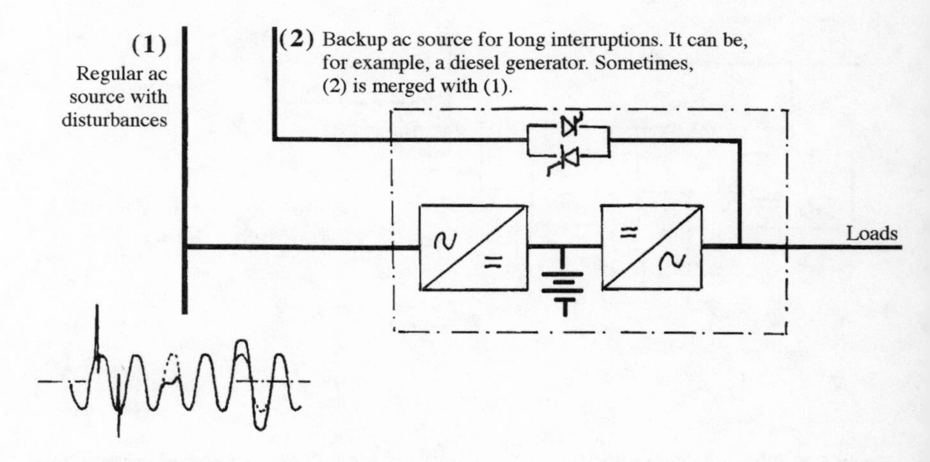

Figure 8.4 Online UPS scheme.

currents or very bad form factor (high peak-to-average ratio) repetitive current will cause a distorted waveform on the "clean" output.

Indications, Prescriptions

The indications and prescriptions are similar to those of the line conditioner, plus the important feature of an interrupt-free supply. However, some of the drawbacks listed above for the online or offline scheme should also be considered as limitations.

Since the choice of installing a UPS to solve a power-line disturbance problem requires a costly investment, a UPS selection form has been provided (Table 8.1).

Table 8.1 UPS Selection Checklist

1. Quality of the regular ac mains delivery:

A. Voltage fluctuations ....

B. History of power outages, dropouts, etc.? ....
   Any existing records from a disturbance monitor? ....
   Is there a possibility of lightning exposure? ....
   Are heavy construction/civil engineering projects in the vicinity? ....

C. Does the user have an existing step-down transformer? ....

D. Are other heavy consumers on the same branch? ....

2. Type of loads:

	I_{avg} per phase	$I_{startup}$ per phase
Equipment #1	______	______
Equipment #2	______	______

Tolerances for voltage and frequency:
Any soft-starts or sequential starts? ______
Power factor (kW/kVA) _____
Any internal backup batteries? _____ If yes, what is the autonomy? _____
Are there auxiliary loads requiring service continuity (e.g., air cond.)? ______

Foresee, in UPS sizing, a possible increase in computer system usage (e.g., a typical criterion for computer rooms is 500–600 W/m^2.

3. Choice of UPS

A. Critical aspect of a power interruption (e.g., loss of data, deterioration of manufacturing processes, medical or safety considerations, etc.).

B. What type of short interruption can be tolerated or handled? (For example, reiteration of the operation underway? Rebooting of entire program? Watchdog with automatic recovery?)

Depending on the above, an online or standby unit may be preferable.

Chapter

9

Signal-Type Isolation Transformers

Isolation transformers for pulse-type (digital or SCR gate-drive) or analog applications operate on the same principle as those used for ac power applications (Fig. 9.1). However the frequency bands to be transmitted are extremely different, and some performances required for processing the useful signal (e.g., distortion, 3 dB-bandwidth, losses, symmetry, impedance, pulse delay, and so on) are quite stringent. These transformers are wideband devices, with F_{max}/F_{min} ratios of several decades.

By interrupting the CM ground loop at the transmitter side or receiving side, they provide a rejection of CM noise while processing the DM signal without alteration. Since the CM voltage stresses the primary-to-secondary barrier, this isolation must have a high level of breakdown voltage—typically 1.5 kV, and up to 10 kV for some applications.

The primary-to-secondary capacitance, C_{1-2}, ranging from 3 pF for a high-quality miniature transformer up to 100 pF for an ordinary audio transformer, will tend to reclose the ground loop across the barrier. In addition, parasitic capacitances C_A and C_B (see Fig. 8.1 for power transformers) between the primary winding inputs and the magnetic core are unequal. This creates imbalance such that a certain percentage of the primary-applied CM voltage appears as DM on the secondary or signal-line side. This transfer, called *mode conversion,* represents 1 to 3 percent up to few megahertz for the best devices.

Such miniature transformers are often used for LAN or ISDN applications, military digital buses such as the 1553, thyristor gate drives, and so forth (Fig. 9.2). In long-haul signal applications, they also can

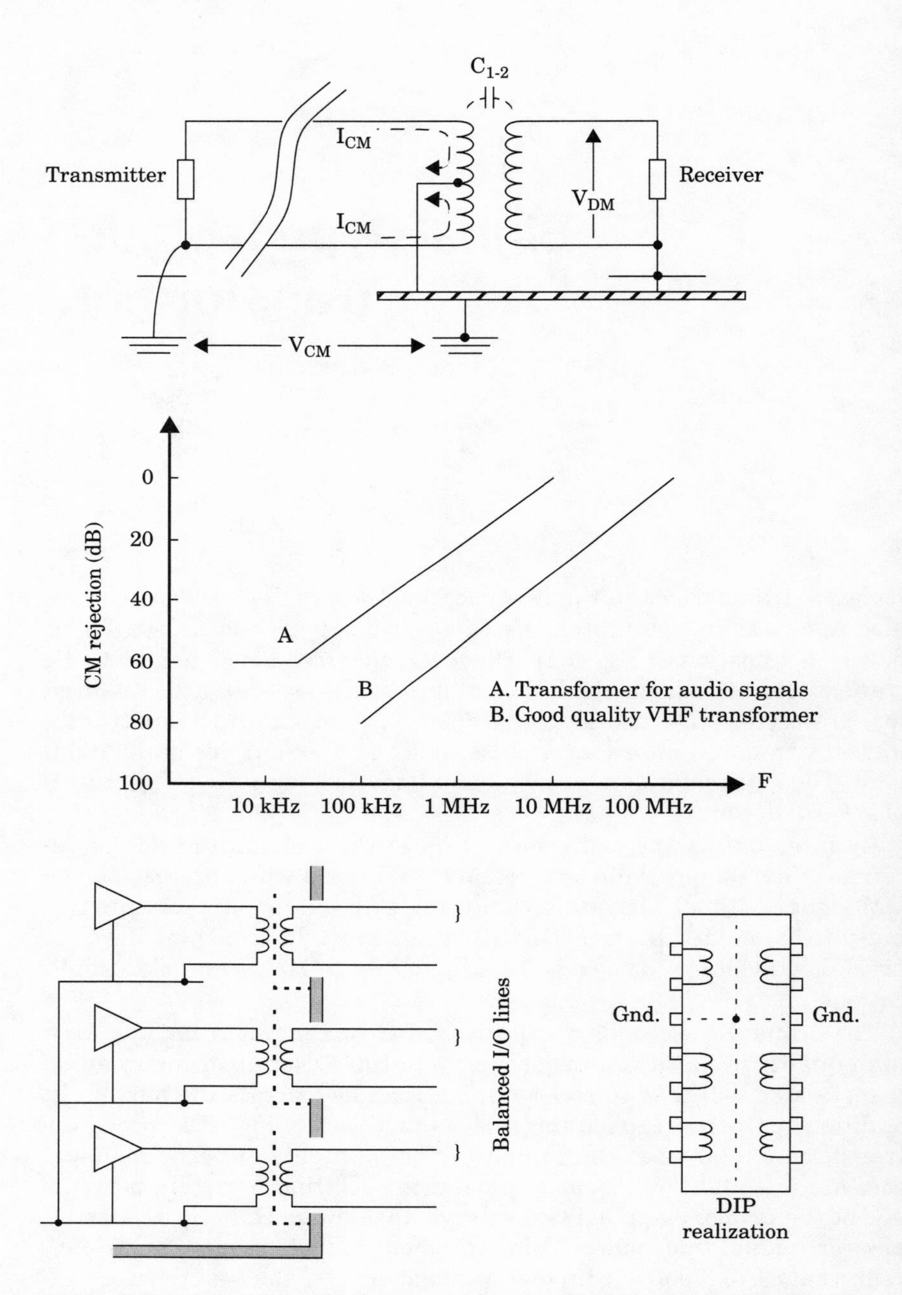

Figure 9.1 Examples of signal isolation transformers. Top: the center tap provides a symmetrical sink for CM current, but it spoils the galvanic isolation between the left and right equipment frames. A center tap on the secondary would save the galvanic isolation, but the receiver must be a floated, differential input. Bottom: a signal transformer with shield, for a fast data bus.

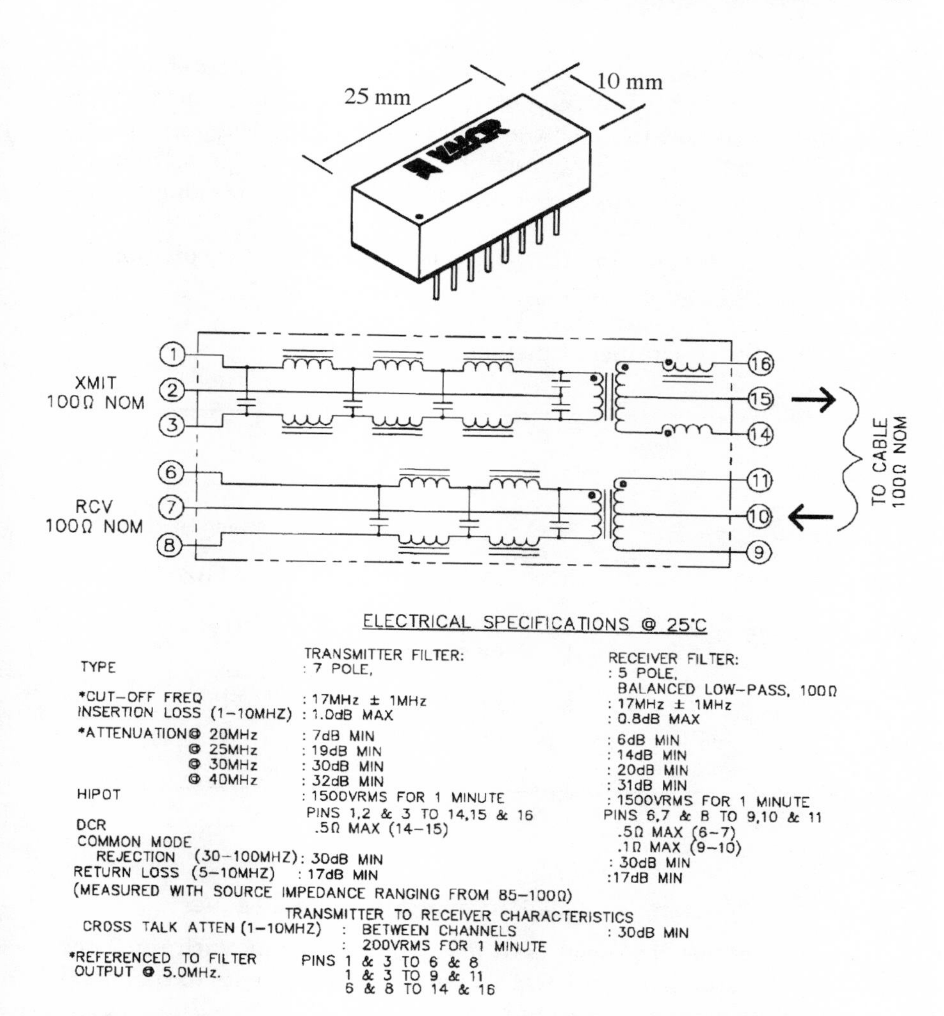

Figure 9.2 Example of a more elaborate signal transformer for a 10Base-T high-speed LAN. It combines balancing, CM filtering, and galvanic isolation such that data transmission on unshielded, twisted pair is possible while meeting most contemporary radiated emission requirements. *Courtesy of Pulse Engineering Inc.*

provide an impedance matching and a balanced-to-unbalanced conversion when driving a wire pair from an unbalanced amplifier. And, like power transformers, their CM rejection can be improved with an internal Faraday shield.

A significant improvement occurs when one of the windings is center tapped, the midpoint being grounded. The two branches of the CM current are drained down to ground via a very low impedance, since their fluxes in the two half-windings are cancelled.

The main advantages of signal transformers are their simplicity, ruggedness, longevity, and linearity, all of which comes at a rather moderate cost. As for their power-line counterpart, EMC performance decreases when frequency increases. Interestingly, their advantages and weaknesses are exactly reciprocal with those of longitudinal transformers (see Sec. 6.6).

The following table summarizes few typical features of common signal transformers. The F_{min}/F_{max} column relates to the longest pulse duration T_d (for F_{min}) and the fastest rise time t_r (for F_{max}) that can be processed without significant distortion.

Type	Bandwidth		$C_{1\text{-}2}$	CM rejection
	F_{min} (T_d max.)	F_{max} (t_r min.)		
Audio	300 Hz	3,400 Hz	60–100 pF	60 dB (F < 30 kHz)
Digital link, DIP	100 Hz (3.5 ms)	100 MHz (3.5 ns)	15–40 pF (unshielded)	40 dB @ 1 MHz
			3–5 pF (shielded)	60 dB @ 1 MHz 30 dB @ 30 MHz
Baseband video	30 Hz (10 ms)	30 MHz (10 ns)	50 pF	120 dB @ 60 Hz 30 dB @ 2 MHz

Indications

- When loop isolation is needed, from dc to tens of megahertz
- For transmitting low-level (≤10 mV) analog signals with minimum noise and distortion, while CM voltages of several volts to thousands of volts may exist on the signal link
- In thyristor applications to isolate the trigger drive from CM voltages
- Generally, as a field fix for breaking a ground loop and creating a balanced link in an otherwise unbalanced transmission

Prescriptions, Installation

- Select a device that has the convenient bandwidth (F_{min}–F_{max}) for the desired signal.
- Apply the same precautions as for filters regarding the clearance between input wires and output wires.

- If the signal transformer has a shield, connect the shield to a nearest CM reference: chassis, metal plate, etc. If the nearest reference is a PCB zero-volt reference, make sure this reference is actually grounded to the chassis close to the transformer location.
- If center tapped, connect the midpoint to a low-impedance ground, as for the shield.

Limitations

- Because of unavoidable primary-to-secondary leakage, this will not block CM noise above a few tens of megahertz. This can be overcome by teaming the isolation transformer with a longitudinal transformer (see Sec. 6.6, Common-Mode Bifilar Chokes), thus getting the best of the two devices (Fig. 9.2).

Representative Signal-Line Isolation Transformer Vendors

Coilcraft, Inc.
1102 Silver lake Rd.
Cary, IL 60013
Tel.: (847) 639-6400
Fax: (847) 639-1469
www.coilcraft.com

Schaffner EMC, Inc.
9B Fadem Rd.
Springfield, NJ 07081
Tel.: (973) 379-7778
Fax: (973) 379-1151
www.schaffner.com

Main office:
Schaffner EMV AG
Nordstr. 11
4542 Luterbach
Switzerland
Tel.: (41) 32 6816 626
Fax: (41) 32 6816 641

Steward
P.O. Box 510
1200 E. 36th St.
Chattanooga, TN 37401
Tel.: (423) 867-4100
Fax: (423) 867-4102
www.steward.com

Vacuumschmelze GmbH (VAC)
Gruner Weg 37
D-63450 Hanau
Germany
Tel.: 49 6181 38-0
Fax: 49 6181 38-2645
www.vacuumschmelze.de

Pulse Engineering, Inc. (formerly Valor Electronics)
12220 World Trade Dr.
San Diego, CA 92118
Tel.: (619) 674-8100
Fax: (619) 674-8262

Pulse Engineering, U.K.
1 and 2 Huxley Rd.
The Surrey Research Park
Guildford
Surrey GU2 5RE
Tel.: 44 1483 401700
Fax: 44 1483 401701

Chapter

10

Transient Suppressors

10.1 Solid State Varistors, Transzorbs®

Varistors and Transzorbs are components with a nonlinear type of V,I curve (smooth for varistors, avalanchelike for Transzorbs). They act as voltage limiters, where the voltage across the device will be clamped to some value equal to or greater than the breakdown voltage, V_{BR}. Their response time is fast, but their energy handling has some limits. Available thresholds (V_{BR}) for PCB or equipment protection typically range from 6 to 600 V (Fig. 10.1). These devices can be bidirectional (against positive and negative pulses) or unipolar. Metal oxide varistors (MOVs) are always bidirectional.

Indications

1. Use on equipment power inputs, telephone lines, or other outdoor wire entries, provided that a primary, first-rate surge suppressor is installed at the building service entrance, telephone MDF, etc., to decrease the energy of the largest pulses (lightning).

2. Add to installed equipment where there seems to be a long-term history of power-line transients on the site (e.g., heavy load switching, UPS transfer from "emergency" to "normal" operation, and so forth) or in the equipment itself (electromechanical or solid state power switches).

3. Use to complement the efficiency of an existing power-line filter. (Ordinary filters do not reduce the amplitude of surges lasting more than a few microseconds.)

4. Use across the coil of relays, solenoids, etc., to protect upstream circuits from the inductive kickoff.

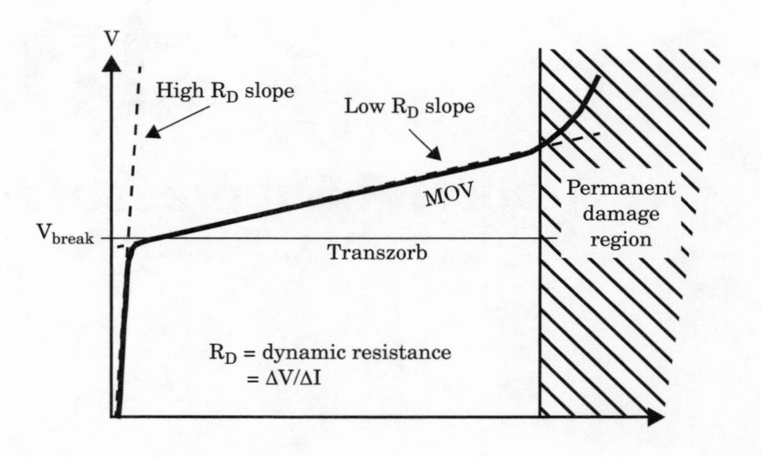

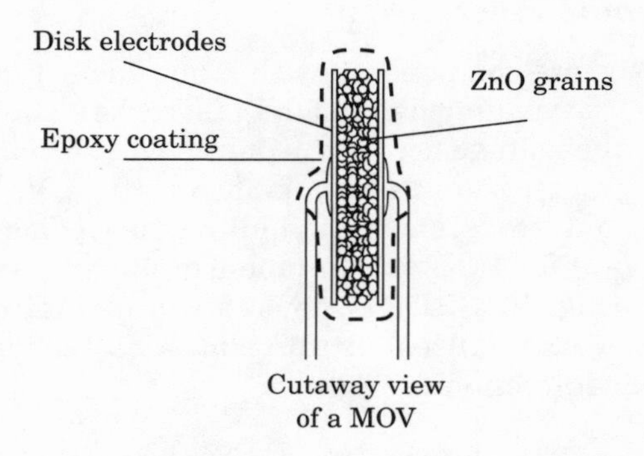

Figure 10.1 General V,I characteristics of metal oxide varistors (MOVs) and Transzorbs®.

5. Use for protecting I/O interface circuits, if the connector area is exposed to ESD (Fig. 10.2).

Prescriptions, Installation

1. Determine whether bipolar or unipolar protection is needed.
2. Select a clamping voltage slightly (8 to 10 percent) above the maximum steady line peak voltage; e.g., 175 V for a 115 Vrms ac mains.
3. Estimate the maximum surge current, I_p, that can be expected during the transient. This can be derived from the surge immunity requirements relevant to the environment where the equipment is installed. Otherwise, it can be approximated by dividing the expected surge open voltage by the power-line dynamic impedance.

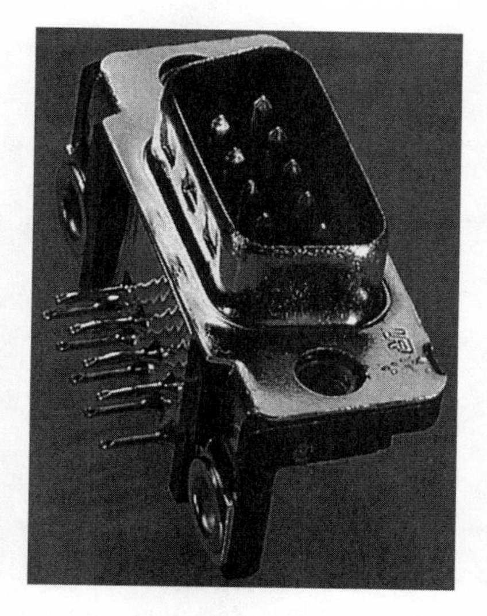

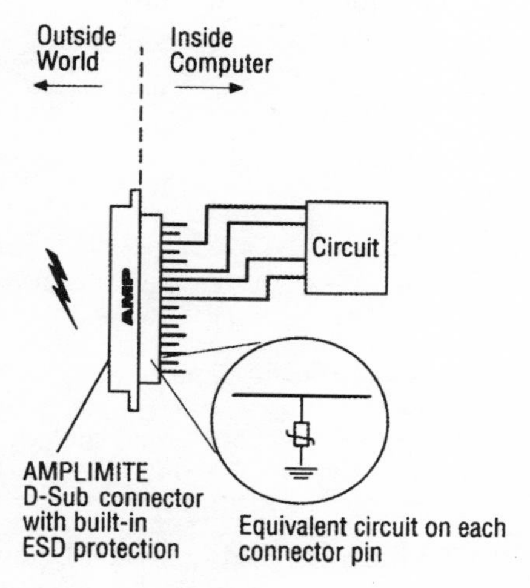

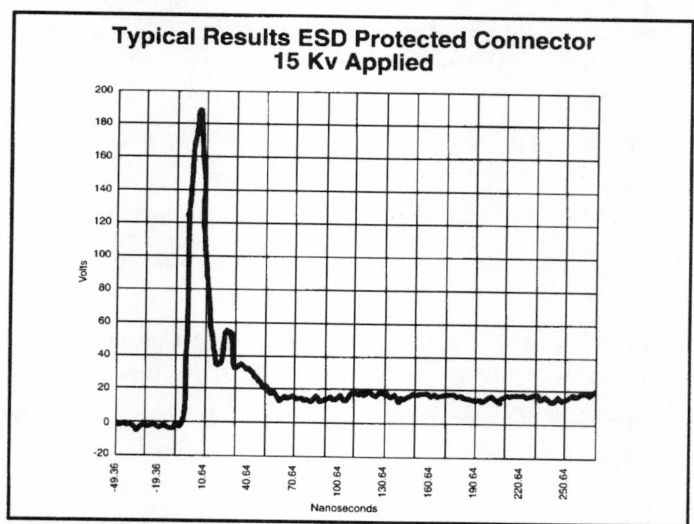

Specifications

Clamping Voltage: less than 30 VDC.

Capacitance: 5pF typical at 1KHz.

Pulse Life: Greater than 1000 pulses @ 15KV.

Leakage Current: less than 25 uA @ ±15V.

Operating Voltage: less than ±25VDC max.

ESD Waveform per IEC 1000-4-2

Figure 10.2 Transient voltage suppressors can be incorporated into standard connector receptacles. They, like EMI filters, can be packaged into male/female adapters for quick-fix applications. *Courtesy of AMP, Inc.*

The following default values are suggested for 230 Vac power mains:

	V_{peak}		I_{peak} max.	
	CM	DM	CM	DM
Lightning, short pulse (50% pulse width= 50 μs)				
Ordinary environment	1.0 kV	0.5 kV	0.5 kA	0.25kA
Severe environment	4.0 kV	2.0 kV	2.0 kA	1.00 kA
Lightning and power switching, energetic pulse (pulse width = 700–1000 μs)	0.75 kV		≈ kA	

4. Make sure that I_p is well below the maximum current that the suppressor can handle for repetitive pulses (single-pulse data can be misleading, since the device's fatigue is not considered). Otherwise, select a larger device. Equipments in lightning-prone areas may have to endure 100 surges in a lifetime of a few years.

5. On the V-I curve corresponding to the suppressor characteristics (Fig. 10.3), draw the load line corresponding to I_p.

6. From the intersection, find the actual voltage V_x across the device.

7. Check that V_x (typically 1.5 to 3 times V_{BR}), is tolerable by the protected equipment or circuit for such short duration.

8. Check that the product of $(V_x) \times (I_p) \times$ (50 percent pulse width) stays below the varistor energy-handling limit, in Joules.

9. To protect power control circuits from inductive kickoff caused by relay coil, solenoid valve, motor, etc., install transient protection across the winding terminals.

10. To protect power-line users from transients caused by utility switching, install suppressors across the line (DM) at the service entrance or power distribution panel.

11. To protect equipment from conducted transients of unknown or out-of-reach sources, install suppressors at equipment entry, typically right after the line filter. They should be mounted DM and CM.

12. To protect PC boards, or more generally discrete low-level components (Fig. 10.4), install discrete or dual-in-line Transzorbs near the edge connector, on the dc supply, and on I/O lines. A DM protec-

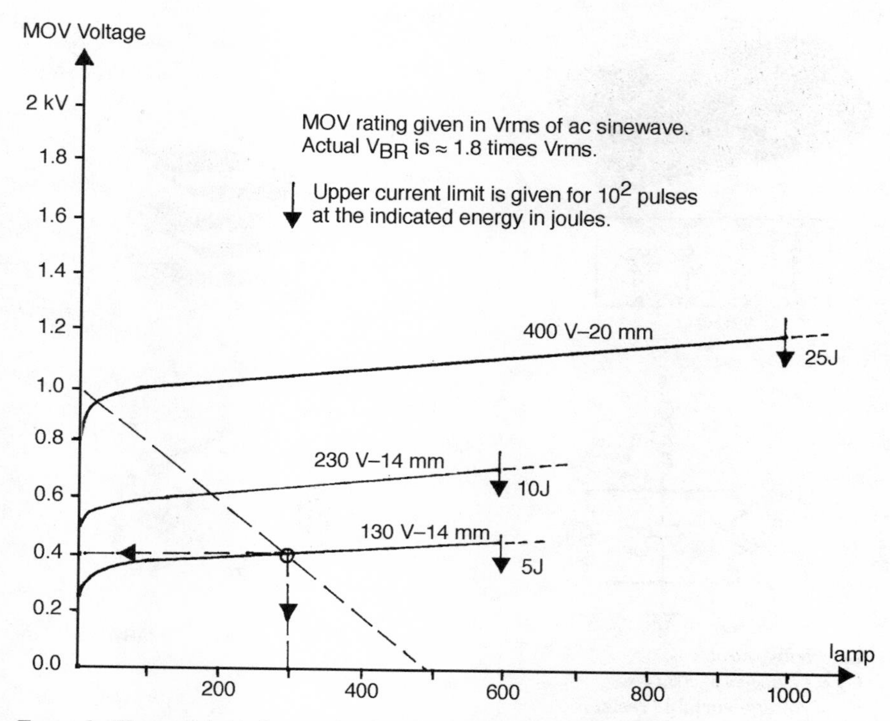

Example: The anticipated surge is 1 kV, with 2 Ω source impedance, resulting in 500 A of theoretical short-circuit current. The MOV selected is a 130 V, 14 mm dia. ceramic element. The 1 kV/500 A load line intersects the MOV characteristic at 400 V/300 A. Assuming a triangular 1 μs/50 μs pulse, the approximate energy to dissipate is

$$U \times I \times T_{50\%} = 400 \times 300 \times 50.10^{-6} = 6\ J$$

This is slightly above the 5 J energy handling for 100 pulses. If a repeated exposure is expected, an MOV with a higher rating must be preferred.

Figure 10.3 A few typical V,I clamping curves for an MOV in power-line applications.

tion is sufficient, if the zero-volt reference is grounded. Some packages provide a "zero-inductance" lead arrangement.

13. In all cases, keep lead lengths short.

Limitations

1. These devices have medium energy-handling capability. This can be overcome by teaming the MOV or Transzorb with a slower but more energy-resistant gas tube.
2. They add a significant shunt capacitance (typically 300 to 3000 pF).

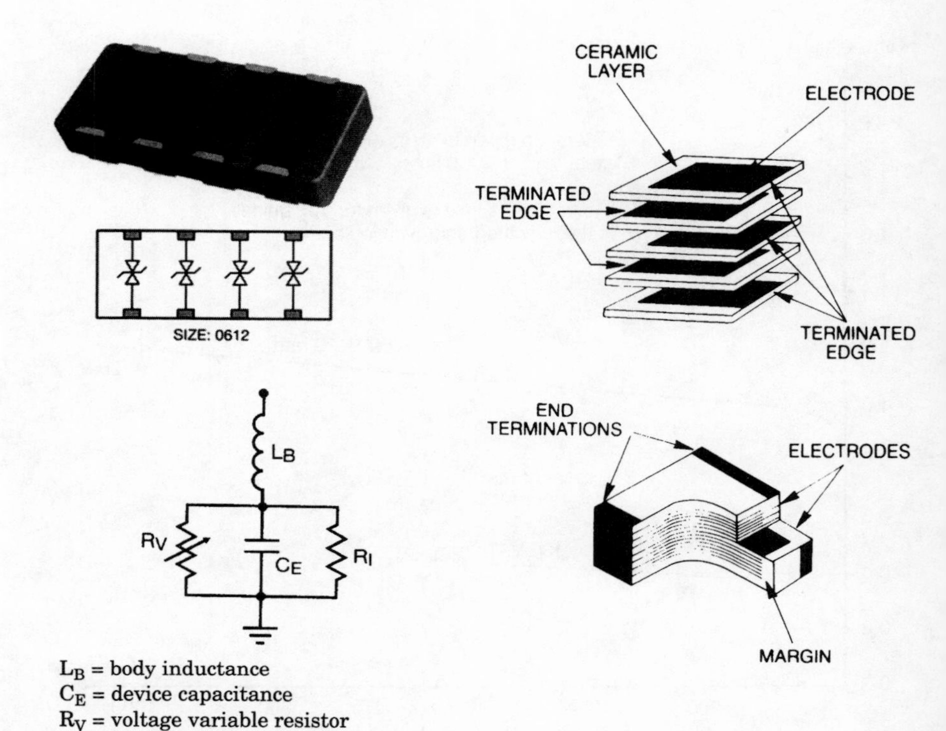

- Multiple electrodes yield a capacitance
- The capacitance can be used in decoupling
- Capacitance can be selected from 30 to 4700 pF

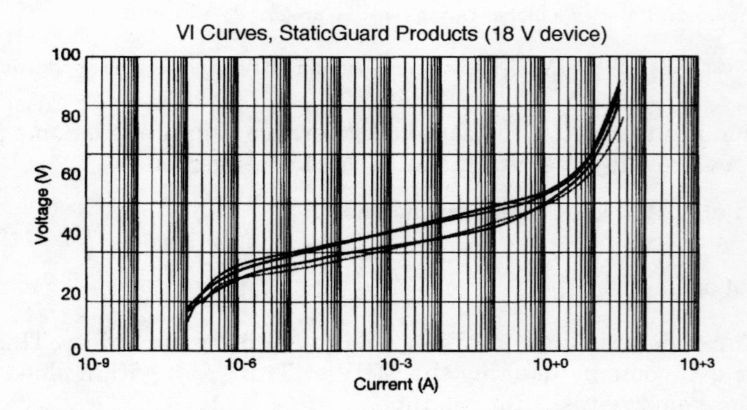

Figure 10.4 Transient voltage suppressor for low-voltage applications. The Transguard® combines the features of a ZnO varistor plus an integrated decoupling capacitor of 50 to 100 pF, compatible with high-speed data. *Courtesy of AVX Corp.*

3. They are easily overstressed when the line voltage remains applied all the time. After many exposures to a crest current close to the I_p max. value, V_{br} tends to shift progressively, leading to destruction by accelerated aging. This can be overcome by employing a generous safety margin.

4. They cannot—and do not pretend to—suppress EMI noise of few millivolts or volts riding over a wanted signal. They protect from damage, but not necessarily from errors.

10.2 Gas Tubes

The gas tube (Fig. 10.5) is the modern version of the ancient air-gap arrestor used on overhead lines. A sealed ceramic capsule contains a calibrated spark gap in an inert, low-pressure gas. When the surge voltage reaches the firing threshold, the device becomes abruptly a short with a very low residual voltage drop and stays like this until the arc extinguishes.

The main advantages are its simplicity and ruggedness, a high current handling of several kiloamps, and a low parasitic capacitance, typically < 3 pF.

Since it is a rather slow-responding device, with a 1 μs typical reaction time for a 1 kV/μs pulse rise, it provides adequate protection for high-voltage power lines, and also for medium- and low-voltage ac power lines, e.g., the power service entrance panels of a commercial or industrial facility. The gas tube will shunt the bulk of the lightning surge energy, leaving the residual voltage to be handled by the secondary protections. They can be mounted CM, DM, or both.

Indications

- Use on power-line and telephone-line entries as a first barrier against incoming surges from lightning, power utility switching transients, and so on. This device is typically used when there is a need for a robust protection that must be resistant to aging and cannot be inspected easily. Its use is restricted to ac lines or current-limited dc lines (telephone, radio) where no follow-on current could occur.

- Use on RF transmitters and receivers antenna cable to protect the sensitive RF circuits from lightning surges arriving by the antenna feeder. Gas tubes are interesting devices for high-frequency transmission lines, because their very low capacitance does not impair the characteristic impedance (Fig. 10.6).

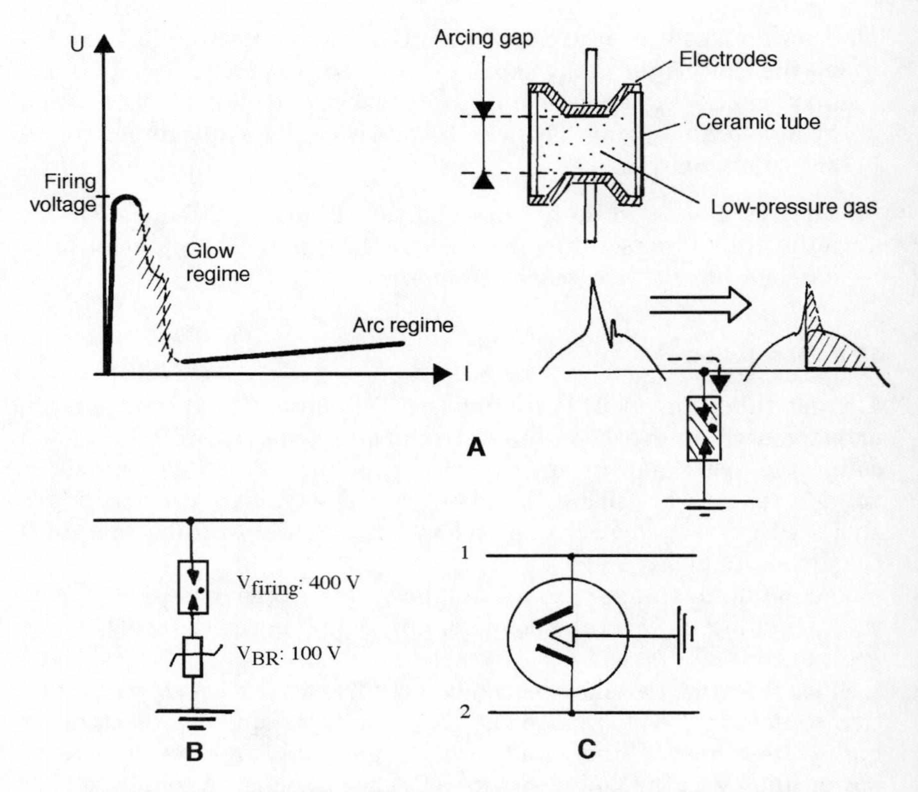

A. The firing takes place after a short delay, and rise fronts shorter than 1 μs pass through. Then, the line is shorted to ground until the arc extinguishes at zero crossing.

B. The combination gas-tube + MOV prevents the follow-on current, forcing the tube to extinguish. Here, the combination of $V_{firing} + V_{BR}$ provides a threshold of ≈ 500 V.

C. Gas tube with integrated triple protection: DM (line 1 to line 2), CM 1 to ground, and CM 2 to ground. Typical application is for telephone/telecommunication lines.

Figure 10.5 Gas tube operation.

Prescriptions, Installation

The rating selection follows the same basic steps as the solid-state devices (see Sec. 10.1). In addition:

- Check that the normal power source will not maintain the gap arcing after the surge is gone. Normally, if the line and source impedance limit the short-circuit current to less than 10 A, follow-on current will extinguish. Otherwise, a fuse should clear it.

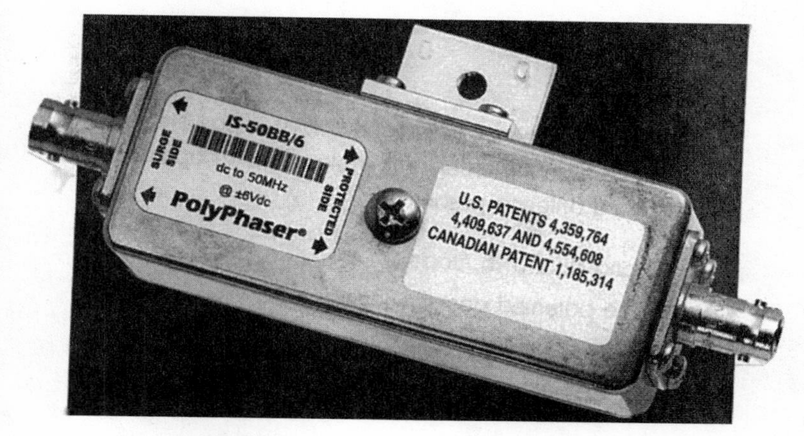

SPECIFICATIONS:

Surge: 18 kA IEC 1000-4-5 8/20 μs waveform 110 Joules

Turn-on Time: 4 ns for 2 kV/ns

VSWR: ≤1.1 to 1 over frequency range

Insertion Loss: ≤0.3 dB over frequency range

User Current: 2.0 Adc continuous

Figure 10.6 Gas-tube lightning protector for 50/75 Ω baseband coaxial protection. These devices have been optimized to preserve good impedance matching for VLF/HF, LAN, and closed-circuit video links. The coaxial electrode gap has also been optimized for faster response. *Courtesy of Polyphaser Corp.*

- Mount the gas tube in DM (between hot and return) and CM (each conductor vs. chassis or local ground). Some gas tubes incorporate such triple protection (Fig. 10.5c).
- If the gas tube is a primary protection device, followed downline by a secondary protection like a varistor, check the coordination of the ratings so that the varistor will not clamp first, preventing the gas tube from ever firing. A good precaution is to ensure that at least 30 m of wiring, or an inductance of ≥ 30 μH, exists between the gas tube and the varistor.
- Some devices (Fig. 10.7) combine the advantages of gas tubes and clamping devices.

Limitations

- Gas tubes are available only for firing thresholds above ≈90 V.
- Their response time is rather slow. They should not be used alone to protect sensitive electronics.

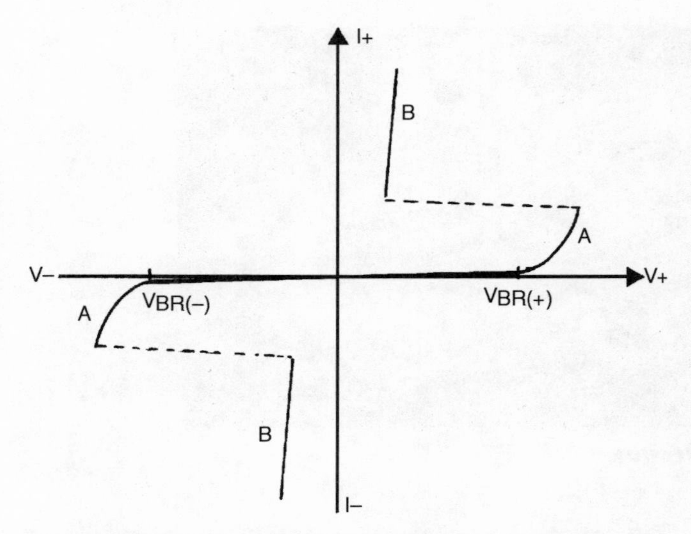

Figure 10.7 Bidirectional three-state transient suppressor. For low values of overvoltages, the device behaves as a clamping diode or MOV (portion A). For higher values, it triggers into a short-circuit, crowbar mode (portion B) like a gas tube.

- The firing causes a very sharp dV/dt collapse, along with a sharp dI/dt rise. This, in turn, can be a disturbance in the proximity of the device.
- Once they fire, they short out any voltage—the surge and the normal voltage. Therefore, the protected equipment loses its power during the time the gas tube stays arcing.
- Follow-on current sometimes may be difficult to control and extinguish. If it lasts, a power input fuse will blow, but the equipment will have to endure a power-down. In the presence of a sustained, strong RF signal, the actual static firing voltage may decrease by 50 percent, causing unexpected triggering.

Representative Transient Voltage Suppressor Commercial Vendors

Varistors

AMP, Inc.
Box 3608
Harrisburg, PA 17105
Tel.: (717) 564-0100
Fax: (717) 986-7575
www.amp.com

AVX Corp.
801 17th St. S.
Myrtle Beach, SC 29577
Tel.: (843) 448-9411
Fax: (843) 448-7139
www.avxcorp.com

Gas Tubes

Citel Inc.
1111 Parkcenter Blvd., Ste 340
Miami, FL 33169
Tel.: (305) 621-0022
Fax: (305) 621-0766
www.citelprotection.com

Citel-2CP
12, Boulevart des Iles–B.P. 18
92132 Issy-les-Moulineaux Cedex
France
Tel.: 33 (0) 1 41 23 50 03
Fax: 33 (0) 1 41 23 50 09

Polyphaser Corp.
2225 Park Place
Minden, NV 89423
Tel.: (702) 782-2511
Fax: (702) 782-4476
www.polyphaser.com

Chapter

11

Bonding, Ground Continuity, and RF Impedance Reduction

The techniques described herein are for improving EMC in equipments, systems, and installations by creating a low-impedance path to the local RF reference, or a low-impedance run along an existing cable. In addition, some techniques create a low-impedance plane, in sites where a real earth plane is impossible to make or reach.

11.1 Grounding Braids and Straps

Wide and flat conductors have less inductance than round conductors of same cross section (Fig. 11.1). In order of preference, try to use:

- flat metal straps
- flat braids with flat-type terminals
- flat braid with crimped ring terminals
- round, stranded-wire jumpers

Indications

1. To decrease CM impedance coupling along structures, chassis, etc.
2. To make a low-impedance bond between cable-tray sections, for upgrading an existing installation
3. To ground a cable shield when no metallic connector bulkhead exists

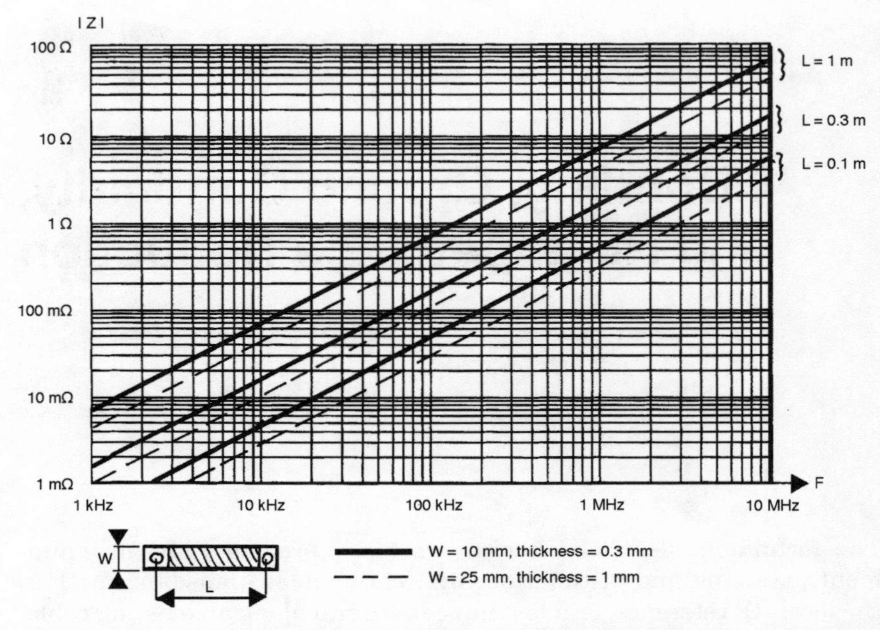

Figure 11.1 Impedances of flat straps of different lengths and widths. The impedance of a flat strap or braid, depending on its L/W form factor, is typically 0.5 to 0.8 times that of a round conductor of the same cross-sectional area. For example, a 0.30 m piece of AWG 14 wire has a self-inductance of ≈0.4 µH. A 25 mm wide strap of the same length will have 0.2 µH, which can reduce a ground shift by a factor of 2.

Prescriptions

For maximum efficiency, grounding conductor length should remain a small portion of the wavelength of the highest EMI frequency where the fix is supposed to work. For instance, if the strap length-to-wavelength ratio, expressed in same units, stays ≤ 1/50, grounding impedance will remain at a low value. Conversely, for a ratio exceeding 0.1, the grounding strap becomes so impedant that its role becomes questionable, reaching an infinite impedance when length/wavelength = 0.25.

11.2 PCB Grounding Spacers

In equipments with metal cabinets, or with metallic bottom plates, PCBs are usually separated from the sheet metal by nylon spacers or stand-offs. For the best efficiency of EMI decoupling capacitors, filtered connectors, cable-shield connections, transient protections, etc., it is important that a chassis ground area be available close to the PCB I/O area. Quite often, the initial PCB layout provides no such commodity. Optimally, the PCB 0 V plane or land is grounded via a long trace that sometimes even must go through a daughter card connector.

To create a more direct low-impedance EMI current sink, grounding spacers are available (Fig. 11.2). A spring-loaded clamp is fitted over a regular resin-type spacer, providing a strong and dependable pressure over the 0 V copper land on one end, and over the PCB mounting chassis on the other end. Owing to the bronze-tin material of the springs, the contact is stable, with a contact resistance in the milliohm range.

Indications

- Use to replace existing isolated stand-offs.
- Use to create or improve a CM current sink from the PCB to the metal chassis or housing.
- Use to reduce the ground-shift between the 0 V references of two or more PCBs stacked over each other.

Prescriptions, Installation

The location of this 0 V-to-chassis tie is extremely crucial. Since this grounding stud will drain all undesirable EMI current, it should not be placed in the middle of the board or far away from cable-entry locations. Typically, the spacer(s) must be in the area where I/O cable connectors or filters are grouped.

Limitations

There are no particular limitations.

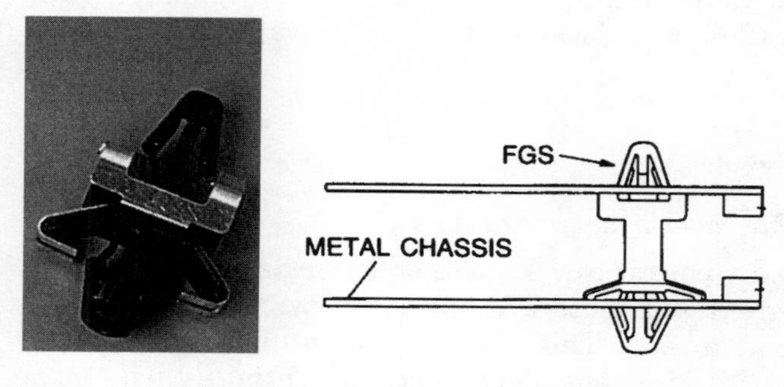

Figure 11.2 PCB-to-frame grounding spacers (FGSs). Contact resistance after repeated snap-on is 4–5 mΩ. *Courtesy of Kitagawa Ltd.*

11.3 Metallic Cable Raceways and Companion Braids

The role of metallic cable trays, or *companion* wires and braids, is to carry some of the ground loop EMI currents (from 50/60 Hz up to tens of megahertz) between several interconnected equipments. Although they might be regarded as *shorting out* the CM voltage between different chassis or earth bars, this explanation does not really apply for long distances, except maybe at dc or 50/60 Hz. The actual mechanism, provided that the companion conductor stays within a very short spacing from the associated, protected cable, is that of mutual inductance $M_{1\text{-}2}$. This strong mutual inductance, created by the nearness of the cable and its companion, generates a canceling current that can amount to 60–95 percent of the initial loop EMI current. The simplicity of the solution is that it does not require the companion structure to totally surround the protected cable (as a shield would do) but simply that they be laid tightly next to each other. Such conductors, if properly connected at both ends to the equipment structure (frames, chassis, and so on), will reduce all susceptibility and emission problems related to CM voltage in large sites (Fig. 11.3).

Indications

Use in computer rooms, factory machine rooms, or larger sites wherein many cables are of the unshielded type, and it is impossible (or difficult) to replace them with shielded cables or tubular conduits. This will

- reduce CM field-to-loop EMI pickup and EMI emissions at any frequency from 50/60Hz up to 30 to 100 MHz,
- reduce CM crosstalk by increasing wire-to-ground capacitances, and
- supplement (or actually create) a reference ground grid underneath the cables' paths when a real conductive floor or wall reference does not exist.

Prescriptions, Installations

1. *For metallic cable raceways (see Fig. 11.4),*
   - Use galvanized steel raceways (solid sheet or perforated). Make sure the different segments are bolted to each other, continuously, from end to end. This includes right-angle turns, wall-throughs, and "Y" junctions. *One single discontinuity will completely ruin* the very mechanism (mutual inductance) upon which the raceway is based (see Fig. 11.5).

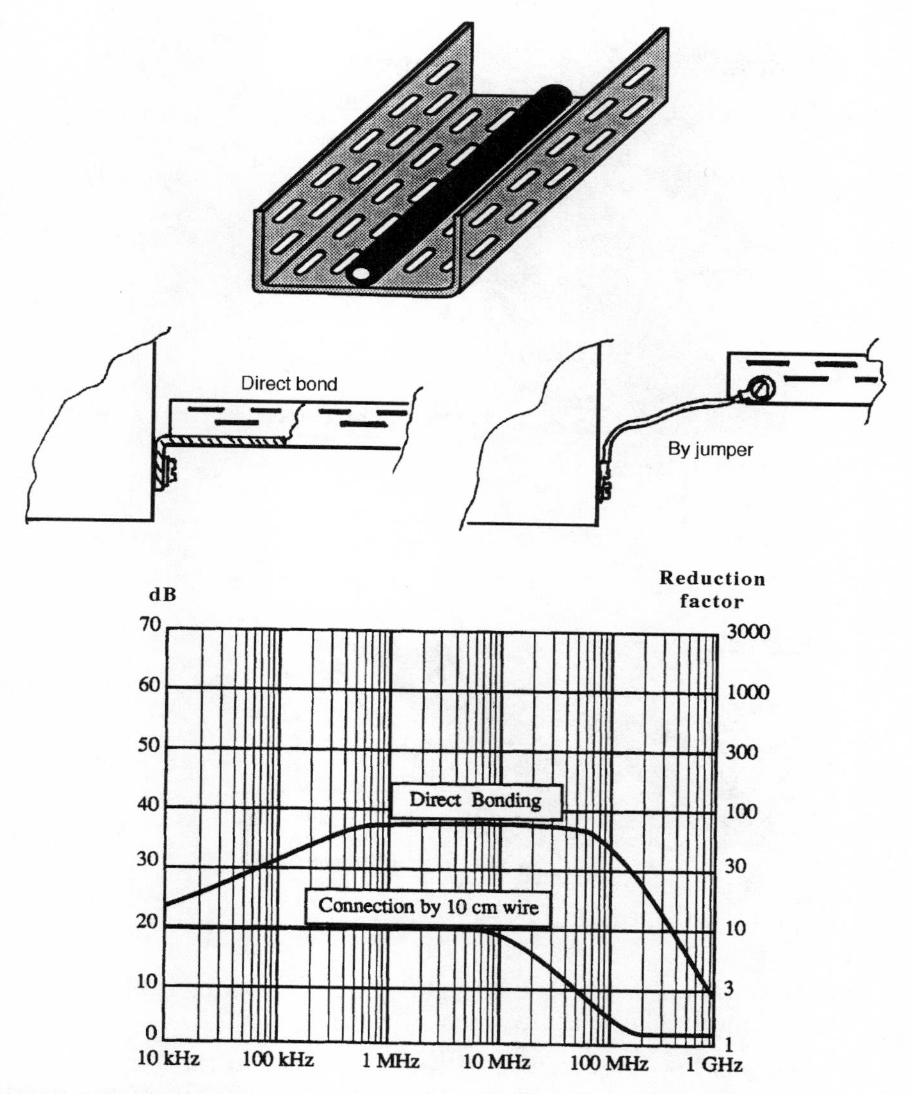

Figure 11.3 Coupling reduction by ordinary, perforated-steel raceway: influence of grounding style. *Courtesy of AEMC.*

- At places where actual piece-to-piece continuity is not feasible, use short straps or braids (L/W ratio ≤ 5). Absolutely prohibit round-wire jumpers.
- Where the raceways terminate near the equipment frames, try to extend at least the bottom part of the raceway up to the frame entry, or

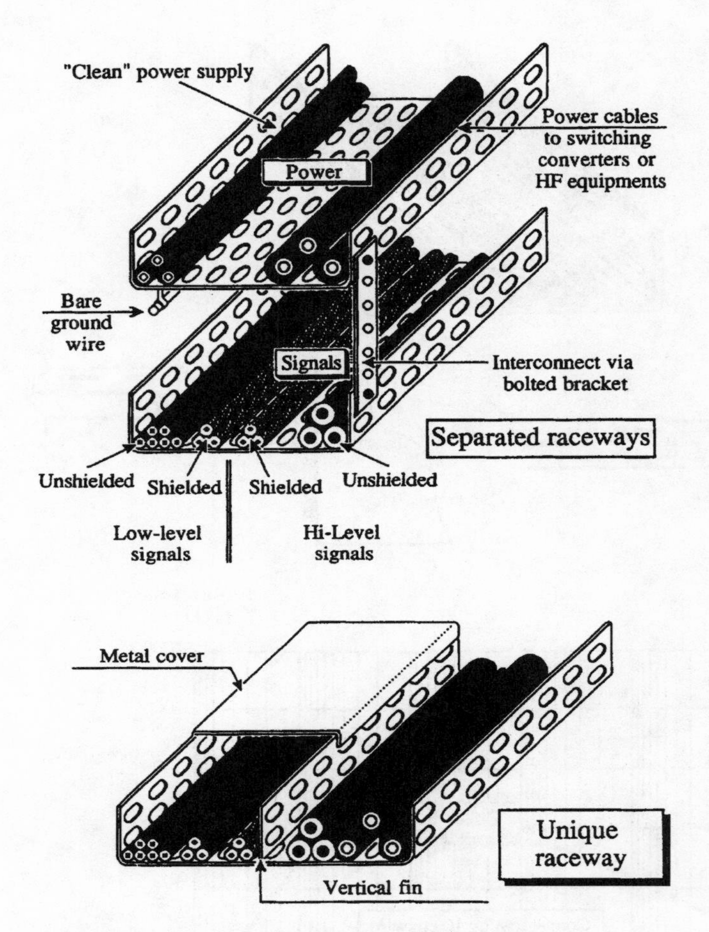

Figure 11.4 Using cable raceways. *Courtesy of AEMC.*

use wide straps or braids held tightly to the incoming cables. Screw these terminations to the equipment chassis.

- Do not mix low-level signals with power cables in the same tray. If the power vs. signal segregation is impossible (i.e., only one raceway is provided), at least try to pack them in opposite inside corners of the "U" shape. A vertical separating wing is preferable.
- If the cable raceways are equipped with covers, try to stagger the cover discontinuities by a half-length of the raceway segments. This will avoid having the cover-to-cover seams in coincidence with the raceway-to-raceway seams. However, the addition of a cover is needed more for mechanical and dust protection than for real EMC improvement. The additional benefit is barely 6 dB.

Figure 11.5 One typical flaw with cable raceways: The installation is generally correct, but the discontinuity at the top ruins most of the EMC benefits of the metallic channel.

- Check that the cables are not bulging over the raceway vertical wings. Try to install the most critical cables (sources or victims) first, in the bottom of the raceway.

Limitations

- Standard electrical practices for buildings consider cable raceways as mechanical entities only. The concept of taking advantage of their electrical continuity is difficult to infuse into long-term construction traditions.
- In old installations, existing raceways tend to be overcrowded, spoiling the EMI reduction effect.
- In existing installations, cables frequently make short detours outside the raceway channel.

- In existing installations, when new cables are installed, they tend to be packed over old ones, making power vs. signal segregation impossible.

2. *For companion braids*

   Companion braid is a fair substitute for metallic raceway on short hauls only, and only inside the same room. When distances are long, with several rooms being crossed, it is not practical to lay a braid nearby, along the entire cable path.

   - Use wide, tinned-copper braid with minimum 25 mm (1 in) width. Hold it tightly against the cable/harness using nylon ties or by a similar method. Whatever detour the cable takes, the braid should stay with it.
   - Connect the braid at each end to the destination equipments using a wide, rigid, metal-to-metal attachment (e.g., large washer, rectangular plate, etc.).

   Where applicable, precautions that have been listed for raceways are also applicable to companion braids.

11.4 Floor Impedance Reduction, RMF Grounding Pads

For reducing the effects of incoming conducted transients or ambient RF fields onto an installed system (e.g., computing or telecommunication center, instrumentation setup, or industrial process supervision room), a room reference plane or grid can bring a substantial improvement, easily reaching 20 dB at up to hundreds of megahertz. Such an equipotential mat also reduces the ground shift between the different equipment frames or bays within the same room.

Erecting a complete Faraday enclosure is an expensive solution, devoted to specific installations where attenuations greater than 60 dB are required. On the other hand, the older philosophy of star grounding is totally inadequate above a few hundred kilohertz: A round wire represents a self-inductance of 1 to 1.5 μH/m, i.e., 6 to 10 Ω/m at 1 MHz. The star branching from the earth rod, a safety requirement, is incapable of providing a low HF impedance between equipments.

When the equipments and their cables are resting closely over a conductive grid, this will provide:

- smaller cable-to-ground loop area
- better decoupling of fast CM transients arriving through power lines
- better sink of HF currents derived by the EMI filters or collected by cable shields

Emissions will decrease and immunity will improve without any quest for a mythical (and generally inaccessible) "clean earth" (Fig. 11.6).

There are two ways of achieving such a low-impedance reference mat close to (generally underneath) a large electronic system.

One technique is to install a net of criss-crossing braids or straps (Fig. 11.7), with a grid size in the meter range. At each intersection, the crossing stripes are electrically connected by any of the following:

- *Tin soldering.* This is a dependable method, requiring almost no height above the resting surface, but it is labor-expensive.
- *U-clamps, Fargo clamps, or Joslyn washers.*
- *Hose-type clamps,* using the raised floor pedestals as attachment points.

The principle of a grid is that it achieves the same ohm/square impedance value, *regardless of the distance between the reference intersection points,* provided that these points are not at the edges of the grid perimeter. By contrast, a serpentine shape of the same braid would produce a linear impedance, reaching 4–8 Ω per meter length at 1 MHz. Therefore, it is crucial for the braids or tapes to be grid-connected rather than just zig-zagging in one way (a practice that is sometimes used).

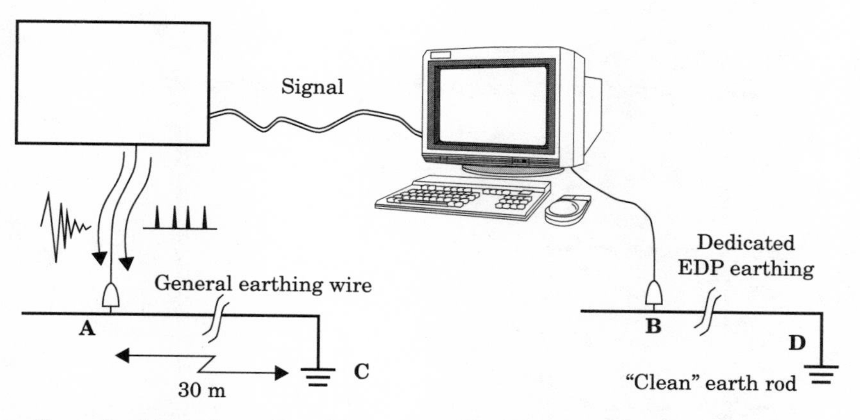

Example of typical noise:

- B and C are *not* isolated.
- $V_{AB} = V_{AC}$
- V_{AB} (repetitive) = 15 V @ 100 kHz (from power converter)
- V_{AB} (ringing transient) = 300 V to 1 kV @ 300 kHz

Figure 11.6 The wrong answer to common-mode ground noise: the mythical "clean" electronic earth rod. By common impedance coupling (see Chap. 1), noise voltages inject CM currents into the signal link.

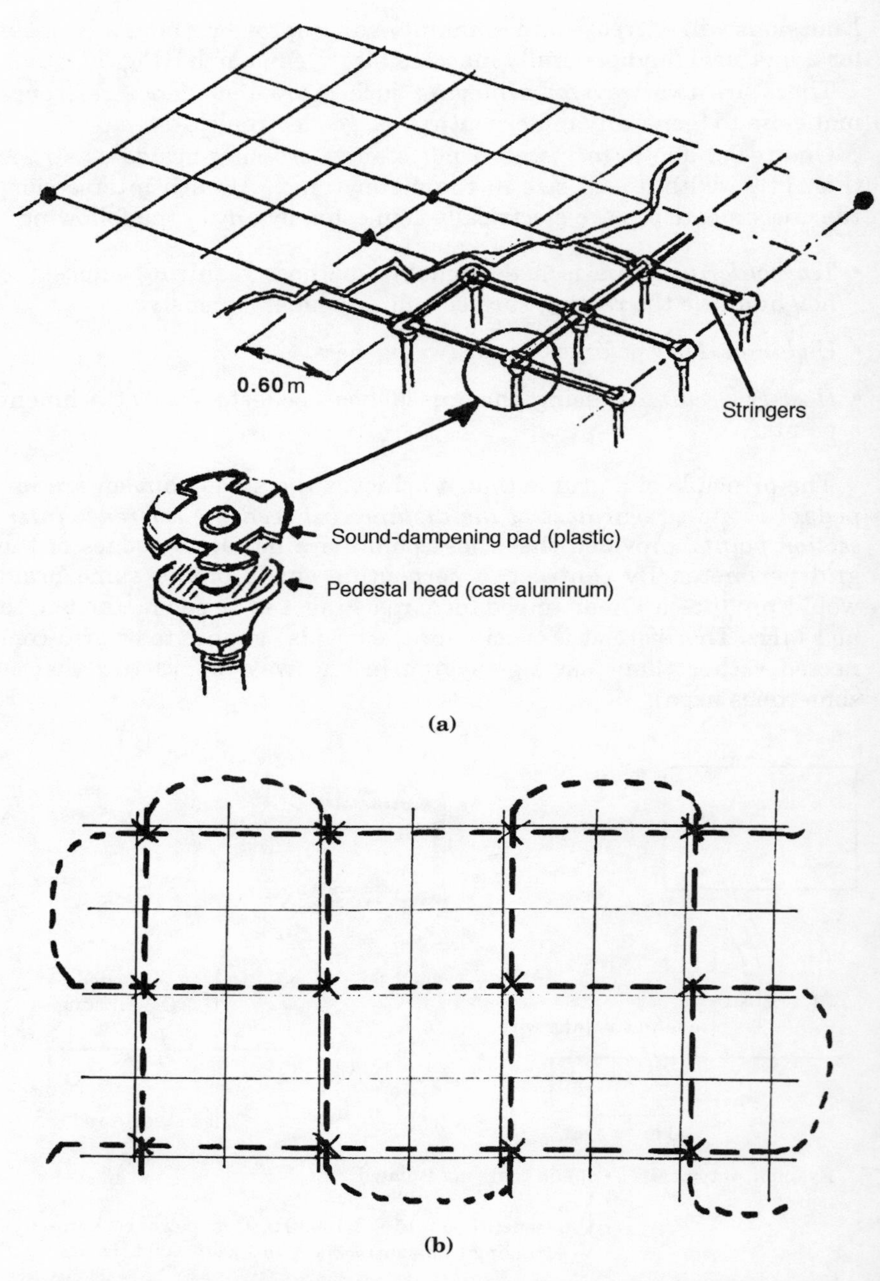

Figure 11.7 Using a raised floor as a ground grid. (a) Typical raised metal floor configuration. This could make an 0.60 × 0.60 m grid, but insulated pedestal heads forbid it. (b) One solution is to use a criss-crossing, serpentine arrangement of braid, connected at every other pedestal to form a 1.20 × 1.20 m grid.

The other technique, which is highly recommended for rooms equipped with a raised-metal floor (RMF), is to use the floor tiles stringers as the reference grid. RMF stringers normally form a 0.60 × 0.60 m grid array. Unfortunately, the electrical contact at their intersection over the pedestal heads is generally poor or nonexistent because of dust, debris, and the plastic damper typically fitted over the head. By using a conductive damping pad instead of the plastic one, an excellent, durable bonding is established in the following areas:

- tile-to-tile
- tile-to-stringers
- stringers (×4) to pedestal

After such a simple treatment, the floor-grid impedance *between any two nodes* drops down to tens of milliohms (dc to kilohertz), rising to less than 10 Ω at 3 MHz (Fig. 11.8).

Indications

RMF gridding is useful in buildings when no reference plane or grid exists in the construction itself that could be easily accessed, and when there is a need for reducing the risk of EMI emissions or susceptibility because of

- older equipments that do not comply with any EMC specification
- poor cabling and grounding practices by unaware contractors
- a specific EMI environment that is more demanding than usual
- certain sensitive equipments that by nature are impossible to harden
- certain equipments that by nature generate strong RF emissions (ISMs, etc.)

This need relates to an average interference reduction in the 10 to 20 dB range without resorting to an expensive six-sided Faraday cage.

Prescriptions, Installation

1. *For the braid, or tape, technique:*
   - Sketch a floor plan to prepare the zig-zagging layout. Divide the floor surface in a grid map of approximately 1 m pitch (0.60 to

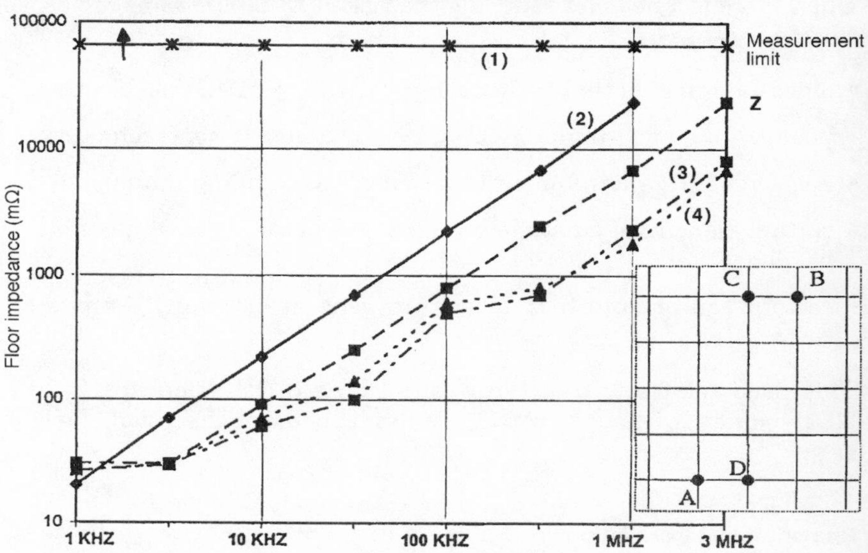

Figure 11.8 Floor impedances vs. frequency. 1 = ordinary raised floor with antistatic pads and tiles (Z is above the 50 Ω measurement limit). 2 = with criss-crossing braids. Z = recommended maximum value for site EMC. 3, 4 = impedance Z_{A-B} and Z_{C-D}, using the stringers + grounding pads technique. *Photo courtesy of J. Dubois.*

1.50 m being an acceptable range). It is better to calculate a grid size that results in an integer number of identical grid cells.

- Install two orthogonal patterns of criss-crossing braids or copper tape (minimum 25 mm wide) on the concrete slab, underneath the carpeting or the raised floor.
- Electrically connect all intersections.
- If the room has a power-distribution panel, connect this panel bottom plate to the grid, using a vertical chute of braid or minimum AWG 8 green wire. This is both for safety (earthing via the ac mains safety wire) and EMC purposes.
- Connect the frame/chassis of each equipment to the grid, using a vertical drop of braid making the shortest possible path. Always make this grounding at a grid crossing—never on the edges.
- If metallic cable raceways, metal pipes, etc., are crossing the room, connect them to the floor grid next to their points of entry/exit.

2. *For RMF stringers technique:*

   Before installing the floor tiles (or after removal, for an existing installation), insert the special cross-shaped grounding pad on the pedestal head plate. The dc contact resistance, under normal tile weight, is less than 5 mΩ.

All further steps are similar to the braid/tape method. Figure 11.9 provides a recapitulation of room grounding practices thus far discussed.

11.5 Transient Plate

This backup solution, initially practiced by IBM installation planning engineers in the 1970s, consists of installing a square sheet of copper or galvanized steel. For cases where the "real" ground is out of reach, this transient plate (TP), by virtue of its large capacitance to the building structure (typically 300 to 1000 pF) provides an efficient sink for EMI filters, transient protectors, and Faraday shields of isolation transformers (Fig. 11.10). This virtual ground is more efficient at HF than the long, green or green-yellow earthing conductor.

Indications

Indications are similar to those for floor grid, in sites where a ground plane or grid cannot be found and where the reinforced concrete rebars cannot even be accessed.

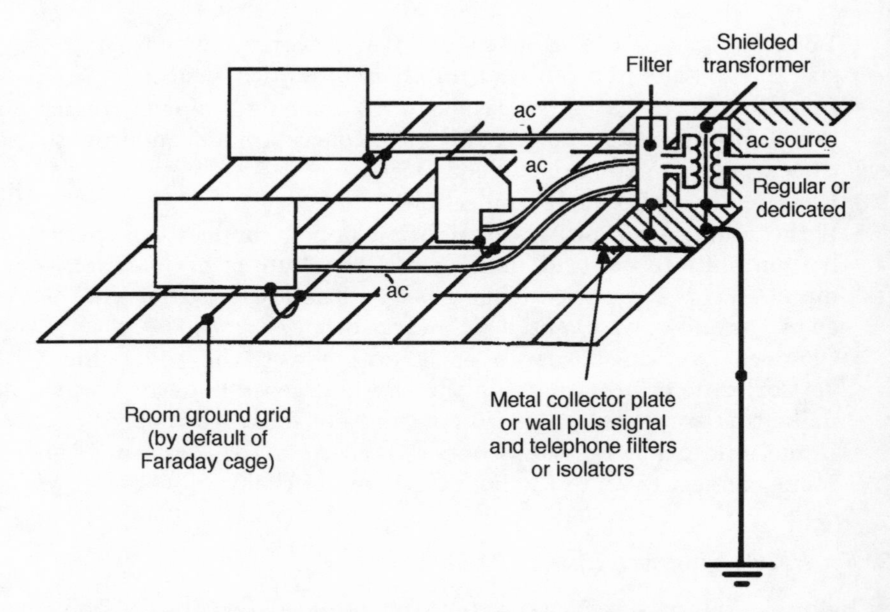

Figure 11.9 Recapitulation of room grounding practices for EMI reduction in large system installations.

Prescriptions, Installation

- The plate should cover an area slightly larger than the footprint of the concerned equipment and, in any case, not less than ≈1.4 × 1.4 m. A rectangular shape is adequate, provided the above minimum area of 2 m^2 is observed, and the length-to-width ratio does not exceed ≈3.
- For equipments that are installed vertically against a wall, the TP can be laid vertically behind the equipment chassis—provided that the wall is of reinforced concrete type.
- Connect the concerned equipment to the plate, using short braid or strap, near the point of entry of the external cables.
- If external EMI filters, surge protectors, or shielded transformers are provided, connect them directly to the TP.

Limitations

- It works only above several hundred kilohertz, at which point capacitive grounding of the TP is effective.
- Even when properly installed, a TP cannot match the performance of a true ground grid.

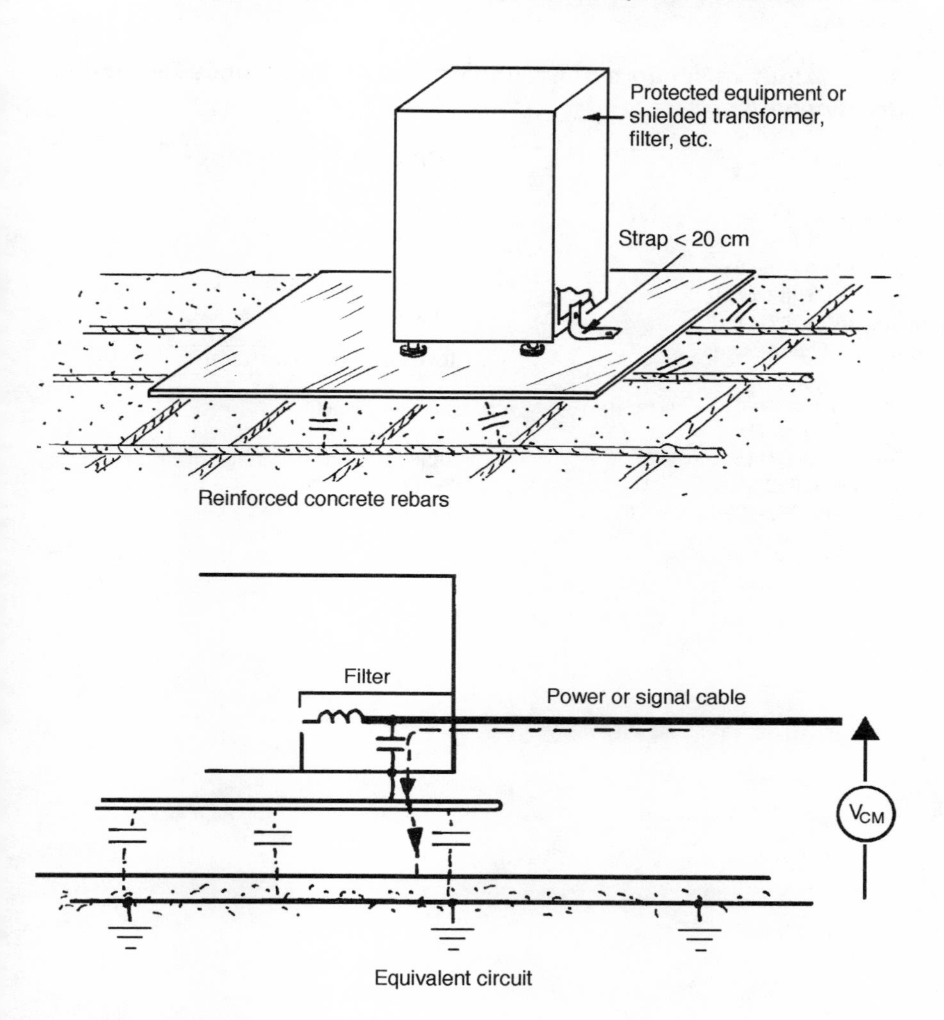

Figure 11.10 Transient plate.

Representative Vendors of Grounding and RF Impedance Reduction Components

Grounding Hardware

Polyphaser Corp.
2225 Park Place
Minden, NV 89423
Tel.: (702) 782-2511
Fax: (702) 782-4476
www.polyphaser.com

Joslyn Electronic Systems Corp.
6868 Cortona Dr.
Goleta, CA 93116
Tel.: (805) 968-3551
Fax: (805) 968-2335
www.jesc.com

PCB Grounding Spacers

Kitagawa
350 Fifth Ave., #704
New York, NY
Tel.: (212) 629-3620
Fax: (212) 967-3948

Raised Floor Grounding

Jacques Dubois
82 r. A. Badin
76360 Barentin
France
Tel.: 33 (0) 2 35 92 32 21
Fax: 33 (0) 2 35 91 42 94

Chapter

12

Radiation-Type Fixes

Acknowledging that, in many cases, a radiated EMI problem can appear and be resolved at the conducted stage, there still are a number of occurrences where the solution can be installed somewhere across the radiation path itself, acting as a field barrier. The efficiency of this barrier, according to shielding theory (Ref. 4) will, of course, depend strongly on the frequency of the EMI source, its nearness to the barrier, and the nature of the EMI field: electric, magnetic, or plane wave.

This chapter describes, for all situations involving a radiated path, what hardware parts to use, and when and how to use them. Using the same routine as for Chap. 4, "Conduction Fixes," each fix is briefly described by a kind of "identity card."

12.1 Conductive Tapes

Copper and aluminum tapes provide an easy and expeditious way to create instant shields and low-impedance straps or busses. They are extremely convenient for either temporary ("band-aid") fixes or more permanent solutions. Thickness is in the 0.035 to 0.1 mm (1.4 to 4 mil) range, and they generally are provided with a conductive adhesive backing for easy installation. For copper tape, the through resistance, once applied, is approximately 20 $m\Omega/cm^2$ (3 $m\Omega/in^2$).

Indications

The applications for conductive tape are countless (Figs. 12.1 and 12.2). The list below shows the most common ones.

1. To electrically seal foil shields
2. To locate leakages during troubleshooting

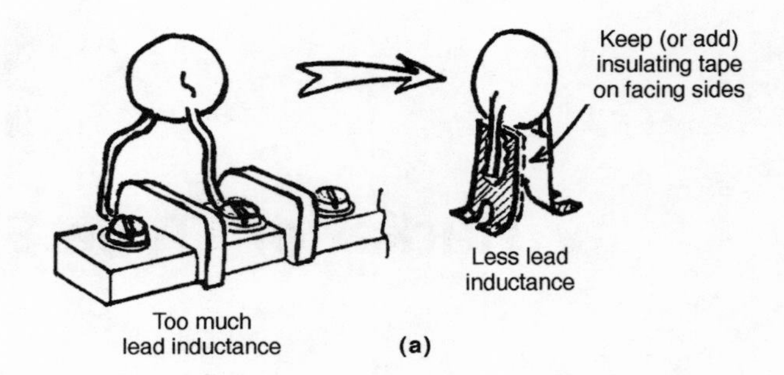

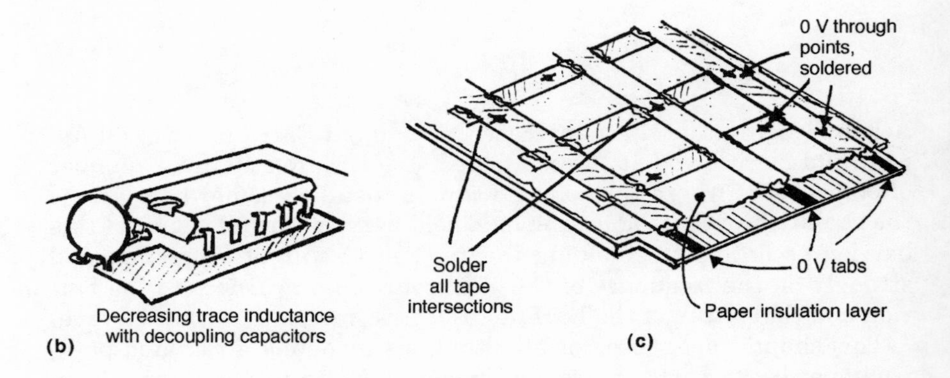

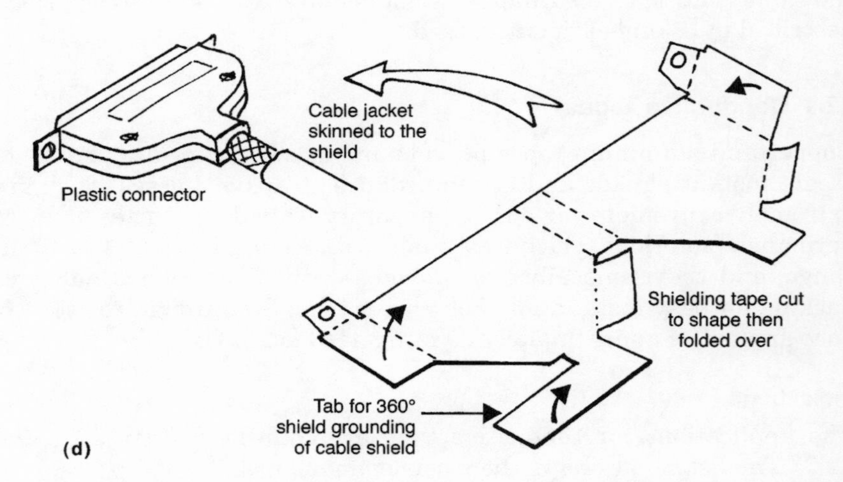

Figure 12.1 A few of the countless applications of conductive tapes, generally used as last-minute fixes. By enlarging component leads (a) or traces (b), the lead inductance is reduced, and the capacitive decoupling remains efficient at higher frequencies. In (c), a grid of wide copper stripes to the 0 V is turning a single-layer PCB into a "poor man's multilayer." In (d), the plastic connector is changed into a shielded one.

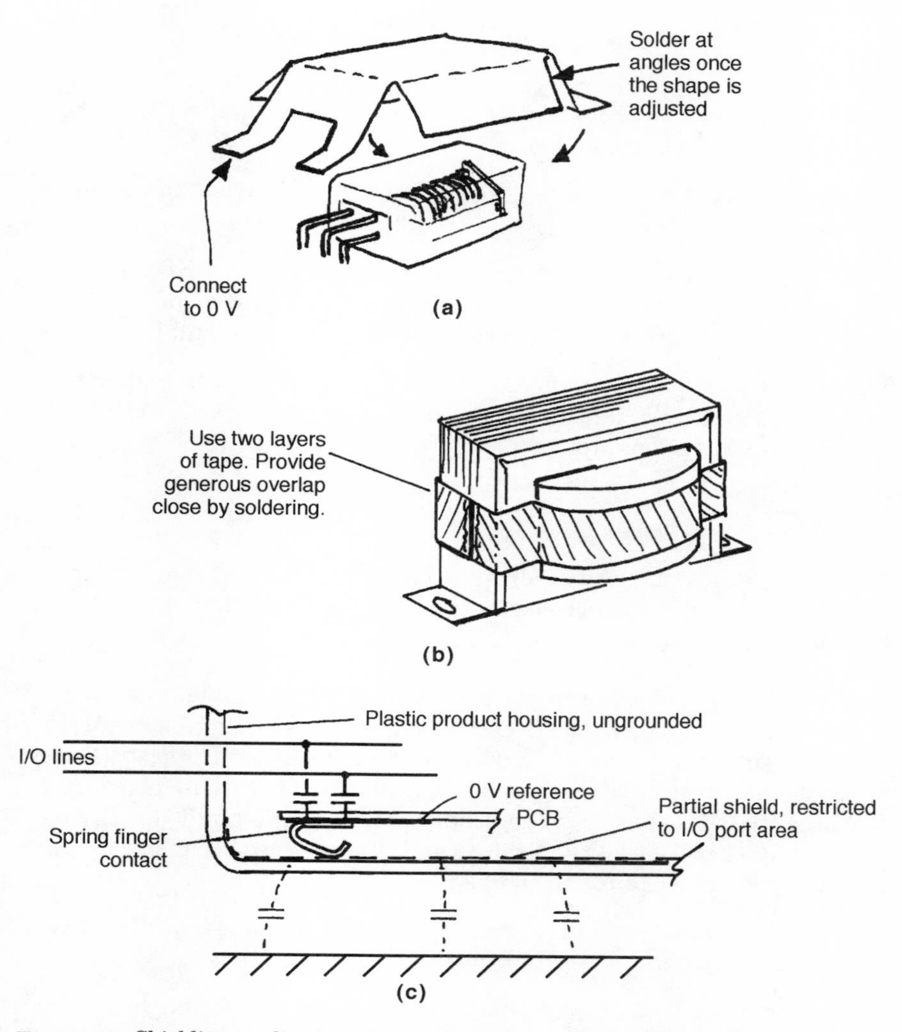

Figure 12.2 Shielding applications for conductive tapes. (a) Shielding a PCB-mounted relay, (b) installing a magnetic-field canceling ring around a transformer, and (c) partial shield inside a floated electronic case. The CM currents derived by the I/O port capacitors are sunk to the actual ground by the shield and stray capacitances.

3. To metallize, as a quick fix, a plastic connector
4. To shield an ordinary flat cable
5. To improve the effectiveness of braided cable shields
6. To create an even, smooth, conductive surface over a rough or poorly conductive zone (a better gasket contact, for instance)

7. To shield small components that are already mounted on a PC board
8. When a Faraday shield needs to be improvised around the coil of an inductor, transformer, etc.
9. When a magnetic canceling loop (shading ring) needs to be improvised around the windings of a transformer
10. For single-layer PC board "patching" against EMI, ESD, or a self-jamming problem, where a temporary ground grid or plane has to be tried for decreasing the zero-volt bus impedance and loop areas
11. When the lead inductance of decoupling components (capacitors, filters, Faraday shields, etc.) needs to be lowered, which relates to conduction fixes as well

Prescriptions, Installation

These tapes generally have a conductive adhesive backing, except for the embossed types, which use ordinary glue. The following hints are helpful when using them.

1. *For shielding applications,* the contact surface must be free of paint, dust, and grease. To clean, use a cloth soaked in solvent. Remove the tape's protective paper progressively and press it firmly onto the desired surface. When the whole piece is in place, reinforce the adhesion by using a hard roller, a tool handle, or the like to emboss it. Due to the limited adhesive properties of the conductive glue, never reuse the same piece of tape.
2. *To improvise a temporary ground grid* over a single-layer PC board without a ground plane, perform the following steps.
   a. First identify as many zero-volt references as possible (via holes) on the solder side of the board.
   b. Affix a kraft paper (or other strong paper overlay) on this solder side.
   c. With a sharp cutter tip, cut away small holes in the paper at the zero-volt connection points.
   d. Install a criss-cross pattern of copper tapes on the paper overlay. Try to make the grid such that the width of the voids is not larger than the tape width. This will ensure a quasi-ground plane behavior.
   e. Connect each intersect with two continuous lines of solder.

f. Perforate the copper tape where the zero-volt connection points occur and solder the tape to these points.

12.2 Shielding Mesh Bands and Zipper Jackets

Mesh. Tin-plated steel mesh tape is primarily used as an easily installed "bandage" type of shield over an already assembled cable harness. The mesh is knitted in the form of a cylindrical stocking, then flattened to form a two-ply tape.

Indications

To reduce H-field emissions or susceptibility of a cable, steel mesh tape is a useful fix. In general, if correctly installed, such tape provides at least 20 dB of LF magnetic field attenuation around a 2.5 cm (1 in) dia. bundle, and more than 20 dB if the diameter is less. This shield also works against high-frequency RF fields, although it will be about 20 dB poorer than a tight (≥90 percent optical coverage) copper braid.

Prescriptions, Installation

1. Wrap very tightly around the bundle of wires (Fig. 12.3).
2. Overlap the steel mesh to produce a four-layer coverage.
3. Do not interrupt at a T or Y junction. Instead, extend the wrapping a little over the fork, then continue the wrapping on one branch. Come back later over the fork to cover the other branch.
4. Terminate to chassis or structure by using a circular clamp or a copper tape (or strap) soldered to the loose end of the mesh. Avoid pigtails whenever possible.

Limitations

1. Multiple holes in the mesh will start leaking significantly at VHF (30 MHz) and above.
2. Electrical contact between adjacent turns will degrade if the cable is subject to motion, bending, etc.

Zipper-Type Shielding Jacket. This consists of aluminum foil, monel, or tinned steel mesh tubing with a PVC coating and a zipper. It provides a quick fix on an existing installation (Fig. 12.3).

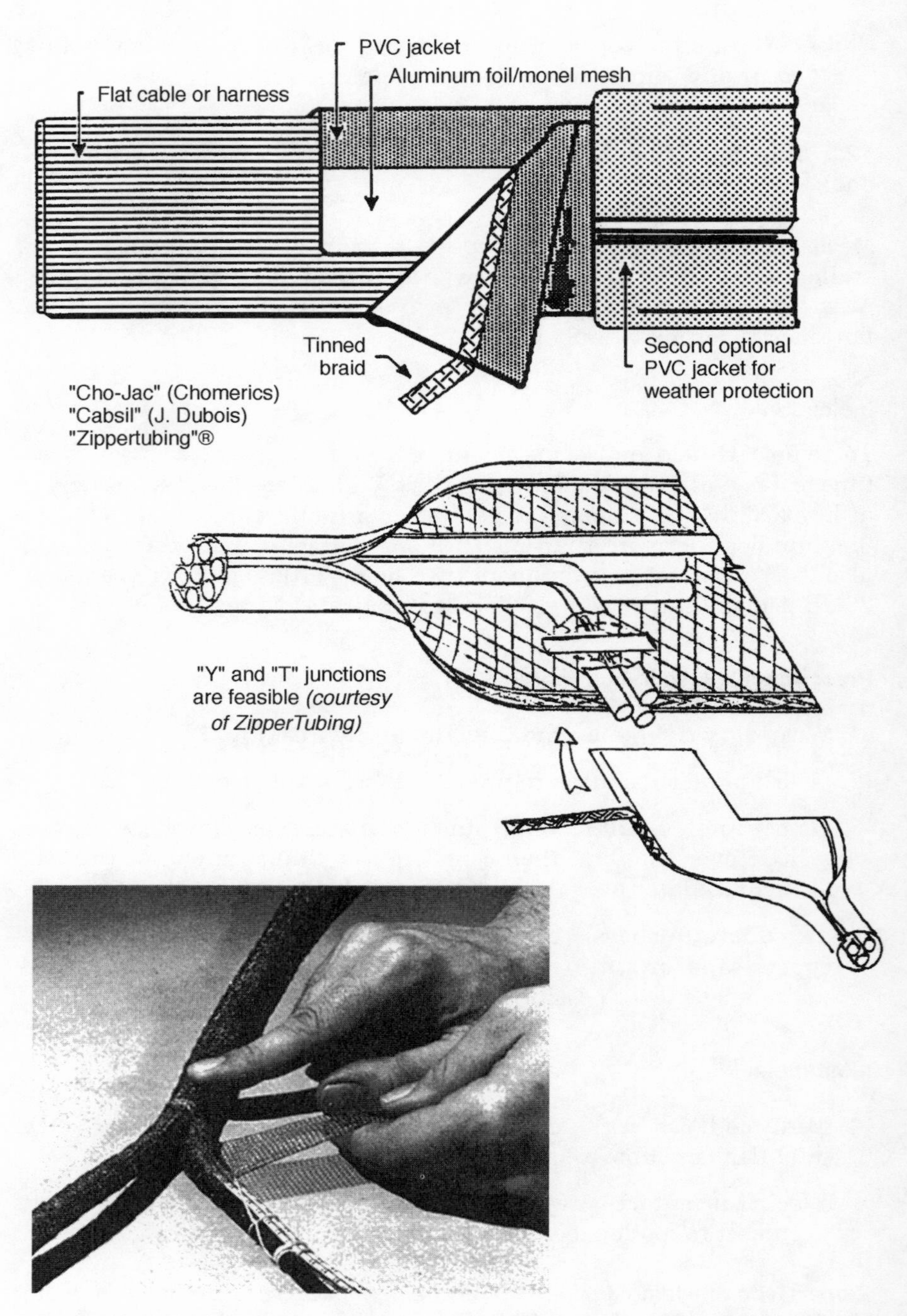

Figure 12.3 On-site overshields for cables and harnesses: (top) zipper type of shielding jacket and (photo) cable shielding by mesh band or tape.

Indications

Use when there are clear signs that cables are the dominant EMI coupling mode.

Prescriptions, Installation

1. Dress the open tube around the cable and close the zipper, making sure the overlap edge extends correctly under the zipper.
2. Terminate at the box's entry points using a short braid connected to chassis.

Limitations

It is difficult to make an ideal 360° bond at shield ends.

12.3 Cable Shields, Connections, and Fittings

The purpose of this section is not to describe the mechanisms by which cable shields perform. This topic is perfectly covered in an abundant literature (Refs. 4 through 6 and 8, to mention a few). The present book being intended to help find and solve EMI problems on a system that probably already has been designed, it is likely that the decision of using a shielded (or nonshielded) cable has already been made by the time you encounter the problem. At the risk of simplification, and exceptions accounted for, we can say that, in most cases, the problem is not in the quality of a cable shield (which generally is sufficient for the task) but in the way it is connected at the ends.

To reinforce this idea, we will just briefly describe how the performance of a cable shield is quantified and, as a result, how a nonideal connection of the shield (and it is never ideal) can deteriorate the expected merits of an otherwise decent shield.

12.3.1 Technical Brief on Cable Shield Effectiveness, Reduction Factor, and Transfer Impedance

Shielding Effectiveness. To grade the merits of a cable shield, until the 1980s, the electronic industry was mostly using a parameter called *shielding effectiveness* (SE), a comparison of the electromagnetic field that would illuminate the wires if there were no shield and the actual field the wires would see once the shield was in place. As an equation, this is

$$SE(dB) = 20\log\frac{\text{E-field (or H-field) before shielding}}{\text{E-field (or H-field) inside the shield}}$$

Although this definition is correct, obtaining a true measurement of the field is not practical, and cable manufacturers instead were using the ratio of the voltage picked up by the field-illuminated wires without the shield, then with the shield.

This has the advantage of deriving a figure in decibels, which seems analogous to the decibels of shielding that one can get from a metallic enclosure, electronic cabinet, etc. However, the rigorous measurement of this SE for a cable shield is less than obvious. It requires generating a strong EM field (i.e., amplifiers, antennas, and a Faraday cage). But, moreover, the figure obtained is not just related to the shield properties *per se* but also to its installation: height above ground, circuit source and load resistances, etc. Flattering SE figures can be obtained with large heights, near E-field exposure, and ideal matching. Conversely, low figures are produced with small heights, near H-field conditions, and mismatched loads. Nevertheless, many vendors still display shielding data in terms of SE (dB).

Reduction Factor (KRF). The reduction factor, KRF, is a conducted measurement whereby a given CM voltage or current is applied to the unshielded cable, and the resulting voltage at the load end is measured. Then, the same CM source is applied to the shielded version of the same cable. The KRF is the ratio of the voltages at load end with and without the shield. This, too, is a relative measure of the shield quality because of the dependency on the test configuration. However, this is a straightforward and rather easy test that immediately informs the user about what the result will be.

Shield Transfer Impedance (Z_t). Transfer impedance relates the voltage, V_i, appearing internally (i.e., between the inner conductor(s) and the shield) to the current flowing on the shield surface (Fig. 12.4).

$$Z_t(\Omega) = \frac{V_i}{I_{shield}}$$

When measured on a sample with length ℓ_m and normalized to a 1 m length,

$$Z_t(\Omega/m) = \frac{V_i}{I_{shield} \times \ell_m}$$

Therefore, Z_t is an intrinsic value of the shield, independent of external variables.

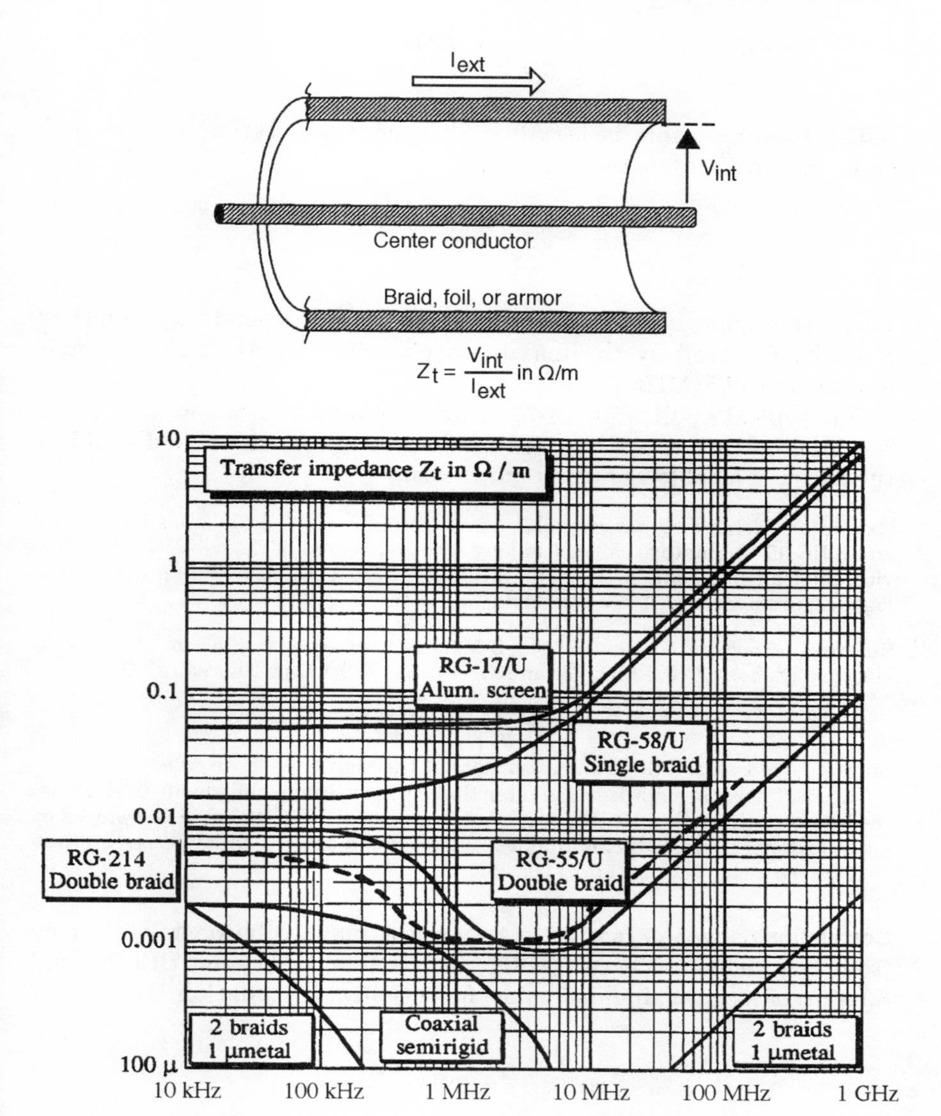

Figure 12.4 Transfer impedances (Z_t) for common cable shields.

Since V_i is caused by the diffusion current through the shield thickness and by the leakage inductance through the braid holes, the better the shield, the lesser the Z_t. Fig. 12.4 shows typical values of Z_t for some popular cables.

Knowing the value of the shield current (I_s), the induced voltage is calculated by

$$V_i = I_s \times Z_t(\Omega/m) \times \ell\,(m)$$

If the shielded cable is terminated by matched resistances at both ends, the terminal voltage is

$$V_l = \frac{V_i}{2} = 0.5 \times I_s \times Z_t \times \ell$$

When the cable becomes electrically long, the geometrical length, ℓ, must be replaced by the maximum electrical length [i.e., 1/2 wavelength, or 150/F(MHz)].

One typical application of Z_t in investigating a problem is to check the likelihood that a measured EMI current over the shield could be the cause of a cable-induced malfunction.

Example Using a clamp-on-probe, a current of 20 mA @ 8 MHz has been found on an RG 174 miniature cable, 4 m long. The source end of the cable is a 50 Ω device, the load side being a 1 kΩ input. Could this be a sufficient threat for an analog measurement with 1 mV threshold, at S = N?

Solution Checking for wavelength, 8 MHz is a wavelength of 37 m. The cable is electrically *short*: Z_t @ 8 MHz for an RG 174 coax is 0.1 Ω/m. Therefore,

$$V_i = 0.1\ \Omega/m \times 4\ m \times 20\ mA = 8\ mV$$

Since Z_{load} = 1 kΩ, all this voltage appears at the high impedance side.

The EMI noise is eight times greater than the sensitivity threshold. To suppress the interference, with a 10 dB margin below the sensitivity, it should be reduced by a 24× factor (i.e., 28 dB). This would call for using a cable with lower Zt, or otherwise filtering the 8 MHz EMI, if possible.

Correlation between Z_t and SE. An approximate relationship, derived from the reduction factor KRF, can be found between SE(dB) and Z_t(Ω/m) of a given shielded cable, for a typical installation:

$$SE(dB) = 20\log\frac{Z_{ext\ loop}}{Z_t \times \ell}$$

Around 1 MHz and above, this simplifies to

$$SE(dB) = 20\log\frac{L_{ext}(H/m)}{L_t(H/m)}$$

where L_{ext} = self-inductance of the cable above ground

L_t = leakage inductance term in Z_t

Typically, for a round cable with 0.5 to 1 cm dia., at a 10 to 30 cm height above ground, $L_{ext} \approx 1\ \mu H/m$. For a good quality single braid shield, $L_t \le 1$ nH/m. Therefore, SE above 1 MHz is in the 60 dB range.

12.3.2 Evaluating the Influence of Shield Ground Connections

The quality of the cable shield is not the only contributor to EMI reduction. The nature of the shield termination plays an important role. In fact, although this may sound exaggerated, a shield termination *always spoils* the effectiveness of a shield (Fig. 12.5). The objective is to minimize this deterioration.

To estimate the contribution that the shield connection (by wires, clamps, or connectors) can add to the overall shield penetration, the EMC community has found it practical to characterize the connection by its transfer impedance, Z_{tc}. As a result, the total coupling impedance between the shield current, I_s, and the internal voltage, V_i, for a cable shield with two identical connectors becomes

$$V_i = I_s[Z_t\ (\Omega/m) \times \ell + 2 \times Z_{tc}]$$

For a negligible contribution, it is clear that Z_{tc} must be an order of magnitude below the term $Z_t\ (\Omega/m) \times \ell$.

The following Z_{tc} data are for the most popular types of shield terminations. All data are for the plug + receptacle (one mating set).

- BNC (new): 1 mΩ + j 0.03 mΩ × F MHz
- BNC (old): 3 mΩ + j 0.03 mΩ × F MHz

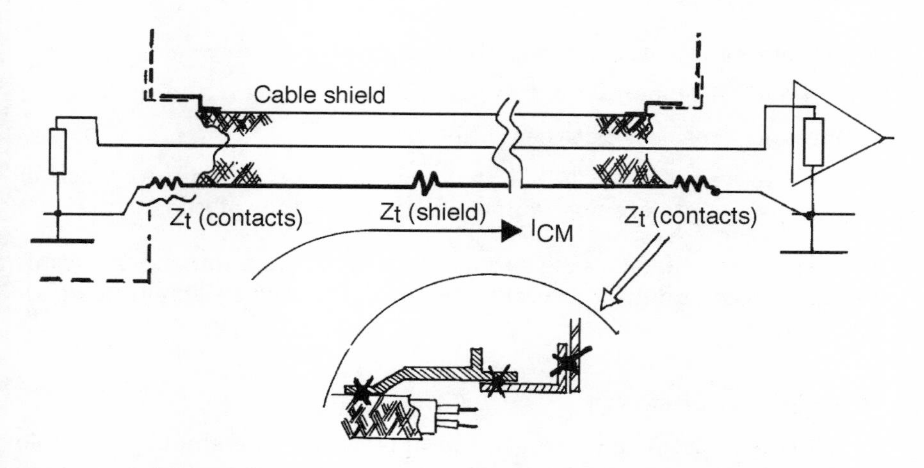

Figure 12.5 Contributions of shield grounding connections to the total shield-induced noise.

- N, TNC, or other threaded type: 0.1 mΩ + j 0.01 mΩ × F MHz
- Multipin with simple slide-on mating: 5 to 30 mΩ + j 1 mΩ × F MHz (e.g., Sub-D, Centronics, etc.)
- One 5 cm (2 in) pigtail: 2.5 mΩ + j 0.3Ω × F MHz

The following example gives a measure of the effect of Z_{tc}.

Example Five meters of RG58 cable are subjected to a 1 MHz EMI. What contribution to the total coupling can be brought by:

- shield terminated by two 5 cm pigtails
- 360° contact by BNC connectors

Solution The transfer impedance of RG 58 at 1 MHz is: –32 dBΩ/m, or 2.5 mΩ/m. For 5 m, $Z_t = 0.025 \times 5 = 0.125\ \Omega$.

1. With the two pigtails adding

$$2 \times (2.5\ \text{m}\Omega + j\ 0.3\ \Omega \times 1\ \text{MHz}) = 0.6\ \Omega$$

Then, the total coupling impedance will be 0.125 + 0.6 = 0.725 Ω, with the pigtails' contribution being five times stronger than the cable alone.

2. With the two BNC (new) adding

$$2 \times (1\ \text{m}\Omega + j\ 0.03\ \text{m}\Omega) \approx 2\ \text{m}\Omega$$

Here, the connector impedance does not practically affect the shield performance.

Indications

This solution is indicated every time a shielded cable is used for the following purposes:

- To reduce susceptibility to radiated EMI on the cable
- To reduce EMI emissions from the signals flowing in the cable
- To reduce crosstalk between different cables
- To reduce the effects of CM impedance coupling (when the chassis of two equipments are not equipotential) above tens of kilohertz

In such cases, the following recommendations regarding shield terminations are mandatory to create, or upgrade, a given shielding effectiveness.

Prescriptions, Installation

In all cases, except for one specific application, the shield of a cable must be connected to the equipment box, or chassis ground, *at both ends*.

For Shielded Cables with a Metallic Connector

1. Make sure that the connector receptacle makes a good metal-to-metal contact with the equipment housing. Avoid or remove paint, oxides, and chemical treatments. Some corrosion protectants, like the military olive-green finish (a bichromate treatment) or anodized aluminum, give a very poor, unreliable contact. Do not rely on a piece of wire for ensuring receptacle-to-baseplate continuity.

2. If the receptacle is mounted on a card edge, make sure the metal housing is positively grounded to the chassis via

   - A connection to an internal PCB ground plane, which itself is grounded to the chassis with a low impedance
   - A mechanical-ground copper land, which is grounded via the card guide rails and spring fingers

3. Check that no weather gasket is interfering with a decent receptacle-to-chassis contact area. If such gaskets are needed, they should be of the O-ring style, saving a sufficient electrical contact footprint. If the gasket is also an EMC gasket, check that the contact area is sufficient. With some MIL-circular receptacles (Fig. 12.6), the contact area is so skinny that the electrical contact is worse with the EMC gaskets than without!

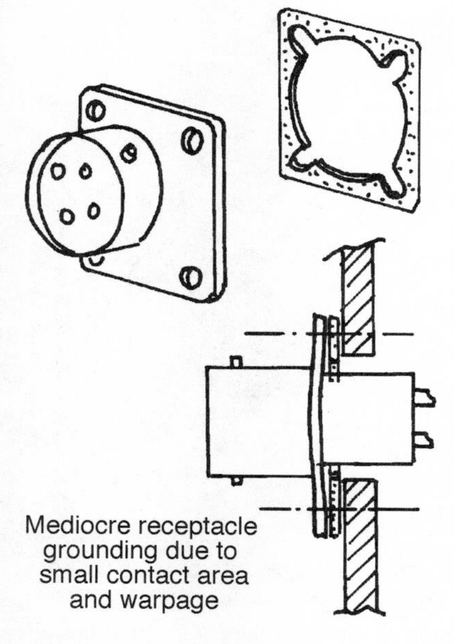

Mediocre receptacle
grounding due to
small contact area
and warpage

Good grounding by
hexagonal nut
and O-ring gasket

Figure 12.6 Connector receptacle grounding.

4. Check the connection of the mating connector plug to the shield. Although it may seem to be a trivial item, this juncture of the shield surface to the connector backshell can be the weakest link in the entire shielding chain. There should be a 360° tight bonding (Fig. 12.7) such that its contact resistance stays $< 1\ m\Omega$. If there is a watertight grommet, it should be set in such a manner that it does not insulate the shield.

Beware of certain connector backshells where the cable clamp is simply a flat piece. This does not make a peripheral contact but a two-point contact (see Fig. 12.8) instead.

Beware also of backshell fittings that are just pressing the shield over the underlying PVC insulation. Because of the long-term fluidity

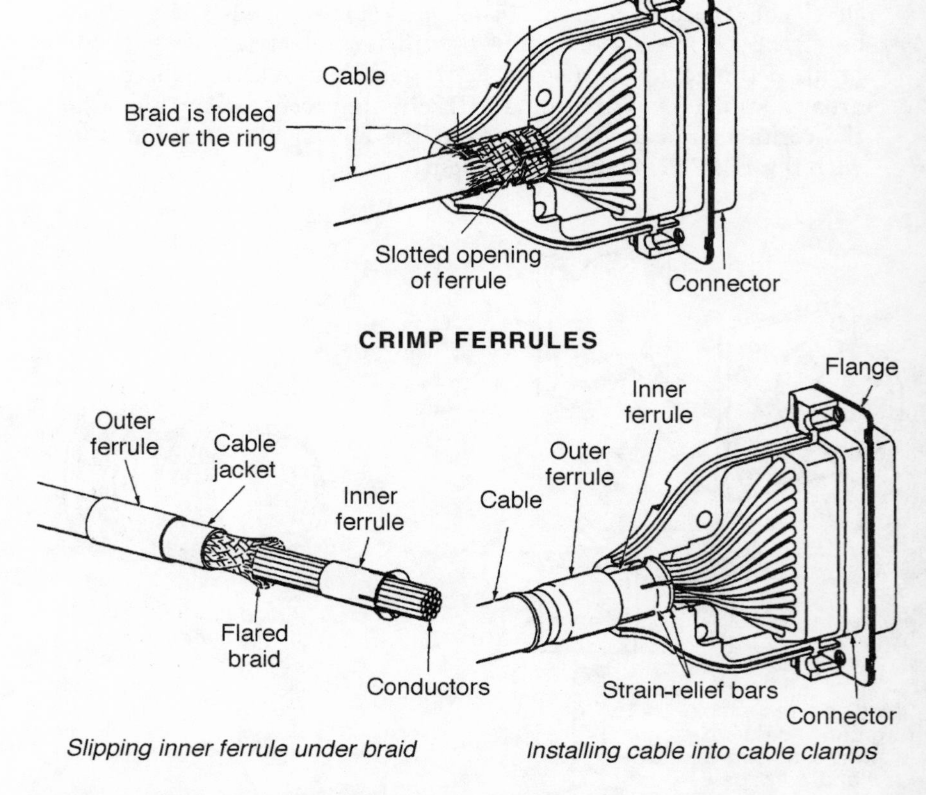

Figure 12.7 Good 360° positive contact by deformable slotted sleeve (top) or slip-and-grip ferrule (bottom). *Source: AMP, Inc.*

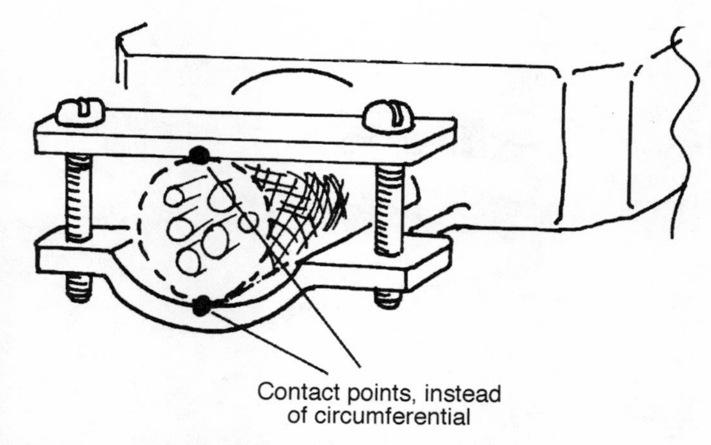

Figure 12.8 Poor shield grounding with some types of backshell connector hardware.

of the PVC, the strain will progressively decrease, leaving the shield almost pressure-free after few months. The solution is to slip a thin slotted sleeve of tinned steel or bronze underneath the shield.

For Individually Shielded Pairs with an Overall Shield

In this case, several techniques are possible.

1. For the overall shield, which is the actual EMC barrier between the inside and outside worlds, strictly apply the recommendations listed above. *Do not* ground this shield via a pigtail then through a connector pin.
2. The individual pair shields (or screen), which are generally made of a lighter braid or foil, must be regarded as pair-to-pair crosstalk protection.

These shields can be grounded collectively by a halo ring or, eventually, by branching short drain wires to some dedicated grounding pins inside the connector. Absolutely avoid daisy-chaining (Fig. 12.9), and try not to branch more than three drain wires to a same grounding pin. If the shielded pairs are carrying signals of very different families (e.g., power, digital signals, and analog signals) the internal pair shields can be connected as shown in Table 12.1 for *optimum crosstalk attenuation* without other detrimental effects.

For Shielded Cables without Metallic Connector

A plastic connector is of little use with a shielded cable, as is a metallic connector plug used on a plastic receptacle. In such cases, the shield-

Connector
I_{sh}
1
2
3
4
Poor
With daisy-chaining, inductive voltage buildups are adding in the remote shields
Ground pin
$I_{gnd} = \Sigma\ I_{sh1} + I_{sh2} + I_{sh3} \cdots$
Better
Still, the length of each individual jumper deteriorates the effectiveness of individual pair screens. It should be restricted to no more than three shields per pin, and to audio or low-frequency (<1 MHz) applications.

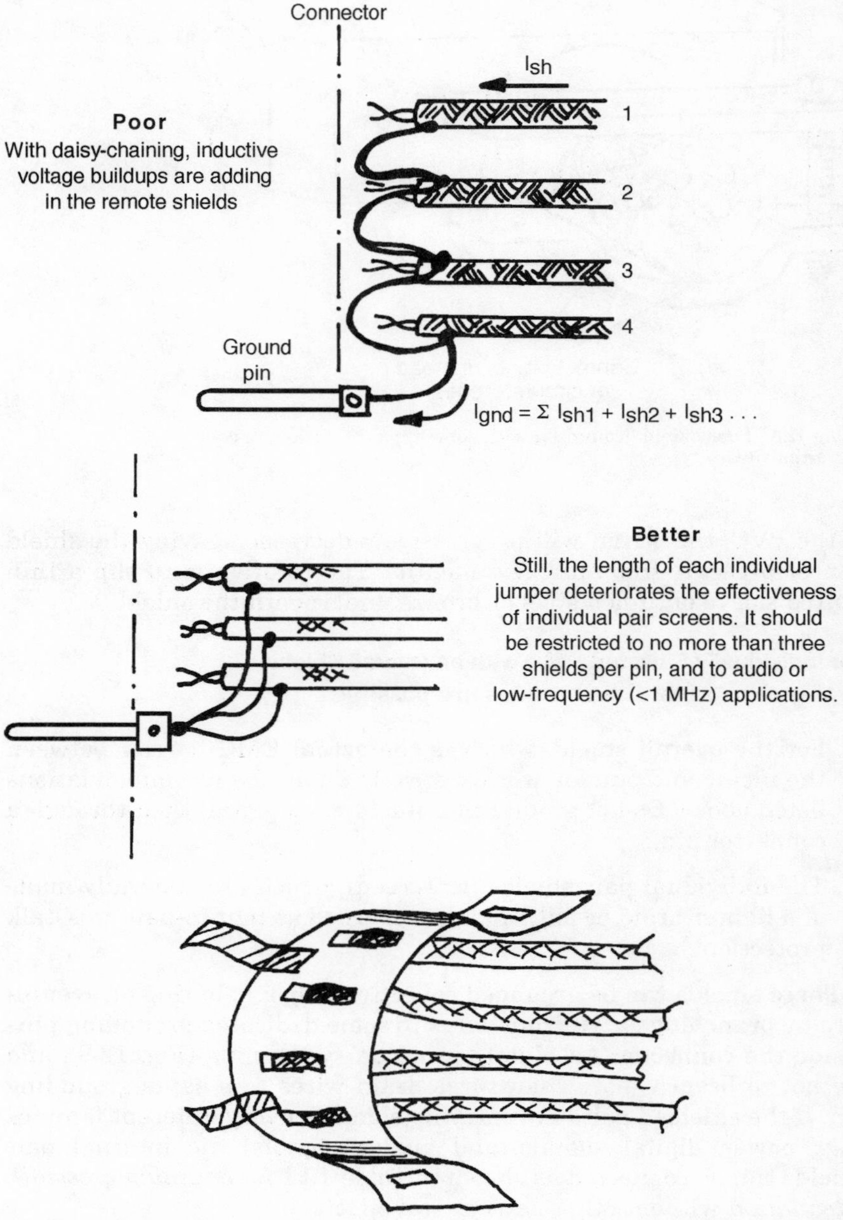

Figure 12.9 One of the many tricks for grounding individual pair shields. Each shield is soldered directly to a tinned copper band provided with punched slots. The band is then folded into a ring to fit inside the connector shell and is soldered to either a connector grounding pin or to a common grounding washer pressed via the connector mounting hardware.

Table 12.1 Connections for Shielded Pairs Carrying Mixed Signals

Culprit pair	Victim pair	Pair shield connection
High dV/dt (typ. > 1 V/μs)	Low-level, low-freq. analog, high-Z input	• Culprit shield at culprit 0 V ref. only, source side • Victim shield at victim's 0 V ref. only, receiver side
High dI/dt (typ. > 10 mA/μs)	Same as above	• Culprit shield at culprit ref. grounds, both ends • Victim shield at victim's 0 V ref. only, receiver side
For all other configurations		• Ground individual pairs, shields at both ends

to-chassis connection must be made, as directly as possible, near the cable point of entry using one of the techniques shown in Figs. 12.10 through 12.14.

Junction Boxes

In large installations, if a shielded cable has to go through junction boxes, the entry and exit must be treated as follows:

1. With metallic boxes, use metal clamps or feedthrough fittings for any chassis or frame penetration.
2. With plastic boxes, keep the shield coverage as far as possible up to the internal terminal board or splicing, then connect the ends of the shields by the shortest piece of braid or wire. Commercial hardware, with self-stripping ferrules, is also available.

In either case, what is important is maintaining the shield continuity while making the smallest possible loop area with the unmasked conductors (Figs. 12.12 and 12.13). Earthing the shield, from an EMC point of view, is not the issue, since the equipment chassis is connected to the safety ground anyway.

Limitations

- With low-frequency, low-level instrumentation cables, grounding the shield at both ends creates a low-frequency ground loop (typically for 50/60 or 400 Hz CM coupling), which can spoil the high CM rejection created by the floating and differential inputs. In this case, and *as an exception to the general rule,* shields are grounded at one end

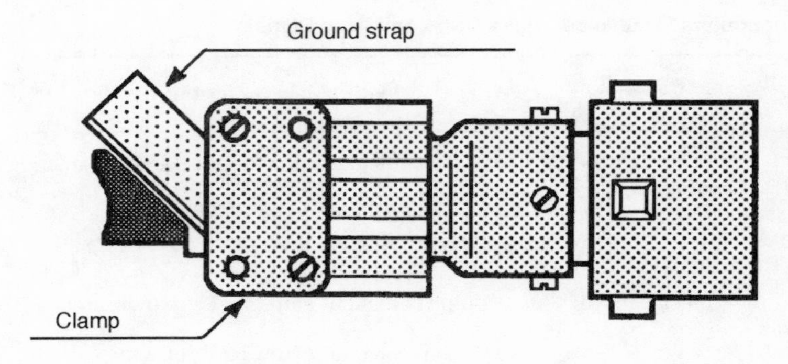

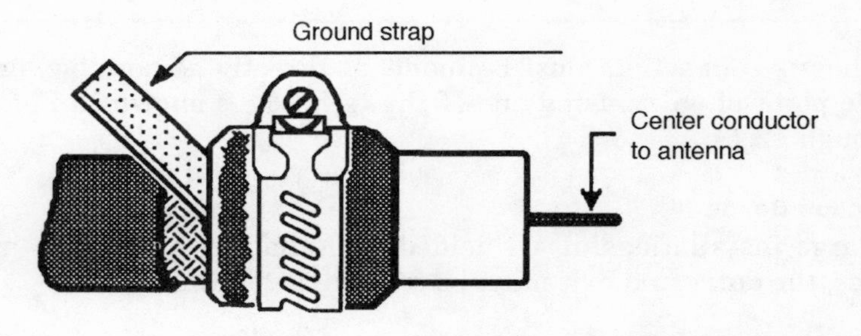

Figure 12.10 Cable-shield connection for armored or radio-communication cables. *Courtesy of AEMC.*

only, generally on the receiver side (Fig. 12.14). This shield is still valuable against electrostatic coupling or crosstalk. However, it is *worthless against any other EMI threat*, for which the circuit must be protected by other means, such as HF filtering.

- In large, complex installations, it is difficult to maintain a high-quality shield continuity throughout the system life, during which it goes through many upgrades, extensions, and various reworks. A rigorous discipline in maintenance operations is necessary.
- When the whole cabling is subject to mechanical abuse, weather, and moisture, the parts of the shielding that suffer the most are the shield connections. A performance decrease by one order of magnitude (20 dB) after five years of harsh service is not unusual.

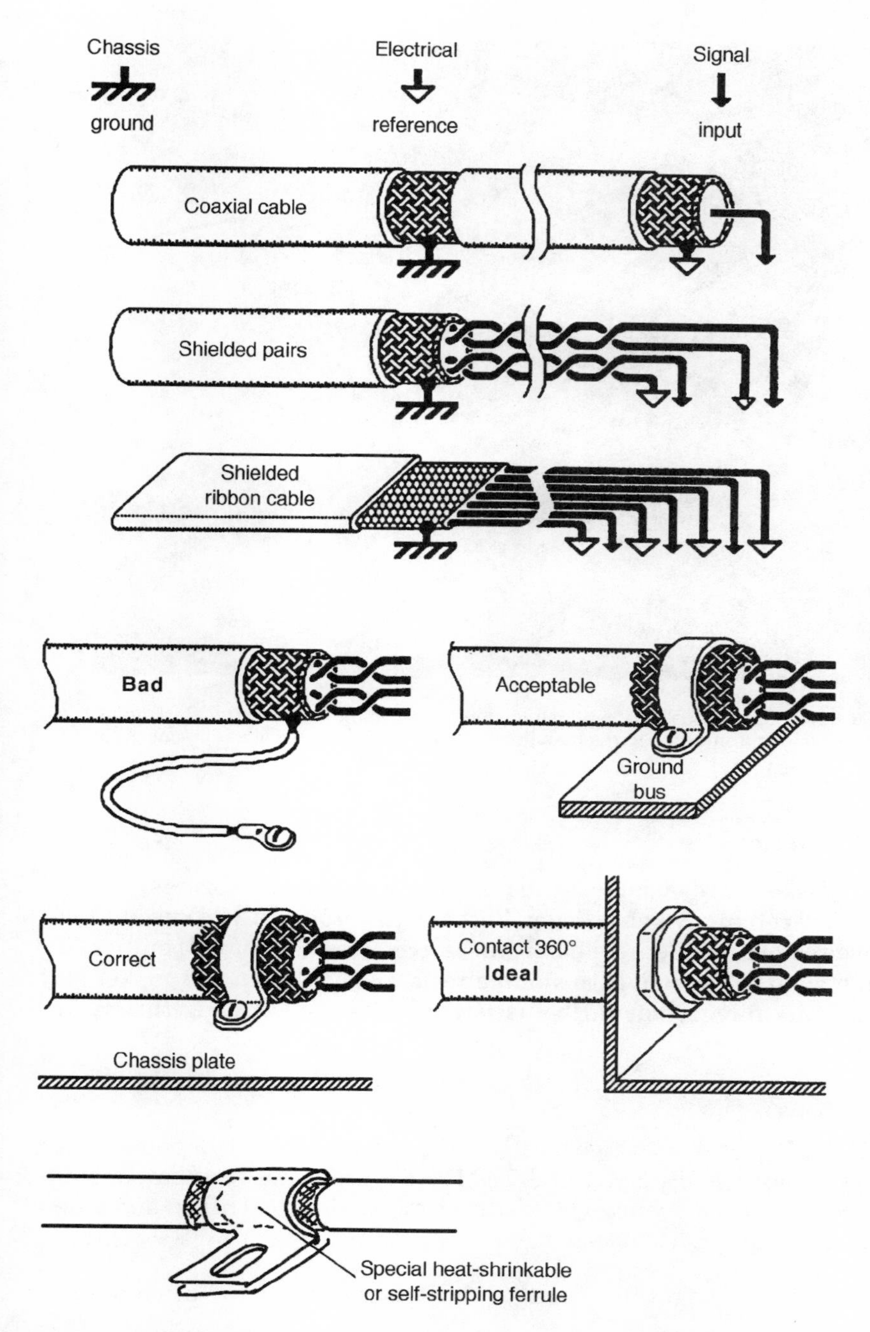

Figure 12.11 Cable-shield grounding when no metallic connector shell is available. *Courtesy of AEMC.*

Figure 12.12 One typical flaw with cable-shield grounding in industrial electronics. The 30 cm long grounding wires make the shields worthless above a few tens of kilohertz.

12.4 EMI Gaskets

EMI gaskets are flexible strips of beryllium-copper springs, knitted mesh, or conductive elastomer (Figs. 12.15 through 12.18) used to improve/restore shielding effectiveness (SE) around box openings. To make a gasket effective, the ohmic resistance between the gasket and its mating flanges should be as low as possible all along the gasket path.

Indications

EMI gaskets are most generally used against radiated emission or susceptibility, ESD, EMP, and TEMPEST problems when a substantial SE is needed (typically ≈ 20 dB or more, above 30 MHz) and when the following conditions exist:

1. The box leakages have been identified as a major radiation path.
2. The mating surfaces are not smooth, flat, or stiff enough to provide a good continuous contact by themselves.

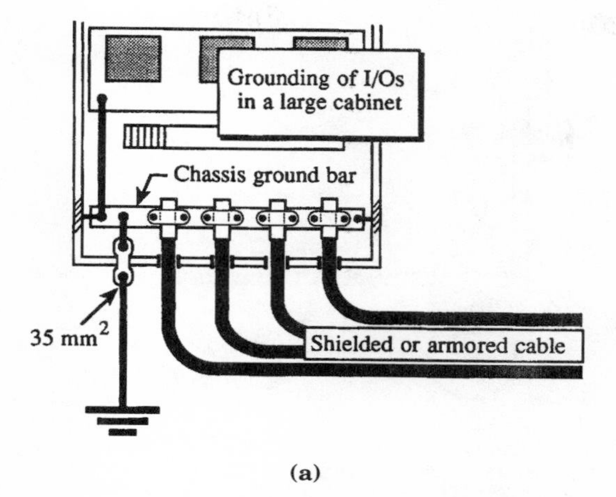

(a)

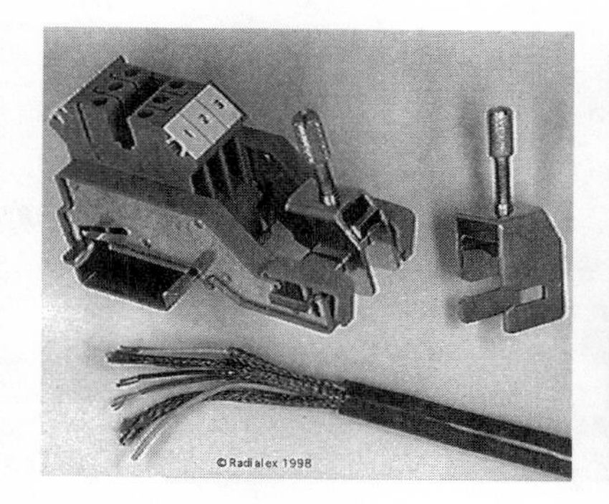

(b)

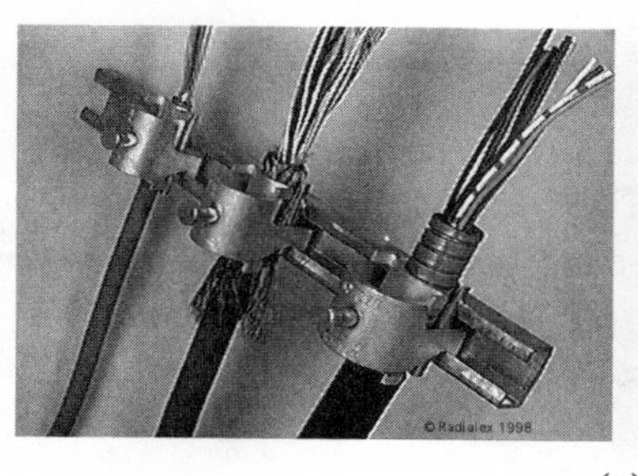

(c)

Figure 12.13 Correct methods for cable-shield grounding in cabinets with many cable entries: (a) regrouping all cables in the same entry zone and using a collecting bar, grounded to chassis, and using fast-clamp devices as provided by (b) Phoenix and (c) Weiland.

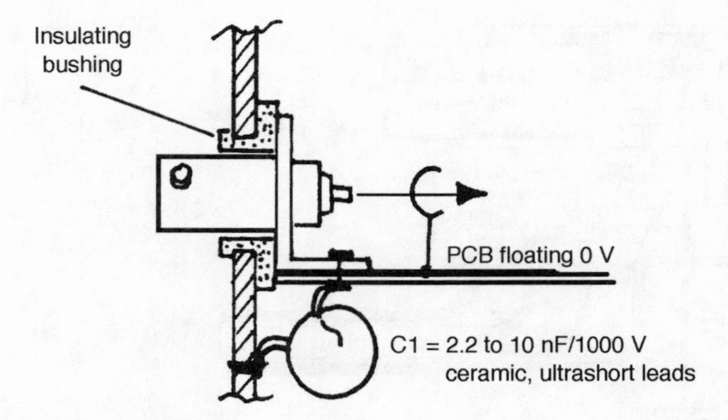

Figure 12.14 Grounding trade-off with some coaxial links. When the 0 V reference must remain isolated, low-frequency ground loops (50/60 or 400 Hz) are avoided, but the shield is grounded to chassis for HF common-mode current sinking via capacitor C1.

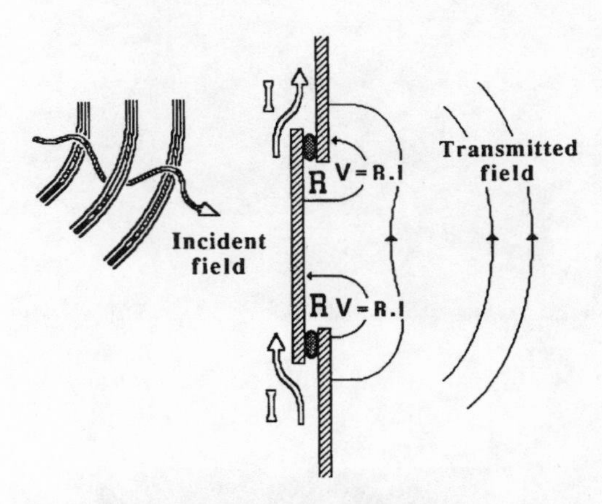

Figure 12.15 Principle of EMC gasketing.

3. The number of screws required to make a perfect metal-to-metal contact would be unacceptably high (e.g., one screw every two inches or less) given the cost, space, or maintenance aspects.

Prescriptions, Installation

1. First try to evaluate the overall SE required for the box to reduce EMI to an acceptable level. For instance, get a measure of the actual EMI emission or susceptibility level without the covers.

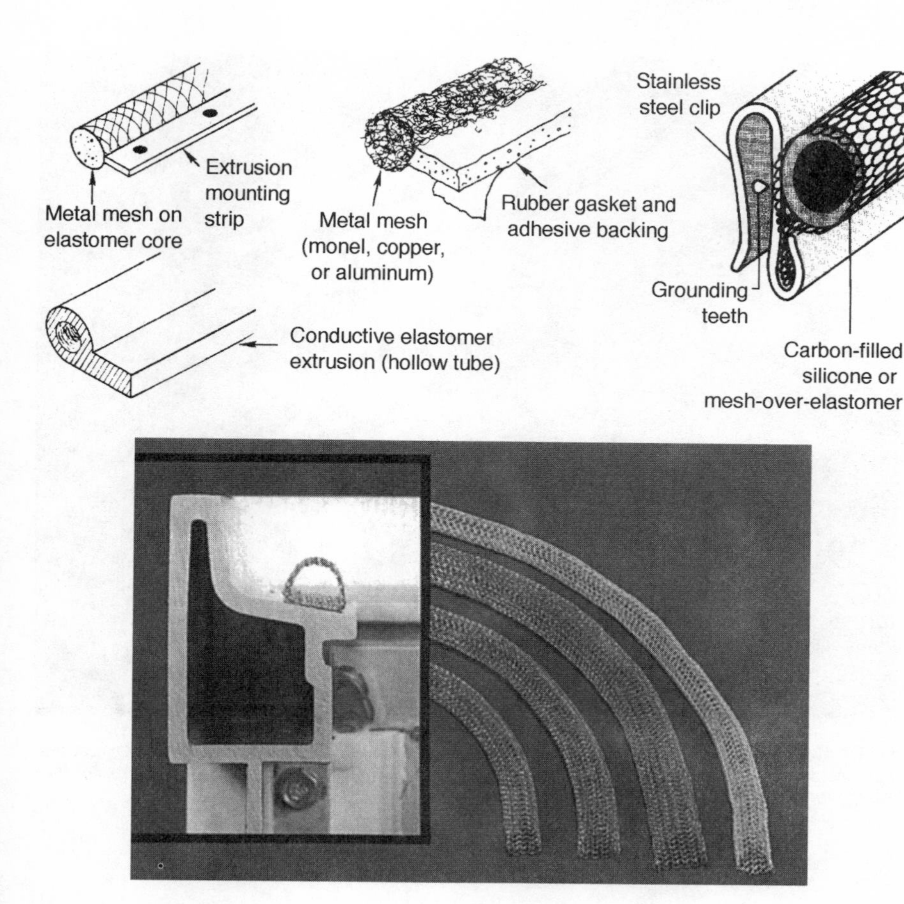

Figure 12.16 Various styles of compressible-core EMC gaskets for SE performances in the 40–80 dB range. The tubular knitted gasket (Ultraflex, by Instrument Specialties) is supplied with a pressure-sensitive adhesive for narrow mounting spaces.

2. Make sure these results are due to box leakages only, and that the contribution of I/O or power cables is below the prescribed level. Otherwise, they need to be sorted out first.

3. Compute the required SE in dB from

$$\begin{gathered}\text{SE(dB)} = \text{open box measurement} \\ - \text{specification limit} + \text{safety margin (typically 10 dB)}\end{gathered}$$

4. Check all seams and other potential leakage points, including permanent covers, access doors, hatches, maintenance panels, air filter flanges, shielded window edges, and so forth.

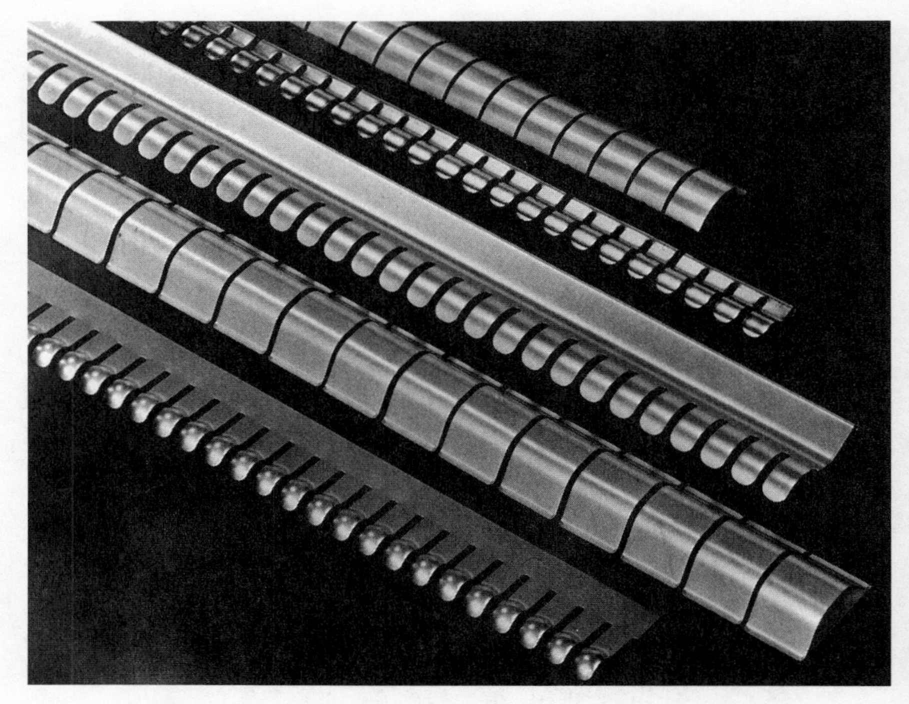

Figure 12.17 Fingerstock spring gasket. These provide 100 percent seam coverage, are very resilient, and offer self-cleaning contact to achieve >100 dB SE. *Courtesy of Instrument Specialties.*

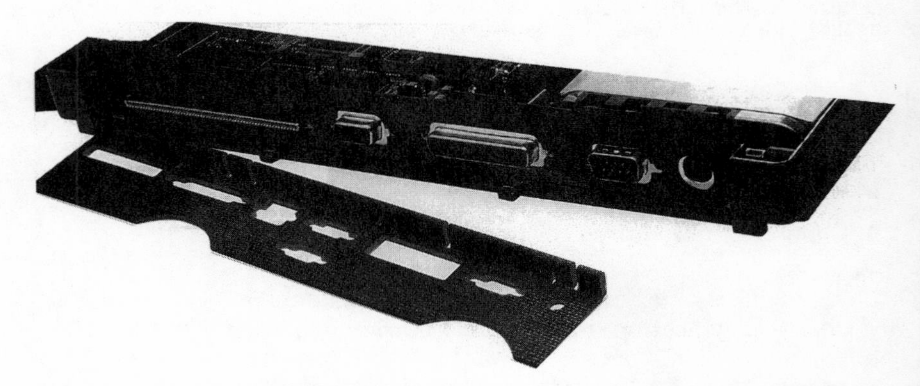

Figure 12.18 Soft-core EMC gasket. The base material is a spongy stock coated with metallized fabric. Performance reaches 60 dB for far-field, plane-wave conditions. It can be custom cut for specific applications. *Courtesy of Schlegel Corp.*

5. For screws/fasteners spacing, use Table 12.2 for a rough approximation. All spacings larger than table values are candidates for gasketing.

Table 12.2 Longest Acceptable Screw or Bond Point Spacing for Ungasketed Joints for a Particular Desired SE

SE	30 MHz	50 MHz	100 MHz	300 MHz	500 MHz	1 GHz
20 dB	30.0 cm	30.0 cm	30.0 cm	20.0 cm	13.0 cm	6.6 cm
40 dB	20.0 cm	13.0 cm	6.6 cm	2.0 cm	1.3 cm	*
60 dB	2.0 cm	1.3 cm	*	*	*	*

*Required spacing is too small to be practical. Gasket is needed.
Note: for F ≤ 100 MHz, although proportionally longer seams could be acceptable, a minimum default value of 30 cm has been assigned to take into account a realistic distance to the circuits behind the seam.

6. Check that the material of the box skin exceeds the desired SE. With metal boxes, this is generally not a problem. With metallized plastics, this is not always true and needs to be verified. Otherwise, you must upgrade the box skin SE first.
7. Using a feeler gauge, estimate the worst tolerance without a gasket for the seam between screws when they are normally tightened (or fasteners are latched). The worst warpage usually occurs midway between screws.
8. Select a gasket such that the tolerance stays within the compression range given (generally 25 percent of gasket diameter or height).
9. With elastomer gaskets, check that the volume resistivity of the gasket material is consistent with the desired SE. A maximum resistivity of 1 Ω·cm is acceptable for an SE objective of ≈ 30 dB overall. For higher SE, seek gasket resistivities much less than 1 Ω·cm.
10. Brush and clean the mating surfaces to remove paint, oxides, and grease.
11. Make sure that the screws and their threads are outboard from the joint, not inboard, as shown in Fig. 12.19.
12. Install the gasket and tighten the screws. If no grooves are provided, install some compression stops in the joints (metal bits or washers at screw locations) to avoid overcompression (Fig. 12.20).
13. Once the screws/fasteners are tightened, check that correct pressure exists all along the gasket. This can be done visually or by

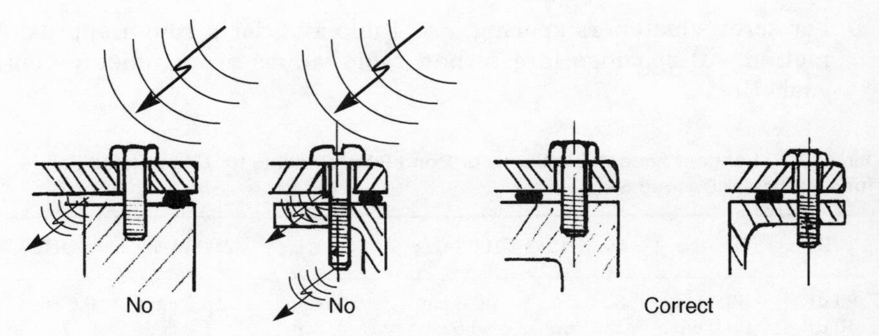

Figure 12.19 Make sure that screws and threads are outboard from the joint.

placing small paper strips between the gasket and the cover before closure, then checking to ensure that they resist a normal tug once the cover is closed.

Limitations

1. Flanges need to be wide enough to house the gasket, and stiff enough to apply an appropriate pressure without deforming.
2. Beware of corrosive/salt spray environments. This may call for an environmental gasket around the EMI gasket.
3. Beware of galvanic coupling between gasket material and box.

12.5 EMI Screens for Windows and Ventilation Panels

Knitted wire mesh provides an SE proportional to the mismatch between the EMI wavelength and the mesh grid size (Fig. 12.21). An approximate model for plane waves (far-field) is

$$SE_{dB} \approx 104 - 20 \log L_{mm} - 20 \log F_{MHz}$$

where L_{mm} = grid/mesh size in millimeters.

For instance, a 1 mm mesh size provides an intrinsic SE at 100 MHz equal to

$$104 - 20 \log 1 - 20 \log 100 = 104 - 0 - 40 = 64 \text{ dB}$$

Of course, as frequency decreases, the upper limit of the mesh SE is that of the base metal itself. In near fields, against H-field sources, the $SE_{(H)}$ becomes independent of frequency and can be approximated by

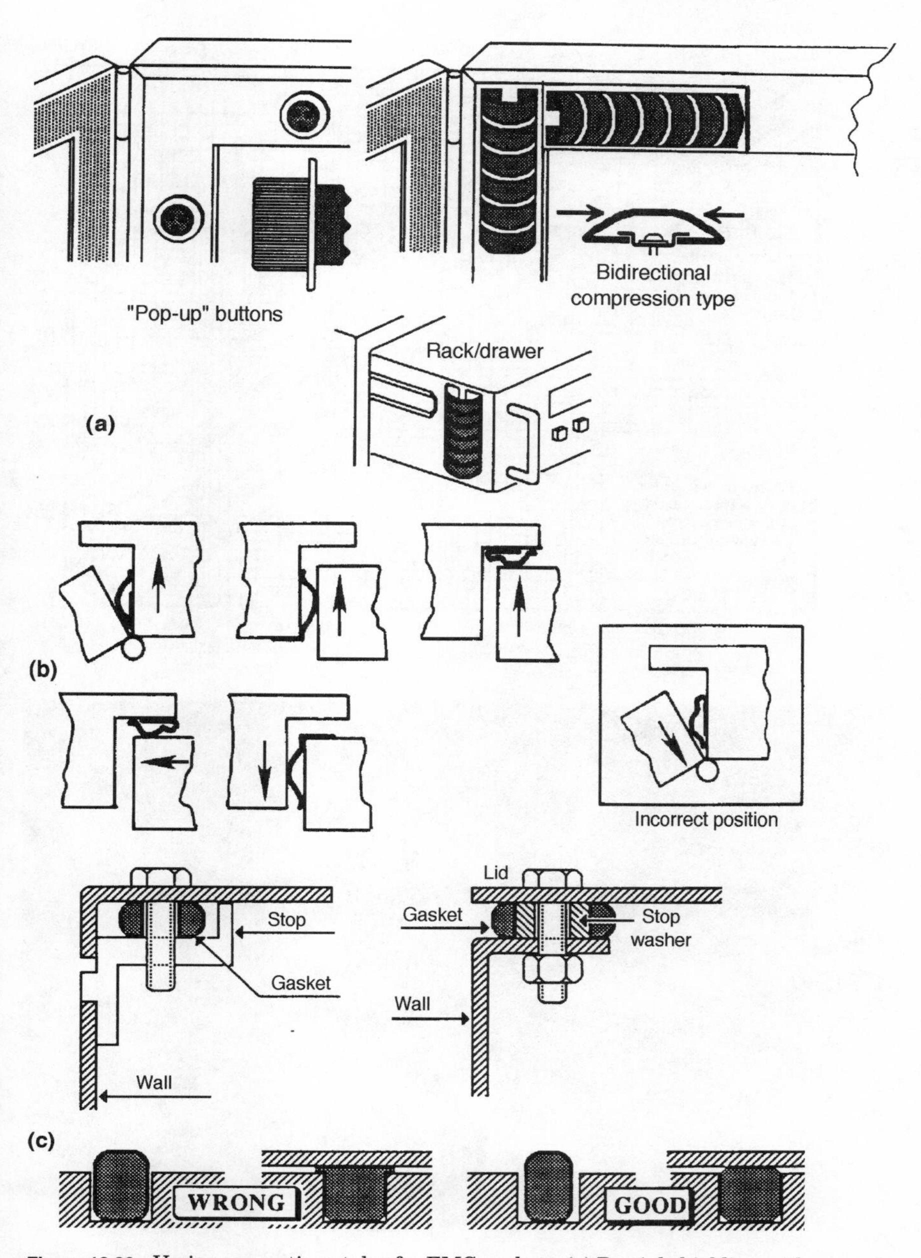

Figure 12.20 Various mounting styles for EMC gaskets. (a) Partial shielding with grounding buttons or bits of finger stock *(courtesy of AEMC)*. Although it provides less than 100 percent coverage, it still accommodates wide mechanical tolerances, making it an attractive solution for industrial electronics cabinets. (b) Fingerstocks offer a self-wiping action that tolerates perpendicular as well as lateral motion, provided the free play of the fingers is not blocked. (c) With deformable-core gaskets, compression stops must be provided.

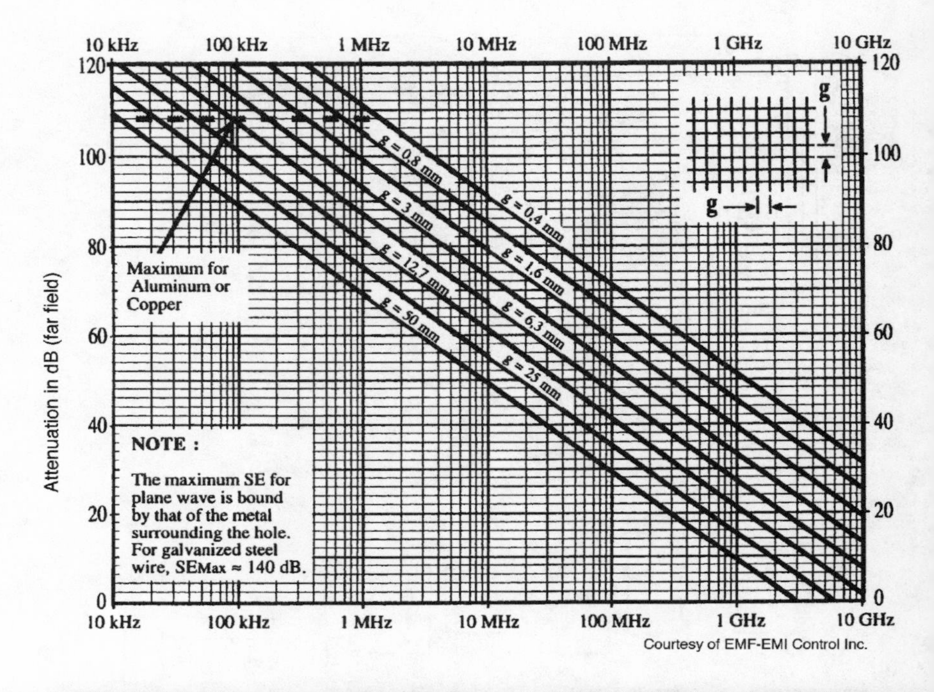

Figure 12.21 Top: shielding effectiveness of screen meshes for ventilations and windows in plane-wave conditions. Bottom: example of commercial shielded air filters with gasketed mounting flange. *Courtesy of Instrument Specialties.*

$$SE_{(H)\,dB} \approx 10 \log (\pi\, r/L)$$

where r = source-to-shield distance
L = grid size

with r and L in the same units.

Mesh can be blackened to eliminate light reflection.

Indications

1. The aperture screen is effective to shield an opening that has to remain transparent (instrument panel, display, etc.).
2. It also may be used to shield an air vent while maintaining some air flow (however, count on a significant aerodynamic resistance).

Prescriptions, Installation

1. Check that the screen meets or exceeds the minimum SE requirements for the whole box.
2. Brush, degrease, and clean the edges of the aperture to be shielded.
3. Install the screen using either (a) or (b) below:
   a. *The preexisting window frame or bezel.* Make sure the edges of the mesh are firmly pressed against the box surface. This generally requires a gasket, several thicknesses of conductive tape, or a sufficient number of screws.
   b. *A rim of conductive tape folded all around the mesh, then another layer of conductive tape to bond the mesh to the box.*

Limitations

1. For the thinner screens, the performance ceiling is 60 to 70 dB at 50 MHz and then rolls off at 20 dB/decade above that level.
2. A good seal with normal pressure is difficult to achieve, especially as a retrofit in the field. As a result, the in-site performance often depends on edge leakage, not mesh SE.
3. The mesh creates a significant airflow resistance.

12.6 Conductive Paint

Conductive paints are made of an acrylic or epoxy binder mixed with small particles of silver, copper, nickel, or graphite. Because their coating thickness, generally 25 to 50 μm (1 to 2 mil), is inferior to one skin depth for the frequency range of most EMI situations, conductive paints work mostly by reflection loss. Therefore, SE (Fig. 12.22) can be predicted by

$$SE_{dB} = 20\log\left(\frac{Z_w}{4 \times Z_b}\right)$$

where Z_w = wave impedance of incident field
Z_b = barrier impedance of conductive coating
≈ surface resistance, R_s, in Ω/sq

In the *far field* (Z_w = 377 Ω),

$$SE_{dB} \approx 20\log\frac{100}{R_s\ \Omega/\text{sq}}$$

In the *near field* (i.e., when source-to-shield distance, r, < λ/2π), the SE will depend on the nature of the field source (predominantly E or H).

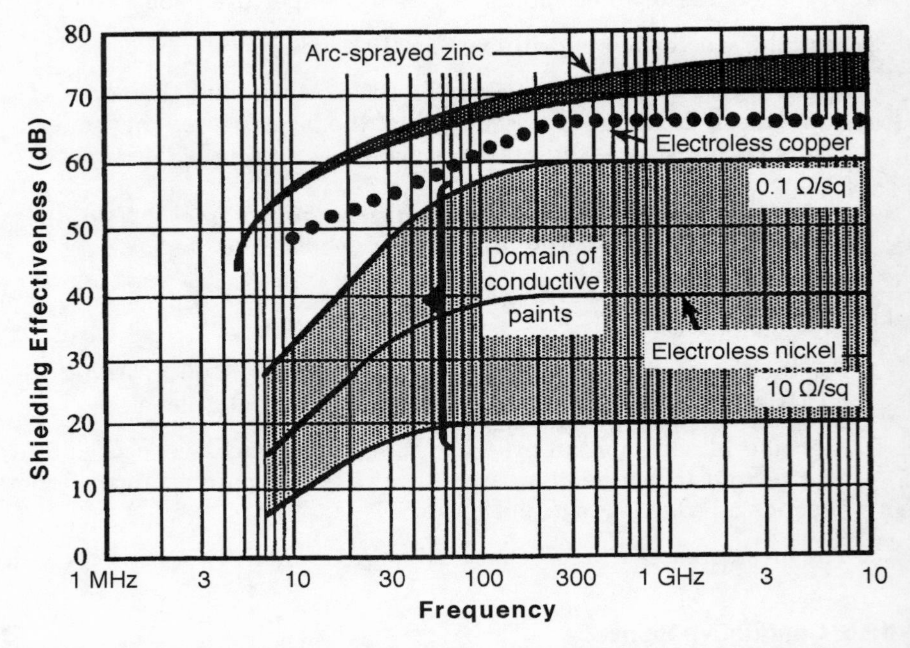

Figure 12.22 Shielding effectiveness of conductive coatings (by standard 30 cm distance test, near-field attenuation given against H-field). For paints, thickness typically is 2 mil (0.05 mm).

For a given distance, r, the boundary frequency below which the shield is in the near-field condition is

$$F_{near-far}\,(\text{MHz}) \approx 50/r\ (\text{meters})$$

Thus, if $F_{EMI} < F_{near-far}$, then

1. *Against H-field sources:*

$$SE_{(H)} = 20 \log(100 \times F_{EMI}/R_s \times F_{near-far})$$

2. *Against E-field sources:*

$$SE_{(E)} = 20 \log(100 \times F_{near-far}/R_s \times F_{EMI})$$

For example, assume a 2 MHz source, 0.3 m from the box:

$$F_{near-far} = 50/0.3 = 150\ \text{MHz}$$

The box is coated with conductive copper paint, providing a conductivity of 0.3 Ω/sq. Since 2 MHz < 150 MHz, we are in near-field conditions. Therefore, if the source is predominantly magnetic (low-impedance circuit with Z load << 377 Ω), then

$$SE = 20 \log(100 \times 2/0.3 \times 150) = 13\ \text{dB}$$

Indications

Conductive paint is useful for the following applications:

1. To create EMI shielding with a plastic housing
2. To improve the SE of an existing mediocre or degraded conductive surface
3. To protect from ESD or to prevent static charge buildup
4. To improve contact area for bonding or gasketing surfaces

Prescriptions, Installation

1. Approximate the SE range needed.
2. Verify that the conductive paint used has an SE figure that *exceeds* the need.
3. Brush and clean the area to be painted to remove dust, grease, etc.

4. Identify and mask with tape the zones that need to remain uncoated (for functional or aesthetic reasons).
5. Shake or stir the paint vigorously.
6. Spray evenly at the prescribed distance, making crossing patterns. (For obvious reasons, the coat is generally applied on the inside face of the boxes.) Do not overcoat a layer until it is at least touch dry. An irregular color is an indication of an uneven layer with "lean" pockets.
7. After a short drying time, quickly check the dc surface resistance (Table 12.3). Although inaccurate, a digital ohmmeter with regular probes firmly applied is an acceptable measuring instrument, provided the meter's zero point is carefully calibrated (with the probes shorted) to make variations of 0.1 Ω readable. Space the probes at intervals of a few centimeters, and do several spot checks over the painted surface. A too-high reading (for instance, several ohms for a copper paint) is a sign that the binder/filler mix was inhomogeneous (not shaken enough).
8. If there are EMI gasket resting zones or bonding areas for ground straps, filters, connectors, receptacles, and so forth, carefully inspect these to make sure they are correctly coated, without voids, cracks, or other imperfections. If necessary, reinforce locally with an extra spray.

Table 12.3 Selector Guide for EMC Coatings (Electrodag Coatings, by Acheson Corp.)

Product	Recommended substrates	Pigment	Carrier liquid	Sheet resistance (Ω/sq/25 μm)	Theoretical coverage (m^2/kg @ 10 μm)
Electrodag® 440AS	Plastic	Ni	Solvents	<0.50	≈17
Electrodag® 442	Metal	Ni	Solvents	<2.0	≈17
Electrodag® SP-010	Plastic	Ag-plated copper	Solvents	<0.125	≈15
Electrodag® SP-012	Plastic	Ag and Ni	Solvents	<0.025	≈11
Electrodag® 6030	Plastic	Ag	Water	<0.025	≈16
Electrodag® 1415M	Plastic	Ag	Solvents	<0.015	≈9

Limitations

1. Coatings have limited longevity (approximately five years) if the surface is exposed to harsh conditions (thermal shock, vibration, abrasion, and so on).
2. They are prone to SE degradation if scratching or other mechanical abuse occurs.
3. They display slight to severe degradation if exposed to salt spray (≈6 dB degradation for passivated copper paint).

12.7 Conductive Foil

Aluminum is a good conductor. A thin aluminum foil, about 25 μm (1 mil) thick, has no absorption loss until about 10 MHz, but it has good reflection loss at any frequency against E-fields. Such thin sheet is easy to tailor cut, form, and wrap.

Indications

Aluminum foil can be a useful fix in the following situations:

1. Against radiated emissions or susceptibility problems, provided the cause is not a low-frequency H-field source close to the shield
2. When the equipment does not have a metal housing and one wants to quickly evaluate the merits of a nonferrous shield
3. When a compartment shield needs to be quickly improvised around a selected component inside a box (however, see Fig. 12.23)
4. When an otherwise mediocre conductive coating needs to be totally or partially upgraded
5. When the shield needs to conform to complicated shapes that could not be quickly obtained from regular sheet metal (Fig. 12.24)
6. When a shield needs to be improvised around a cable harness
7. To create temporarily an RF equipotential plane underneath a system made of several cabinets and interconnecting cables
8. To improvise a ground plane over a workbench, table, or floor for FCC/MIL-STD evaluation on a non-EMC test site, or for ESD testing
9. To shield the walls of an ordinary room, which makes it a "poor man's Faraday cage," realizing that, due to installation constraints, openings, etc., one cannot expect more than 40 dB of

Figure 12.23 When radiating components (e.g., high-speed digital modules) need to be shielded on the circuit board itself, aluminum foil cannot provide sufficient attenuation of such near-field sources. As an alternative, tin-plated steel covers with fenced sides are available in many standard sizes and pin-grid patterns. *Courtesy of Leader Tech.*

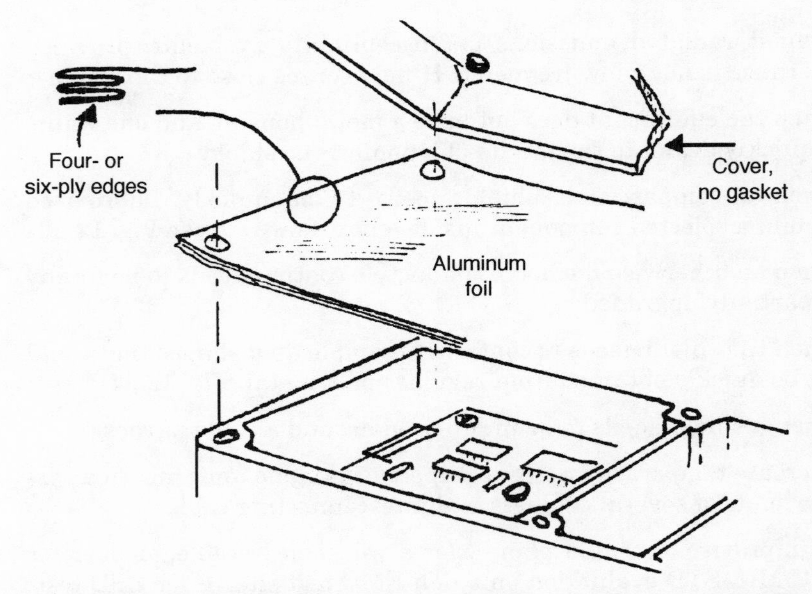

Figure 12.24 One example of conductive foil application. The electronic box needs a shielding cover, but the edges have a complex shape that would require a special die-cut gasket. A temporary shield is made by aluminum foil covering the top side, with edges (four or six plies) being pressed under the cover. There is no need for electrical contact between the cover and the foil—only the foil to box edges.

plane wave SE with such an installation (see also Sec. 12.8, Conductive Fabrics)

Prescriptions, Installation

In general, prefer the heavy-duty type of foil, with 50 μm (2 mil) or greater thickness.

1. First, prepare a template of the shield by cutting a piece of paper tailored to the shape of the box. Cut the aluminum foil to this template, extending about 1.25 cm (0.5 in) beyond the edges. Fold the excess edges to fit the exact contour while giving some strength to the borders.
2. If several side-by-side widths of foil are necessary, provide generous overlaps and apply conductive aluminum tape all over the seams.
3. Double or triple the foil thickness if more stiffness, mechanical strength, or performance is needed.
4. If used as a temporary cable shield, make sure that the wrapping is perfectly sealed at the seams with conductive tape. At the point of cable entry into the equipment box, flare the aluminum sleeve at the ends and bond tightly around the cable-entry hole using conductive tape. *Do not use jumpers.*

Limitations

1. Fragility: it is easily peeled, scratched, and torn, for thicknesses of ≤25 μm (1 mil).
2. Aluminum foil is useless against low-frequency H-fields or, in general extreme near fields.
3. With foil, it is difficult to achieve a low-resistance, durable contact with other materials.

12.8 Conductive Fabric

Metallized textiles are available that incorporate nickel, passivated copper, or other thin metallic fibers. Surface resistances of <1 Ω/sq are obtained, corresponding to ≥40 dB of shielding effectiveness in far-field conditions. Such material, being thin (0.1 mm thickness or less) and light (≤100 g/m^2) is relatively easy to apply on the walls of an ordinary room or to wrap around any three-dimensional shape.

Indications

Conductive fabric is indicated by any need for volumetric shielding with 30 to 50 dB attenuation from 100 kHz to GHz,

- To "faradize" an ordinary, unshielded facility
- To dress-up unshielded windows in a metallic shelter
- To rapidly erect a temporary shield (In less than a day, one can build a shielded hut of a respectable size.)
- To make RF protective clothing

Prescriptions, Installation

Other applications being rather straightforward, the following summarizes the basic steps for realizing a Faraday shielded room (Fig. 12.25):

1. Cover all the inside corners with ≈30 to 50 cm wide stripes. Provide generous overlaps at three-plane angles.
2. Cover all ceiling, walls, and pillar angles with 30 to 50 cm wide stripes.
3. Grind or otherwise abrade sharp protrusions and wall irregularities that could scratch the material.
4. Lay full-width fabric strips in the same manner as wallpaper. Provide ≈7.5 cm (3 in) overlap at the edges. Fabric can be glued or stapled over plasterboard or other existing wall material.
5. Cover, over the full length, all overlapping edges with copper or aluminum tape, with conductive adhesive backing.
6. For floor treatment, use floor gridding as described in Chap. 11 or use same textile or aluminum foil laid under the carpeting or tiles.
7. Treat carefully all doors, windows, and penetrations with mesh and gasketing as for any shielded box.

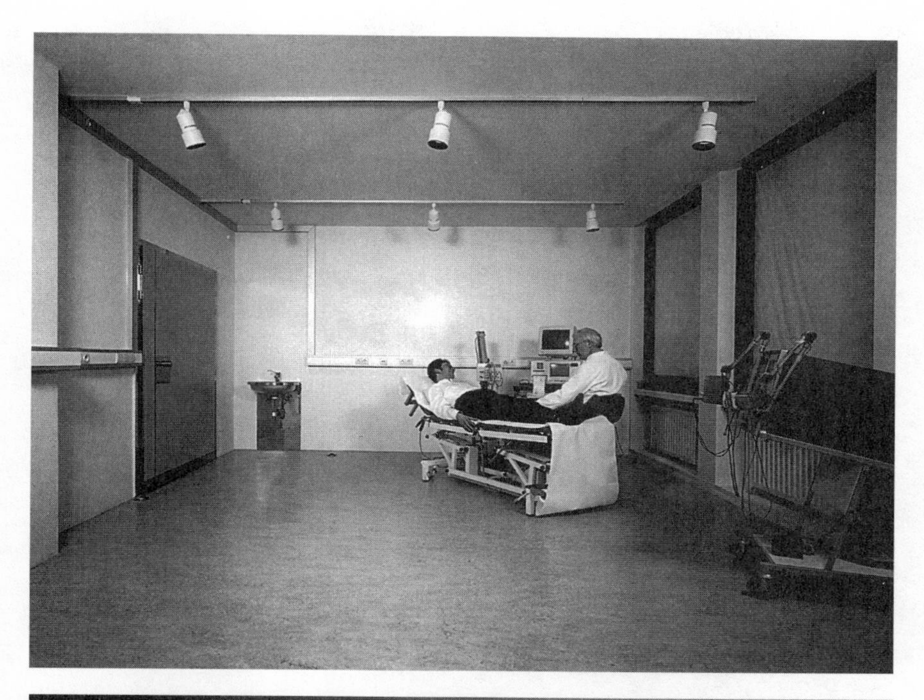

Figure 12.25 Shielding a room with conductive fabric: (top) before and (bottom) after. *Courtesy of Schlegel Corp., Belgium.*

Representative Vendors of Commercial Shielding Products

Conductive Tapes

3M Electrical Products Division
6801 River Place Blvd.
Austin, TX 78726-9000
Tel.: (512) 984-2666
Fax: (512) 984-2211
www.3m.com/elpd

Chomerics
77 Dragon Court
Woburn, MA 01888-4014
Tel.: (781) 935-4850
Fax: (781) 933-4318
www.chomerics.com

Mesh Bands and Zippers

Chomerics
(see above)

Tecknit
129 Dermody St.
Cranford, NJ 07016
Tel.: (908) 272-5500
Fax: (908) 272-2741

The Zippertubing Co.
13000 S. Broadway
P.O. Box 61129
Los Angeles, CA 90061
Tel.: (310) 527-0488
Fax: (310) 767-1714
www.zippertubing.com

Gaskets and Fingers

Instrument Specialties
One Shielding Way
Delaware Water Gap, PA 18327
Tel.: (717) 424-8510
Fax: (717) 424-6213
www.instrumentspecialties.com

Chomerics
(see above)

Tecknit
(see above)

Schlegel Systems Inc.
1555 Jefferson Rd.
Rochester, NY14623
Tel.: (716) 427-7200
Fax: (716) 427-9993
www.schlegel.com

Schlegel, Europe
Rochester Laan
8470 Gistel
Belgium
Tel.: (32) 59 27 03 56
Fax: (32) 59 27 03 45

Jacques Dubois
82 r. A. Badin
76360 Barentin
France
Tel.: 33 (0) 2 35 92 32 21
Fax: 33 (0) 2 35 91 42 94

Screened Windows and Vents

Chomerics
(see above)

Tecknit
(see above)

Dontech
700 Airport Blvd.
Doylestown, PA 18901
Tel.: (215) 348-5010
Fax: (215) 348-9959
www.dontechinc.com

PCB Component Shields

Instrument Specialties
(see above)

Leader Tech
14100 McCormick Dr.
Tampa, FL 33626
Tel.: (813) 855-6921
Fax: (813) 855-3291
www.leadertechinc.com

Connor Formed Metal Products
275 Shoreline Dr., Ste. 530
Redwood City, CA 94065
Tel.: (650) 591-2026
Fax: (650) 620-0850
www.cfmp.com

CONDUCTIVE PAINT

Acheson Colloids Co.
1600 Washington Ave.
P.O. Box 611747
Port Huron, MI 48061-1747
Tel.: (810) 984-5581
Fax: (810) 984-1446
www.achesonindustries.com

Spraylat Corp.
716 S. Columbus Ave.
Mt. Vernon, NY 100550
Tel.: (914) 699-3030
Fax: (914) 699-3035

CONDUCTIVE FABRICS

Schlegel
(see above)

Chapter

13

EMI Problems During EMC Tests: Practical Hints

EMC measurements themselves are not exempt from EMI problems. No matter if you are making a development, formal, or field type of measurement, it is not uncommon to spend a few hours to half a day on chasing EMI problems within the test setup itself. This section describes the frequent pitfalls and causes of errors in EMC testing.

13.1 Validation of the Measurements

EMC tests, especially in non-ideal conditions, are prone to many errors and mistakes. Always be your own devil's advocate, and critique your measurement raw data while you are on site. (Back at your desk, when formatting data for a final report, it will be too late for a second check.)

- Did we really measure what we pretended to measure?
- Did we check that the ambient noise (EUT *off*) was at least 6 dB below the limit or criteria (E- or H-field, voltage, current, etc.) we are checking against?
- Did the radiated emissions that seem to correlate with EUT *on* really emanate from the EUT? (This can be checked with small sniffer probe.)
- Are the measured levels, after applying the proper transducer factor (antenna, loop, current probe, etc.) in the likely range? (A single faulty connector can skew all results.)
- Is the dynamic range sufficient for our expectations?

Modern automated instrumentation tends to prevent such errors, but a good EMC engineer or technician never blindly trusts automated acquisition/analysis: A gross hardware mistake can be *undetected and the wrong data displayed, even by automation software with superb accuracy and reproducibility.*

13.2 Instrument-Related Errors

Some measurement errors are inherent to the instrument and its settings. This is especially true with spectrum analysers. The following is a list of common instrument-related errors. Those set in italic are exacerbated by on-site, non-ideal conditions.

1. Improper bandwidth selection when changing frequency bands
2. *Receiver front-end overload by out-of-band (nonvisible) signals, including 50 Hz components with conducted emission test*
3. Improper detector selection
4. Too fast a scan-rate (normally impossible with modern analyzers)
5. Video filter affecting readout (inaccurate amplitude)
6. Vertical linearity affected:
   - Always check linearity by switching attenuator steps.
   - Always work with a sufficient reserve below top reference.
7. *EMI signal and ambient signals adding up in RBW*
8. Broadband EMI contents "undercarrying" the narrowband EMI signal, combining to produce an overestimate of the amplitude
9. *Insufficient sensitivity with regard to ambient background* and sensors being used (the opposite of problem No. 6)

All these mistakes should be reviewed and quantified by an error analysis.

13.3 Problems with Weak Signals and Strong Background Noise

Strong background noise, or insufficient sensitivity of the measurement chain, results in a measured EMI level that either:

- barely emerges over a broadband clutter (incoherent addition of background + signal) or
- rides over an existing narrowband ambient (≈ coherent addition).

In either case, if the EMI signal does not exceed the ambient by at least 12 dB, a pessimistic error (≥0.3 dB for incoherent, 2.5 dB for coherent) is introduced. Figure 13.1 shows correction curves to apply in either situation, with a real case example (Fig. 13.2).

13.4 Setup- and Accessory-Related Errors

When radiated fields are measured, either for EUT or ambient assessment, one frequent cause of error is the parasitic pickup and asymmetric conditions of the dipole, biconical, or log-periodic antenna. For instance (Fig. 13.3), the capture area of the single braid RG58 coaxial is causing ≈ 1dB pessimistic error. This can be reduced by bringing the antenna cable down, close to the ground level, and using a better coaxial cable.

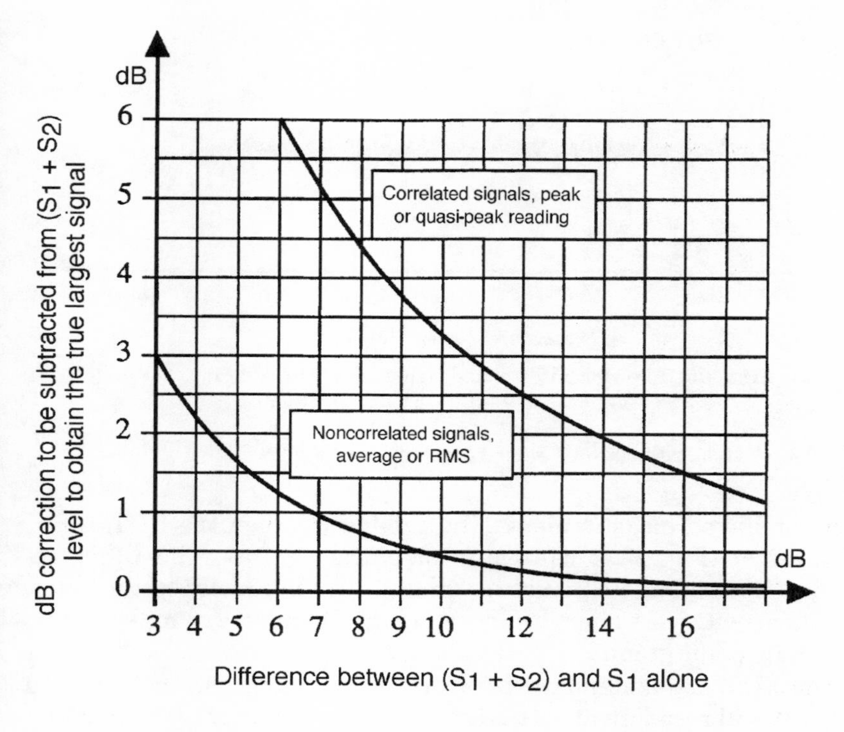

Correlated signals: $S_{tot} = S_1 + S_2$

Uncorrelated signals: $S_{tot} = \sqrt{S_1^2 + S_2^2}$

Figure 13.1 Combined readings of two signals.

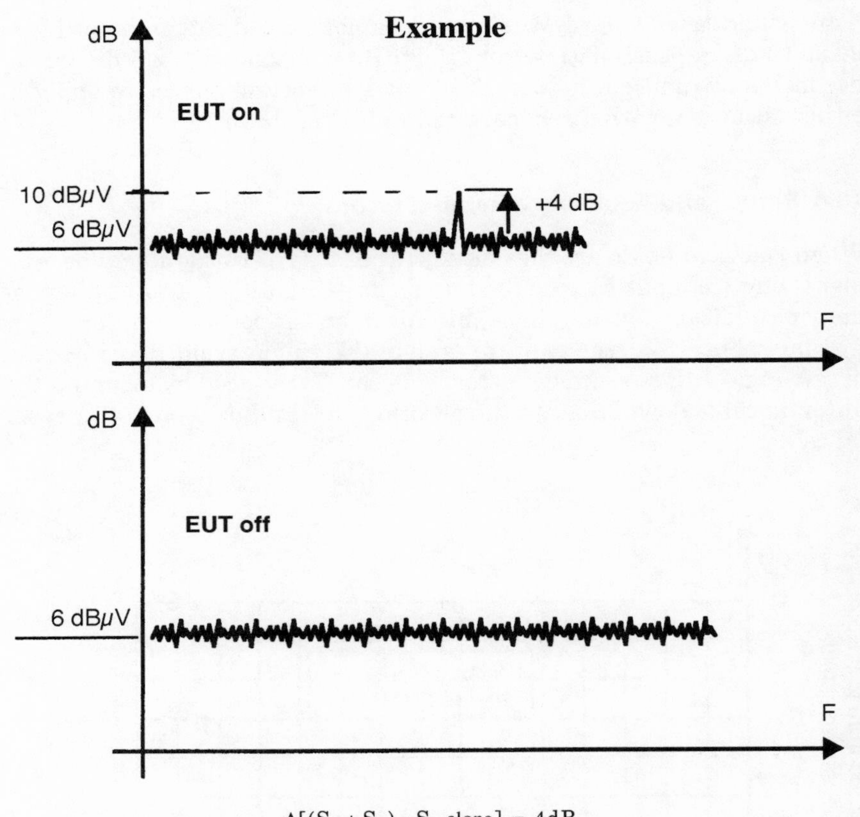

$$\Delta[(S_1 + S_2) - S_1 \text{ alone}] = 4\text{dB}$$

From the curve, for uncorrelated signals, the error = 2 dB, so
S2 (unknown) = 10 dBμV – 2 dB = 8 dBμV

Figure 13.2 Dealing with low-level signals, near sensitivity level.

For conducted measurements, the cabling between the EUT, LISN, and receiver (Fig. 13.4) can cause significant errors. The LISN, receiver, preamplifier, etc., must be grounded individually to the reference plane, with very short braids or straps. Do not try to make a star-type single-point ground.

If the EUT has a standard power cord (including hot, neutral, and safety), it will resonate at certain frequencies (typically, 10 to 30 MHz) with the EUT chassis-to-ground capacitance, C_p. To maintain reproducibility, always use the same, minimum length of power cord at a constant height.

Figure 13.4 recapitulates some other classic pitfalls of a conducted (or radiated substitution) EMC test setup.

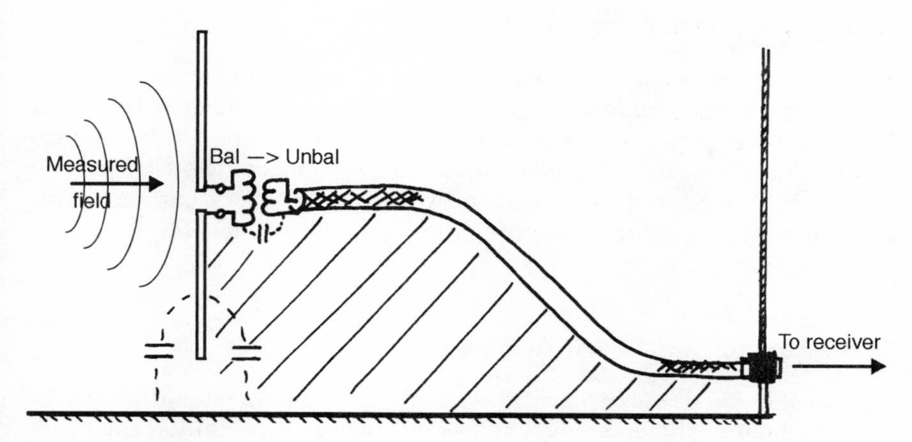

Figure 13.3 Inherent error caused by the asymmetrical pickup loop formed by the coaxial cable, the lower arm of the balanced dipole, and the ground plane.

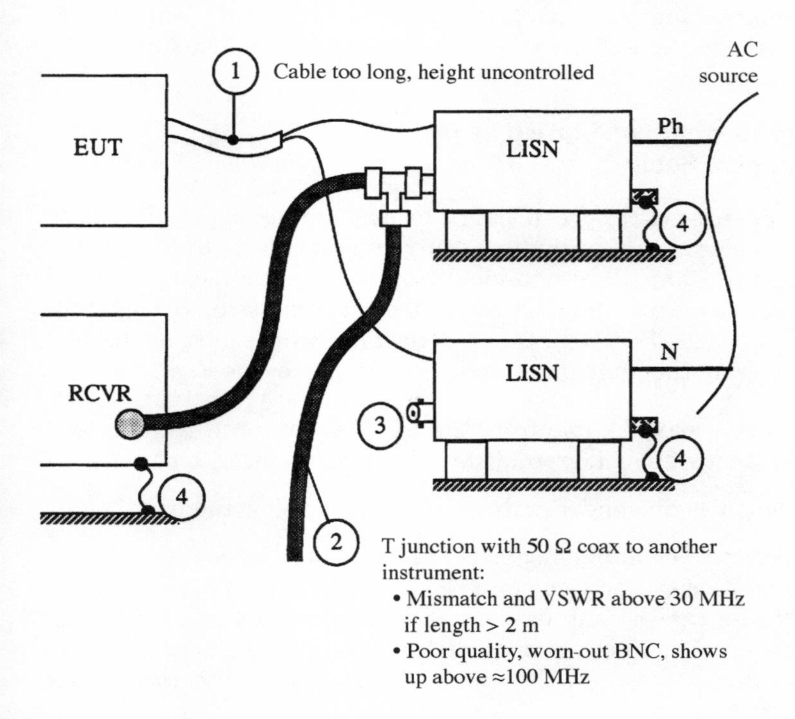

(3) Open-ended coaxial port (not loaded by 50 Ω)

(4) Too-long grounding straps, causing common impedance errors

Figure 13.4 Typical errors in LISN test setup.

13.5 LISN-Related Errors

Some commercial LISNs, due to internal construction elements, show a very irregular profile of impedance vs. frequency above ≈10 MHz (Fig. 13.5). Before acquiring a LISN, or when using an unknown model, it is prudent to check the impedance seen from the EUT power port when the RF output is loaded by 50 Ω. This check can easily be made using a spectrum analyzer plus a tracking generator, or a network analyzer.

13.6 Mismatch and VSWR-Related Errors

When cable lengths approach or exceed a quarter wavelength, significant (and not easily detected) errors may appear, especially if the measuring instrument falls into a null of the standing waves' distribution. Figure 13.6 shows the risks associated with T junctions and other cascaded mismatches if the derivation branch is sufficiently long. Figure 13.7 shows the use of resistive dividers to avoid impedance mismatch.

13.7 Background Noise Carried by the Instrumentation Setup

This type of noise pickup can be quite misleading, especially when measuring emissions (e.g., conducted or radiated substitution methods as discussed in Chap. 3) in non-ideal test sites. If it appears that the background noise, with the EUT *off*, is above the criterion for a decent test credibility (see Sec. 3.2), try to determine which part of the test setup is bringing this unwanted noise. For instance (see Fig. 13.8),

1. Remove the coaxial cable from the LISN output and terminate it by a 50 Ω coaxial load, grounded to the bench reference plane.
2. Does the noise disappear or drop sufficiently below the limit?
3. If the answer is *no*, this means that some CM noise is picked up by the coaxial + receiver + ac mains loop. Check and shorten the routing of these cables, making sure that they are resting completely on the ground plane, far enough from the edges. Install ferrites (two or three turns) on the coaxial cable and on the receiver ac power cord.
4. If answer is *yes*, it means that the noise is coupling via the ac mains down to the EUT and auxiliary equipments. Reconnect the coaxial cable. Disconnect all the EUT external cables (except the power cord).

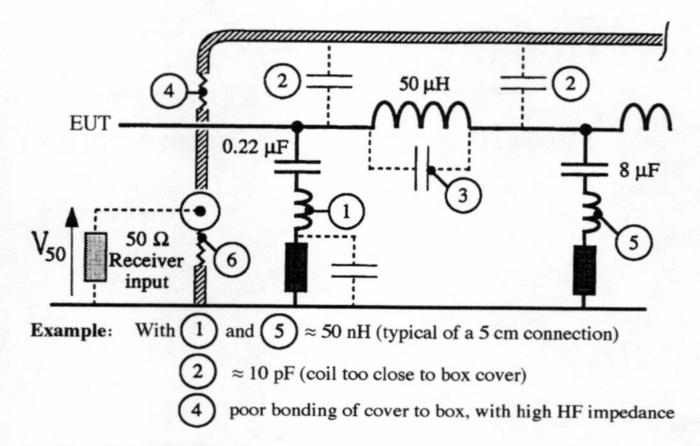

F_{MHz}	50	100	150	300
$\lvert Z_1 \rvert$	16 Ω	32 Ω	50 Ω	100 Ω
$\lvert Z_2 \rvert$	300 Ω	150 Ω	100 Ω	50 Ω
Error $V_{(50)}$ vs V_{actual}	- 1.8 dB	- 4 dB	- 7 dB	- 10 dB

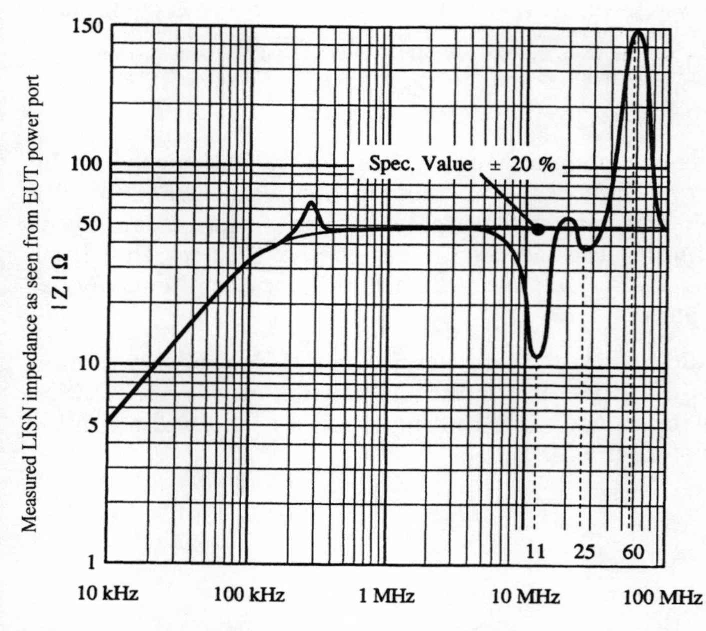

Figure 13.5 Construction defects of some commercial LISNs, and deviation from the standard impedance curve.

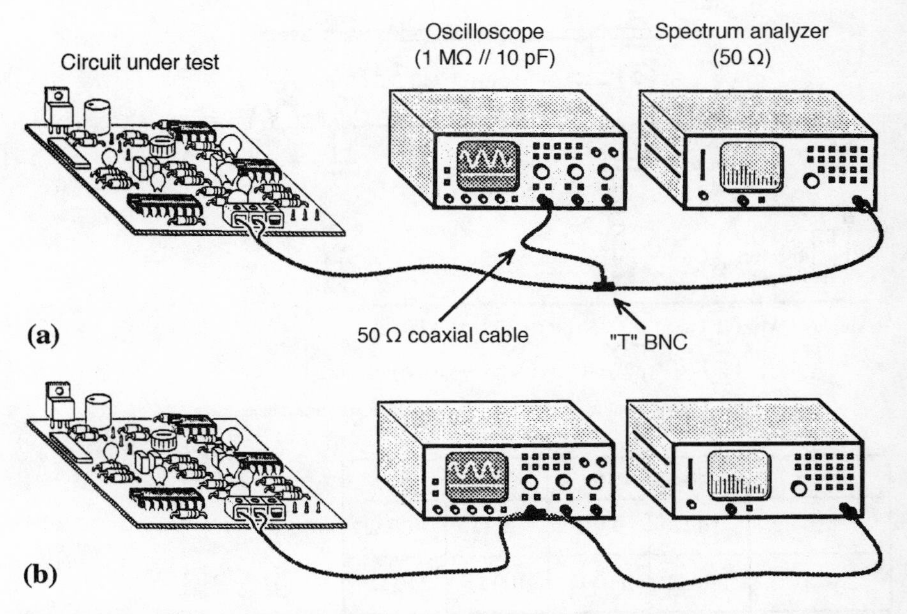

Figure 13.6 Errors related to mismatch and voltage standing wave ratio (VSWR). (a) Three 50 Ω cables, 2 m long. The mismatch caused by the T branching (50 Ω in parallel on the PCB-to-analyzer link) will cause a sizable VSWR error when the line length reaches 1/10 of the wavelength. At 7 MHz, the error exceeds 10 percent. (b) Improvement: the T BNC has been connected directly to the high-impedance input of the oscilloscope. The setup now can be used up to ≈150 MHz, where the 12 pF oscilloscope input will start to be a significant shunt load (12 pF ≈ 100 Ω @ 150 MHz). *Courtesy of AEMC.*

5. Does the noise drop significantly? If so, the EMI is picked up by the auxiliary equipments, even though they were turned off. Try filtering their power input or supplying them from a clean power source—preferably the same source as the EUT. Check that their I/O cables are laid closely over the ground plane, without any external loose path.

6. If the noise did not drop at question 5, the EMI is probably coming from the mains and is not sufficiently attenuated by the LISN. Check and shorten the cable routing upstream and add a T filter (C + L + C) before the LISN.

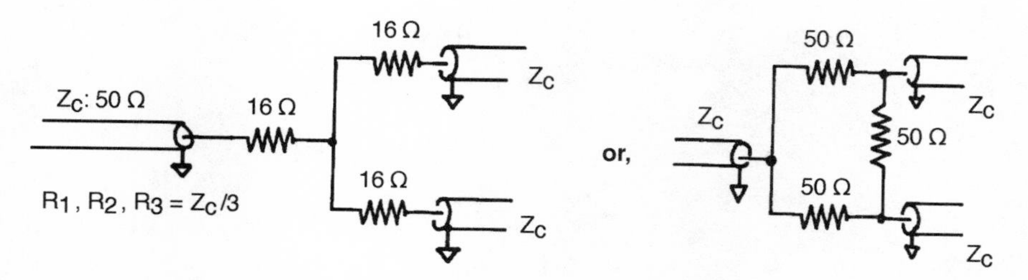

"Y" branching: one- to two-way splitter, 6 dB conversion loss, usable up to gigahertz range.

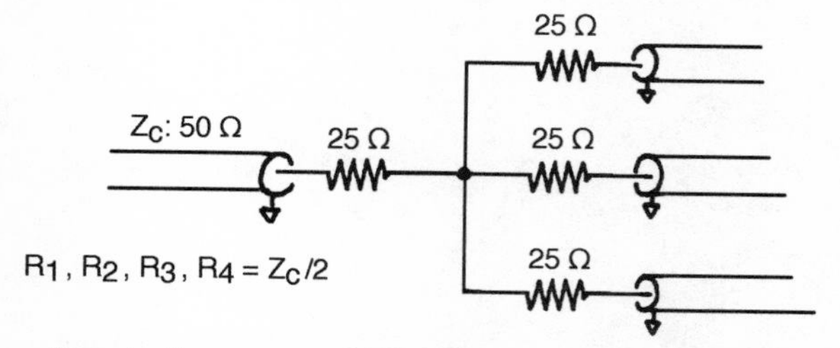

One- to three-way splitter, 9.5 dB conversion loss

Figure 13.7 Simple resistive dividers to avoid impedance mismatch when branching several instruments in parallel.

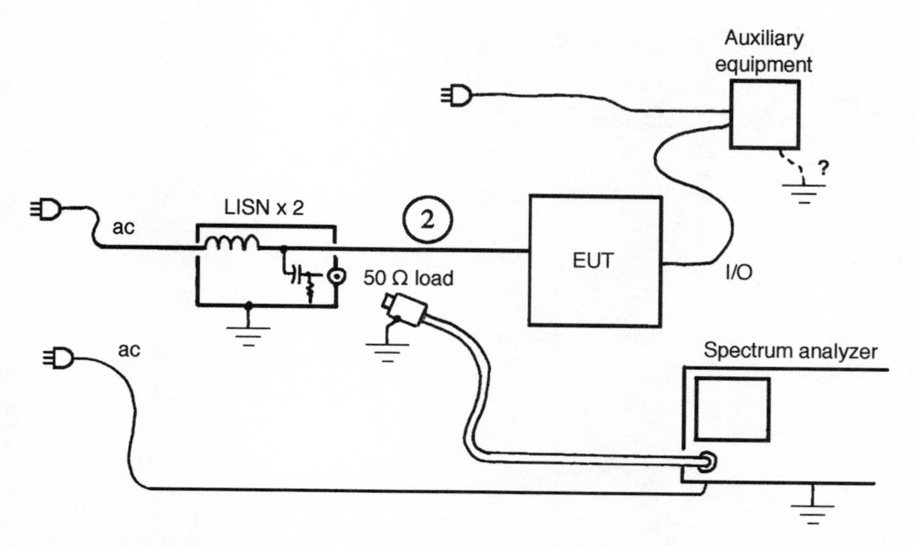

Figure 13.8 Dealing with ambient noise during EMI emission measurements.

Appendix

A

Receiver Sensitivity versus Bandwidth and Noise Figure

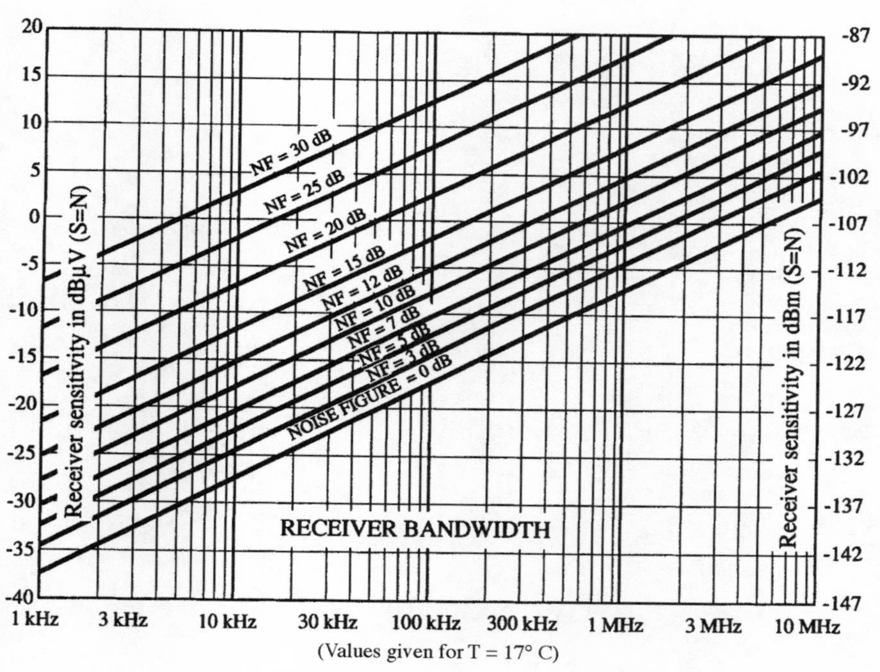

Courtesy of EMF-EMI Control

Application Examples

Question: A receiver with a 12 dB noise figure is set on 100 kHz bandwidth. What is its sensitivity?

Answer: The 100 kHz vertical line intersects the 12 dB noise figure for –5 dBμV (left scale) or –112 dBm (right scale). This "smallest discernible signal" (S = N) will emerge at 3 dB above the receiver noise because

$$S + N = \sqrt{S^2 + N^2} = N \cdot \sqrt{2}$$

Question: A spectrum analyzer exhibits a background noise of –110 dBm for a 9 kHz IF bandwidth. What is its noise figure?

Answer: The –110 dBm (i.e., –110 + 107 = –3 dBμV) line intersects the 9 kHz vertical line for F = 25 dB.

Appendix

B

EMI from Switched-Mode Power Converters

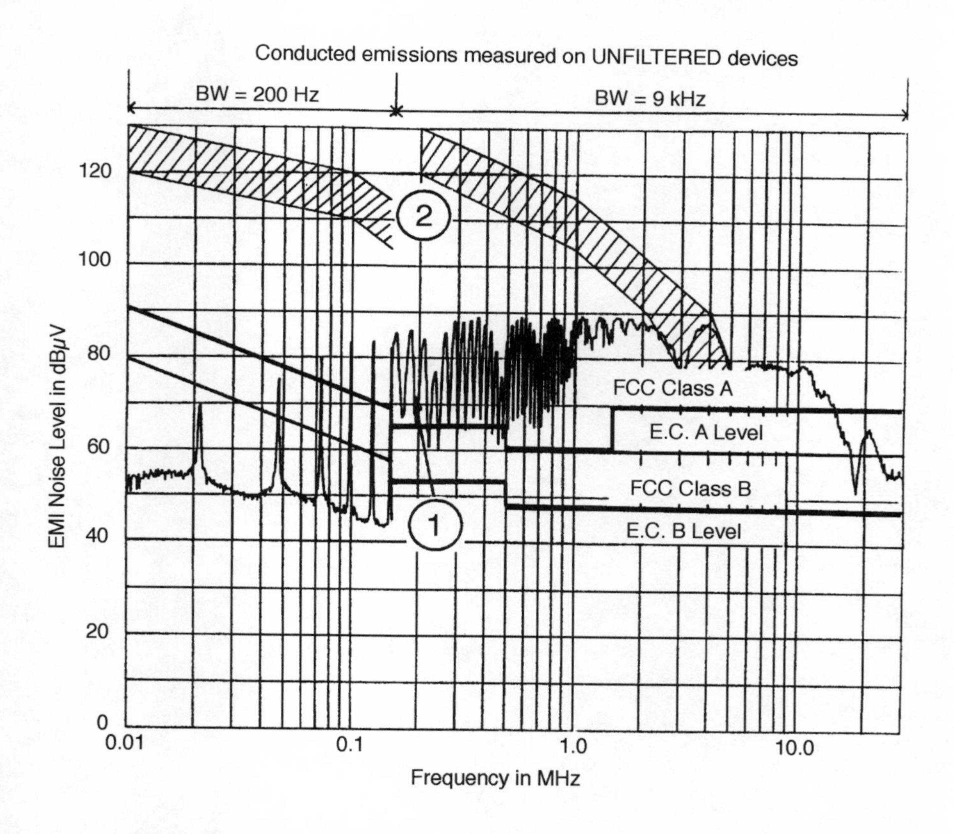

B.1 Conducted EMI (CM and DM) into 50 Ω/ 50 μH LISN (or Translated Equivalent) with CISPR Receiver

1. 500 VA, 20 kHz switch-mode power supply (typical)
2. Merged envelope of six variable speed drives, 30 kVA to 600 kVA switching rate is in the kHz range. Therefore, EMI appears as

a. NB in the 200 Hz BW (receiver BW is $<$ EMI rep. frequency)
b. BB in the 9 kHz BW (receiver BW is $>$ EMI rep. frequency)

Note: Switch-mode power converters and speed drives are very common sources of EMI problems in commercial and industrial installations.

Appendix

C

Ambient Fields from Radio Transmitters

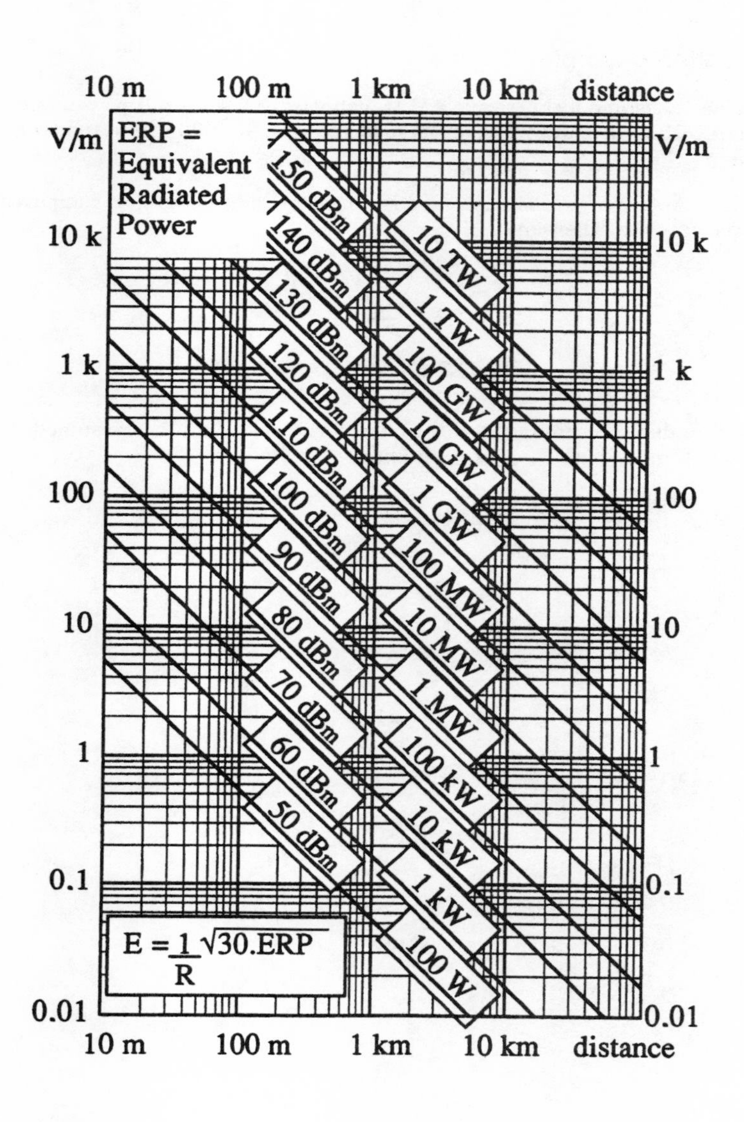

- Equivalent radiated power = transmitter output power (W) × numerical antenna gain.
- Conditions are assumed as free space, lossless propagation (dry air) and far field [distance (m) > $48/F_{MHz}$).
- In actual conditions, the local field could be 3 to 6 dB stronger than the above values in the case of

  a. proximity to conductive objects (ground, walls, etc.)
  b. high percentage of AM content.

Application Example

Question: Assume a shortwave AM transmitter, 6 kW output power, and a whip antenna with 2 dB gain. What is the field received at 100 m from the transmitter antenna?

Answer: Antenna gain is expressed in terms of radiated power, compared to an isotropic antenna. Therefore,

$$G(\text{num}) = (10)^{\frac{2\text{ dB}}{10}} = 1.6$$

$$\text{ERP} = 6\text{ kW} \times 1.6 = 10\text{ kW}$$

The 100 m distance intersects the 10 kW curve at 5 V/m. It is assumed that far-field conditions ($d > 48/F_{MHz}$) are satisfied.

Appendix

D

Impedance of Copper and Steel Planes

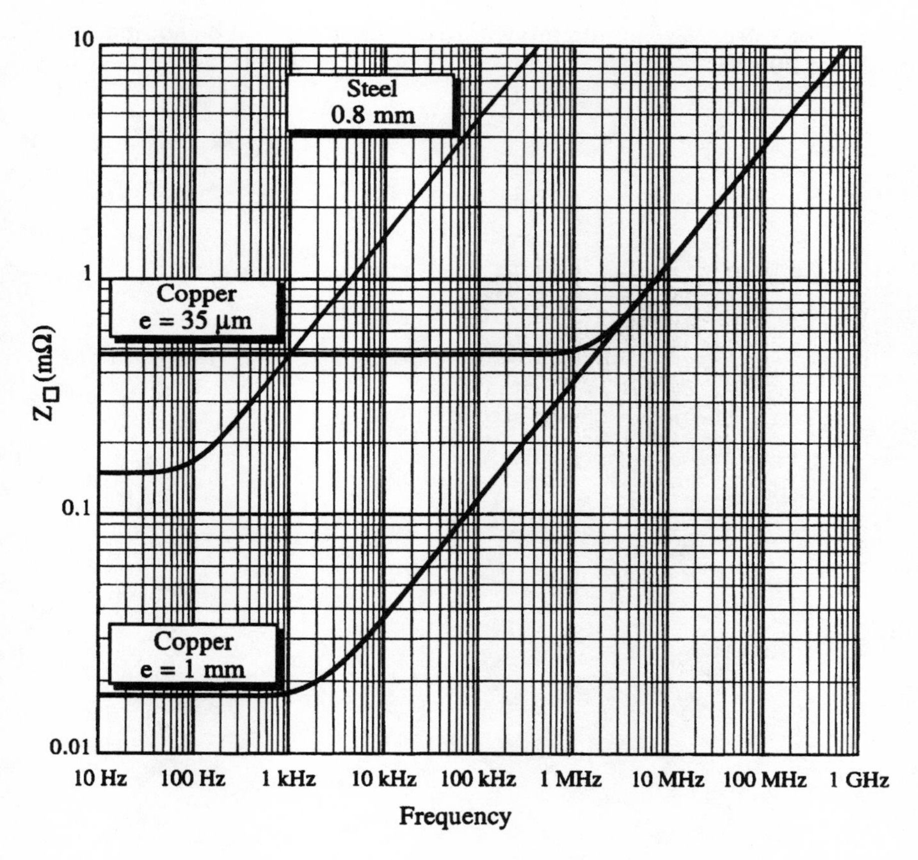

$$\text{DC and LF:} \quad Z/\text{sq}\ (\mu\Omega) = \frac{17.2}{e(\text{mm}) \times \sigma_r}$$

$$\text{HF:} \quad Z/\text{sq}\ (\mu\Omega) = 370 \times \sqrt{\frac{F(\text{MHz})\mu_r}{\sigma_r}}$$

where σ_r, μ_r = relative conductivity and permeability, as referred to copper

Application Example

A 1 mm thick copper sheet, for 300 kHz, shows an impedance of 0.2 mΩ/sq. Provided it takes place far from the sheet edges, a uniform flow of 100 mA HF current will cause a ground reference noise of

$$100 \times 10^{-3}\ \text{A} \times 0.2 \times 10^{-3}\ \Omega = 20 \times 10^{-6}\ \text{V or } 20\ \mu\text{V/sq area}$$

In other words, two points anywhere on the plane will be equipotential within 20 µV.

Appendix

E Voltage Induced by a Field into a Loop

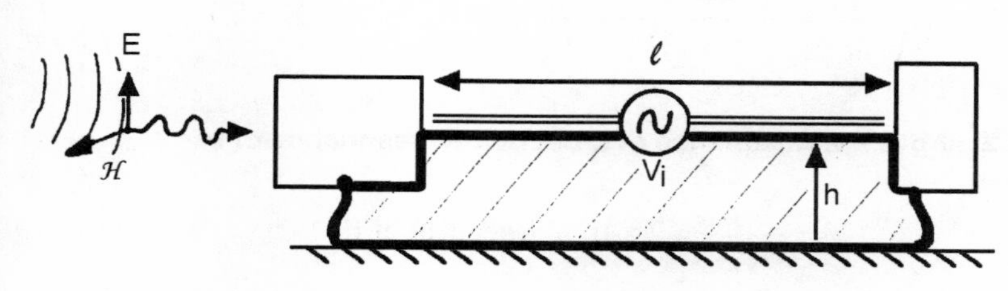

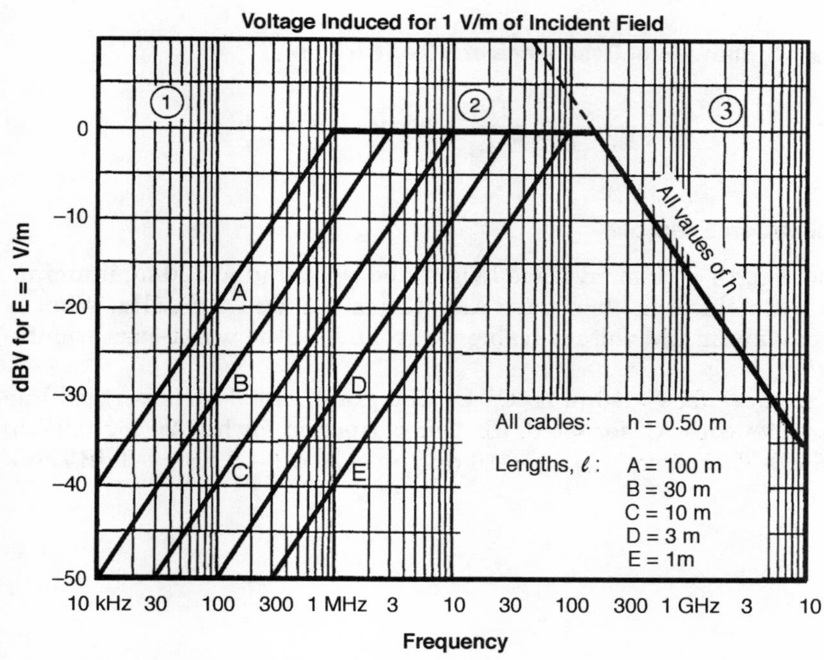

Correction for Other Values of h:

If h ≠ 0.50 m (a typical height, chosen as the default value), the only correction to apply in regions 1 and 2 is as follows:

$$\text{dBV (curve)} + 20\log\frac{h(m)}{0.5}$$

V_i is given for worst-case polarization. Dimension ℓ is assumed to be the largest loop dimension (i.e., $\ell >$ height).

The graph can be used for estimating open-circuit voltage, V_i, in any loop. However, a frequent application is for calculating CM voltage into ground loops.

Curves for $V_{i(max)}$ have been plotted from the following models.*

Region 1, below First Resonance of Cable Length

$$\frac{V_{max(V)}}{E_{V/m}} = \frac{\ell_m \times h_m \times F_{MHz}}{48} \text{ (exact up to} \approx F = 75/\ell)$$

Region 2, above λ/2 Resonance of ℓ, but below Resonance of h

$$\frac{V_{max}}{E_{V/m}} \cong 2 \times h, \text{ independent of F}$$

Region 3, above First Resonance of h (i.e., F ≥ 75/h)

$$\frac{V_{max}}{E_{V/m}} = \frac{150}{F_{MHz}}, \text{ independent of } \ell, h$$

Application Example

Cable length = 10 m, average height above ground = 0.20 m, illumination = E-field from shortwave AM station = 3 V/m @ 3 MHz. What is the maximum CM voltage (cable vs. ground) in the worst-case orientation?

The curve for $\ell = 10$ m is "C." On the graph, the 3 MHz vertical line intersects with "C" for ≈ –10 dB. The correction for h is 20 log(0.2/0.5) = –8 dB. Therefore, $V_{induced}$ = 10 dBV/m – 10 dB – 8 dB = –8 dBV or ≈ 0.4 V.

* White, Donald R.J., *EMI Control Methodology and Procedures,* Interference Control Technologies, Inc.

Appendix

Wire-to-Wire Capacitance and Mutual Inductance for Crosstalk Estimation

F.1 Wire-to-Wire Capacitance for Crosstalk Estimation

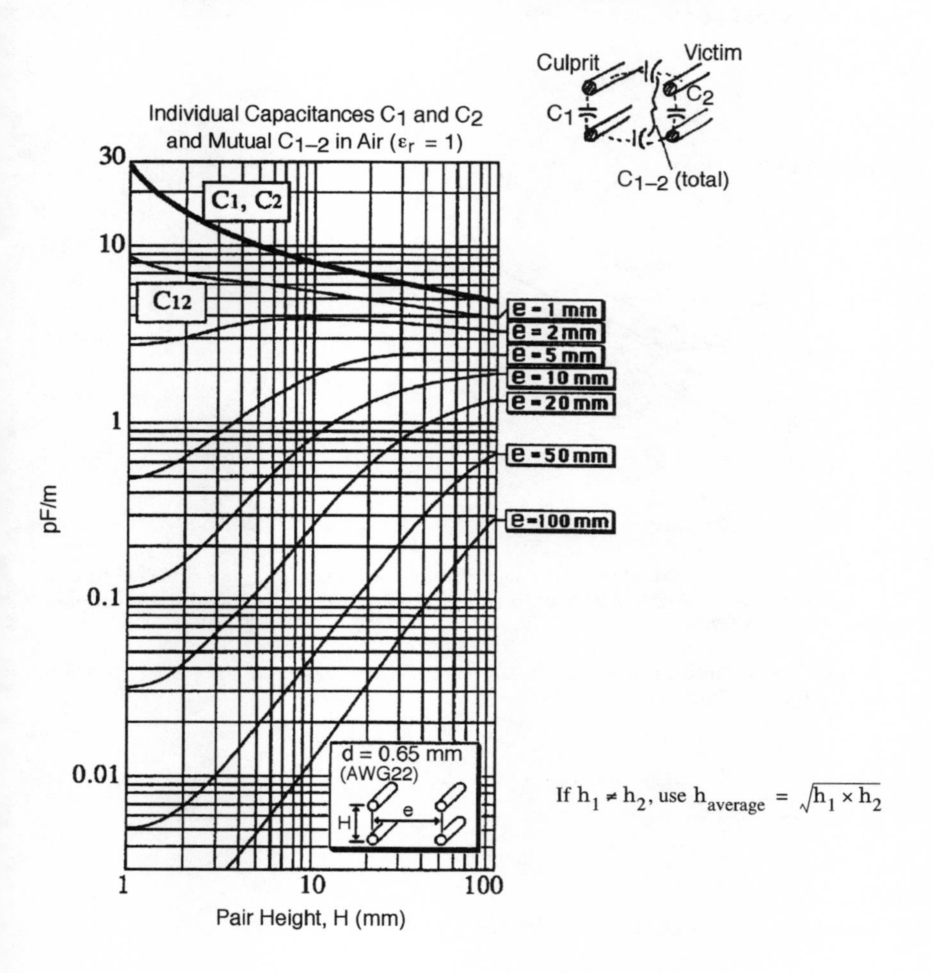

- If culprit + victim wires are *totally embedded* in a dielectric, the values of C must be adjusted for ε_r as follows:

Dielectric	ε_r
Polyethylene	2.3
PVC and nylon	3.5
Teflon®	2.2

- For wire size AWG22, C_{1-2} does not differ significantly from the curves while e, h ≥ 3 × d.

Correction for Length and ε_r

$$C_{1-2} = C_{1-2(\text{curves})} + 20\log \ell_m + 20\log \varepsilon_r$$

Application Example

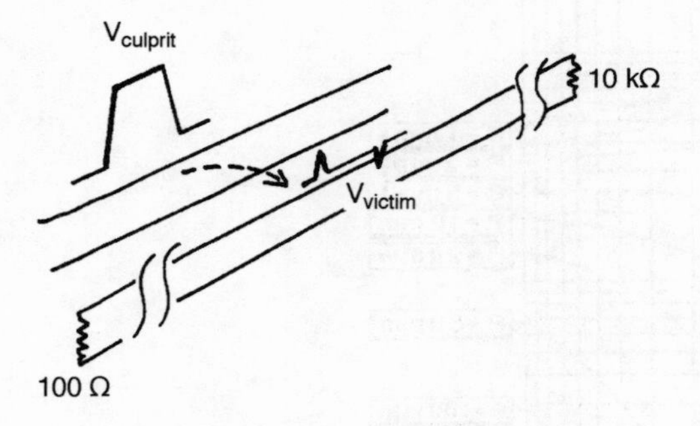

Question: In the figure above, $h_1 = h_2 = 3$ mm, e = 5 mm, $V_{culprit} = 5$ V, rise time = 20 ns, and parallel length = 2 m with coupling in air. What is the voltage coupled on the victim pair?

Answer: Total victim impedance = 100 Ω//10 kΩ ≈ 100 Ω. $C_{1-2} = 0.9$ pF/m (from curve) × 2 m = 1.8 pF.

$$V_{\text{victim (time domain)}} \cong R\,C\,\frac{\Delta V}{\Delta t}$$

$$= 100 \times 1.8{\times}10^{-12} \times \frac{5\text{ V}}{20{\times}10^{-9}}$$

$$= 45\text{ mV}$$

F.2 Wire-to-Wire Mutual Inductance for Magnetic Crosstalk Estimation *(Courtesy of AEMC)*

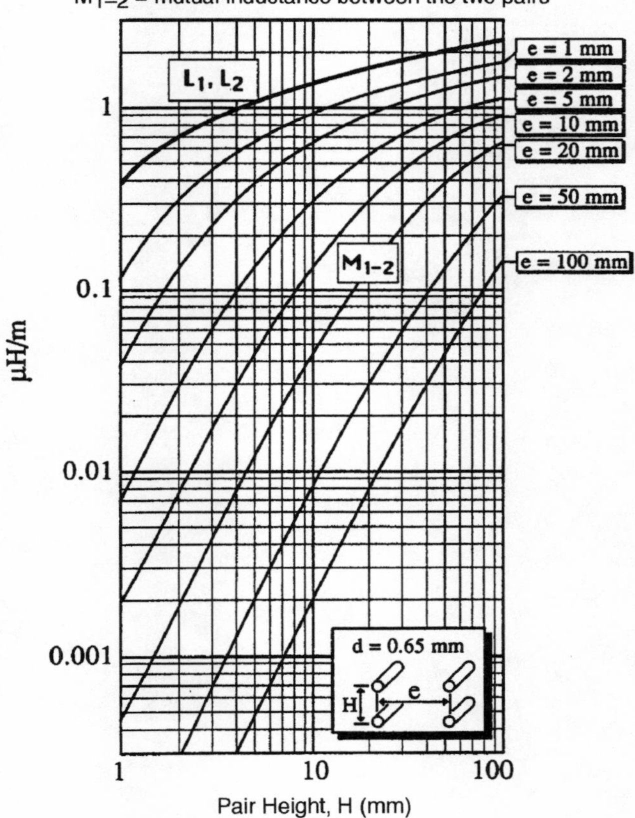

Application

- From "H" and "e," determine M_{1-2} μH/m.
- Multiply by coupling length in meters to get M_{1-2} (μH).

Victim pair voltage is equal to

$$V_{\text{volt}} = 2\pi F_{(\text{MHz})} \times M_{1-2\,(\mu\text{H})} \times I_{\text{culprit (A)}} \quad \text{(frequency domain)}$$

$$V_{\text{volt}} = M_{1-2\,(\mu\text{H})} \times \frac{\Delta I_{\text{culprit (A)}}}{\Delta t_{(\mu s)}} \quad \text{(time domain)}$$

Appendix

G

Conversion of Radiated Emission Limits into CM Current Limits

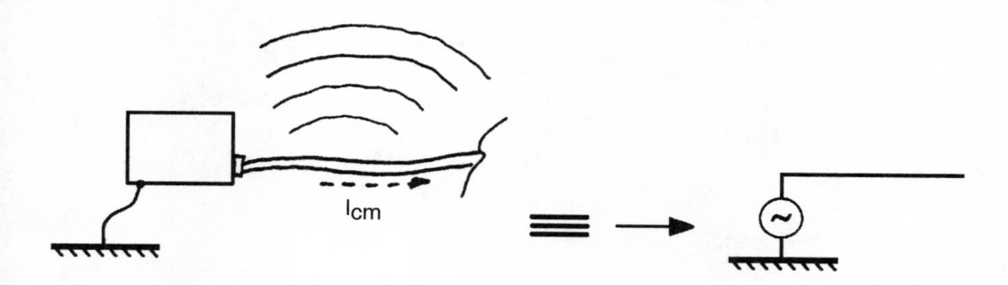

A typical I/O cable exiting the EUT can be regarded as a base-driven monopole. Its radiated field can be calculated by a rather simple model when

1. cable length, ℓ, exceeds 1/4 wavelength (i.e., $\ell_m \geq 75/F_{MHz}$), and
2. cable height above ground exceeds $\approx$ 0.1 wavelength (i.e., $h_m \geq 30/F_{MHz}$) in actual EUT application or radiated emission test setup.

Then, the maximum E-field amplitude becomes independent of frequency and is approximately calculated by

$$E(\mu V/m) = 60 \times \frac{I_{cm}}{D} \text{ for far-field conditions}$$

where I_{cm} = common-mode current (in microamperes), and D = distance to the receiving antenna (in meters).

Therefore, in practical FCC or CISPR/EC standard test configurations, provided that

- $F \geq 50$ MHz
- $D \geq 1$ m
- $\ell > 1.5$ m, and
- $h > 0.60$ m

a simple pass/fail criterion can be set, based only on CM current measurements:

$$I_{cm(\mu A)} < E_{limit} \times \frac{D}{60}$$

For Mil-Std-461 radiated emission testing (RE02), the cables are laid 5 cm above the ground plane, which reduces the radiated emissions by a factor equal to $10 \times h/\lambda$, hence relaxing the I_{cm} criterion by a corresponding amount, from 50 MHz < F < 600 MHz.

Appendix

Filter and Input Circuit Universal Rejection Curves

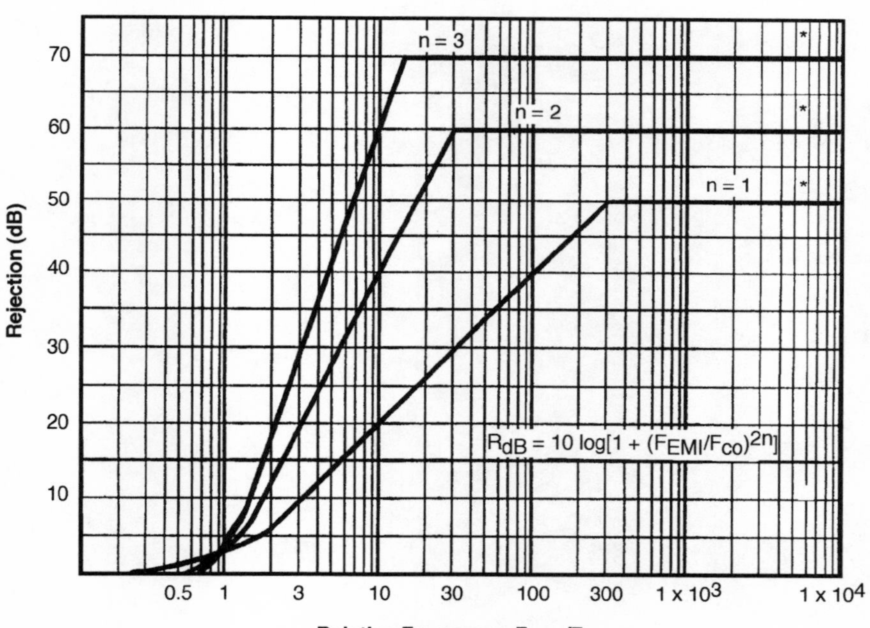

Notes:

n = number of poles (elements) of the filter circuit

F_{co} = cutoff frequency of the victim's input or filter network in your application; i.e., the frequency at which ωRC (capacitive filter) or ωL/R (inductive filter) = 1

* = maximum rejections have been clamped to realistic values, due to component parasitic effects

In the 3 dB region around $F_{EMI}/F_{co} = 1$, rejection is shown for ideal damping conditions with resistive loads on both sides. Actual circuits may behave differently at this peculiar frequency, with low resistance or reactive loads causing resonances and negative rejection (RdB < 0).

Application Example

Question: An L,C filter (n = 2) has been designed such that F_{co} = 1.5 MHz. What rejection does it offer against a 27 MHz noise?

Answer: $\frac{F_{EMI}}{F_{co}} = \frac{27.5}{1.5} = 18$. RdB for n = 2 is 50 dB.

Appendix

I

EMI Fixes Toolbox: Recommended Parts List

I.1 Instruments and Accessories

- Oscilloscope (minimum BW 100 MHz, 200 MHz preferred)
- Digital multimeter (preferred 20,000 pts)
- Hand-held capacitance meter (0.1 pF resolution)
- H-field loop, shielded
- Current probe(s)
- 1 m rod antenna with banana terminal
- 10/1 oscilloscope probe, 100 MHz
- Two 1.50 m coaxial cables, RG58 + BNC
- Ordinary instrument cables with stackable banana plugs
- Complete set of coaxial adapters:
  - Banana-BNC, male and female
  - BNC, female to female
  - "T" BNC
  - N to BNC, BNC to N
  - SMA to BNC
  - BNC to TNC
- 50 Ω/1 W coaxial load, through-type
- AC cord extender and two-to-three-prong adapters
- Wire hooks and crocodile clips
- Soldering iron and solder

- Wire stripper and crimping tool
- Minimal, usual tool set (screwdrivers, hex keys, pliers, a thin file, emery paper, etc.)

I.2 Components

- Roll of heavy-duty aluminum foil
- Copper tape, minimum of 25 mm (1 in) wide
- Steel wool pad
- Insulating tape
- Cable shielding mesh tape
- Set of fingerstocks and compressible gaskets
- Capacitors:
  - 0.22 µF and 2.2 µF, 500 V (polypropylene preferred; by default metallized polyester
  - Feedthrough filter capacitor set (1 nF to 1 µF, 100 V)
  - Ceramic capacitor set (22 to 2,200 pF)
  - Ceramic capacitor + ferrites three-lead combination
- Ferrites:
  - Large split ferrites for cable bundles, inside dia. 12.7 mm (0.5 in), impedance ≥ 300 Ω @ 30 to 300 MHz
  - Ferrites for flat cables
  - Assortment of small ferrites for PCBs
- Filters:
  - Power-line filters, 1 and 10 A min. attenuation @ 500 kHz, 50 dB (CM), 40 dB (DM)
  - Filter connector adapter (e.g., sub-D, 9 position and 25 position)
  - 0.5 mH CM chokes, two- and four-wire
- Varistors (MOVs), 130 Vac (US) or 250 Vac (Europe) rated

Appendix

J

Test Data Report Forms for Conducted and Radiated Emissions, ESD, and EFT

J.1 Conducted EMI Test Log Sheet

Test configuration/EUT: ____________________ Date: ________

Conducted EMI Probe location: ____________________

Type of instrumentation
Receiver: ____________________
Probe: ____________ or LISN: ____________
Input attenuator (if not accounted for in readout: ____________
____
Low-noise preamp: ____________________

Frequency	Receiver BW	Reading (dBm)	dBµV = dBm + 107 (if 50 Ω)	Corrections (Atten., LNA)	Probe factor	dBµV or dBµA	Comments

J.2 Radiated EMI Test Log Sheet

Test configuration/EUT: ______________________ Date: ______

Radiated E- or H-field Antenna/probe location: ______________

Type of instrumentation
Receiver: ______________________________________
Probe/antenna: __________________________________
Input attenuator (if not accounted for in readout: ______________
____
Low-noise preamp: _______________________________

Frequency	Receiver BW	Reading (dBm)	dBμV = dBm + 107 (if 50 Ω)	Corrections (Atten., LNA, coax loss)	Antenna factor	dBμV/m or dBμA/m	Comments
		Pol. H/V	H/V			H/V	

J.3 ESD Test Log Sheet

EUT: ______________________________ Test date: ________

Test type: ____

Prototype: ________________ Release or QC: ______________

EUT software used: ________ Auxiliary units: _____________

Simulator: _________________ Disch. resistor value: _________

Last date checked: ______________________________________

Sketch of EUT showing discharge points:

Test point no.	V_{run}/V_{fail} No. of pulses	Observations: Direct contact, air disch., or indirect, failure mode, fixes, etc.

J.4 Electrical Fast Transient (EFT) Test Log Sheet

EUT: ______________________________ Test date: ________
Test type: ____
Prototype: __________________ Release or QC: ______________
EUT software used: ___________ Auxiliary units: _____________
Simulator: __________________ Last date checked: ___________
Duration of each test (default 1 min): ______________________

	Level				
	0.25 kV	0.5 kV	1 kV	2 kV	4 kV
Polarity	+/–	+/–	+/–	+/–	+/–
1. Power input, direct injection					
Line 1					
Line 2					
Lines 1 + 2					
Others					
Observations (failure mode, fixes, etc.)					
2. Signal lines, capacitive injection					
Observations (failure mode, fixes, etc.)					

Appendix

K

HF Losses with Coaxial Cables

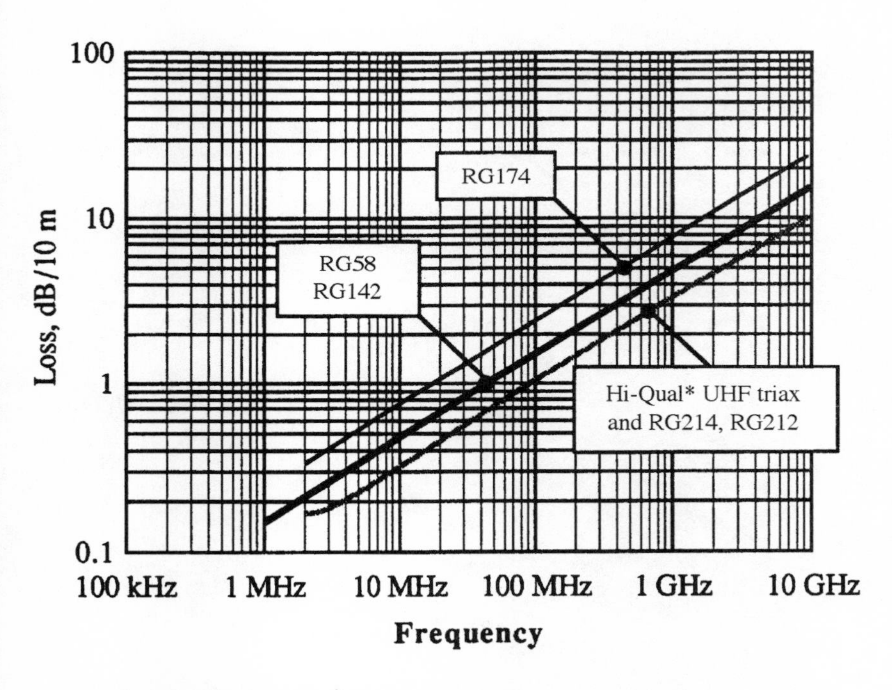

RG58: single-braid coax, 0.8 mm dia., good coverage.
*Huber+Suhner "SUCOFLEX 104" type.
RG142: light construction, double braid

- Losses may vary from one manufacturer to another
- Figures shown are typical for 10 m cable fitted with high-quality threaded connectors (N-type or similar)
- N-type connector loss is ≈ 0.07 dB $\times \sqrt{F_{GHz}}$

Application Example

A receiver is displaying an amplitude of 24 dBμV @ 350 MHz. The antenna (or probe) is connected by 5 m of RG58. The cable loss is calculated as follows:

$$3 \text{ dB (for 10 m)} \times \frac{5 \text{ m}}{10 \text{ m}} = 1.5 \text{ dB}$$

The actual signal delivered by the antenna is

$$24 + 1.5 = 25.5 \text{ dB}\mu\text{V}$$

References

1. Mardiguian, M., *Controlling Radiated Emissions by Design.* New York: Van Nostrand Reinhold, 1992. (Title acquired by Kluwer Academic Publishers, Norwell, MA., in 1999.)
2. IEC Immunity Standards 1000/4-2, -3, -4, -5, and -6, and Mil-Std-461, 462.
3. Mardiguian, M., *Electrostatic Discharge: Understand, Simulate and Fix.* Gainesville, VA: Interference Control Technologies, Inc.
4. White, D.R.J., and M. Mardiguian, *Electromagetic Shielding.* Gainesville, VA: Interference Control Technologies, Inc.
5. Vance, E., *Coupling to Shielded Cables.* New York: John Wiley & Sons.
6. White, D.R.J., and M. Mardiguian, *EMI Control Methodology and Procedures.* Gainesville, VA: Interference Control Technologies, Inc.
7. Keenan, K., *Digital Design for Specification Compliance.* Pinellas Park, FL: The Keenan Corporation.
8. Tsaliovich, A., *Shielding Electronic Cables for Electromagnetic Compatibility.* New York: Van Nostrand Reinhold.
9. Goedbloed, J., Aspects of EMC at Equipment Level, *Proc. 1997 IEEE EMC Symposium,* Austin, TX (tutorial notes).

Index